T0321061

Processing Networks

This state-of-the-art account unifies material developed in journal articles over the last 35 years, with two central thrusts: It describes a broad class of system models that the authors call "stochastic processing networks" (SPNs), which include queueing networks and bandwidth sharing networks as prominent special cases; and in that context, it explains and illustrates a method for stability analysis based on fluid models.

The central mathematical result is a theorem that can be paraphrased as follows: if the fluid model derived from an SPN is stable, then the SPN itself is stable. Two topics discussed in detail are (a) the derivation of fluid models by means of fluid limit analysis and (b) stability analysis for fluid models using Lyapunov functions. With regard to applications, there are chapters devoted to max-weight and back-pressure control, proportionally fair resource allocation, data center operations, and flow management in packet networks. Geared toward researchers and graduate students in engineering and applied mathematics, especially in electrical engineering and computer science, this compact text gives readers full command of the methods.

J. G. DAI received his PhD in mathematics from Stanford University. He is currently Presidential Chair Professor in the School of Data Science at the Chinese University of Hong Kong, Shenzhen. He is also the Leon C. Welch Professor of Engineering in the School of Operations Research and Information Engineering at Cornell University. He was honored by the Applied Probability Society of the Institute for Operations Research and the Management Sciences (INFORMS) with its Erlang Prize (1998) and with two Best Publication Awards (1997 and 2017). In 2018, he received the Achievement Award from the Association for Computing Machinery (ACM) SIGMETRICS. Professor Dai served as editor-in-chief of *Mathematics of Operations Research* from 2012 to 2018.

J. MICHAEL HARRISON earned degrees in industrial engineering and operations research before joining the faculty of Stanford University's Graduate School of Business, where he served for 43 years. His research concerns stochastic models in business and engineering, including mathematical finance and processing network theory. His previous books include *Brownian Models of Performance and Control* (2013). Professor Harrison has been honored by INFORMS with its Expository Writing Award (1998), the Lanchester Prize for best research publication (2001), and the John von Neumann Theory Prize (2004); he was elected to the U.S. National Academy of Engineering in 2008.

Processing Networks

Fluid Models and Stability

J. G. DAI

*The Chinese University of Hong Kong, Shenzhen
and Cornell University*

J. MICHAEL HARRISON

Stanford University

CAMBRIDGE
UNIVERSITY PRESS

University Printing House, Cambridge CB2 8BS, United Kingdom

One Liberty Plaza, 20th Floor, New York, NY 10006, USA

477 Williamstown Road, Port Melbourne, VIC 3207, Australia

314–321, 3rd Floor, Plot 3, Splendor Forum, Jasola District Centre,
New Delhi – 110025, India

79 Anson Road, #06–04/06, Singapore 079906

Cambridge University Press is part of the University of Cambridge.

It furthers the University's mission by disseminating knowledge in the pursuit of
education, learning, and research at the highest international levels of excellence.

www.cambridge.org
Information on this title: www.cambridge.org/9781108488891
DOI: 10.1017/9781108772662

© J. G. Dai & J. Michael Harrison 2020

First published 2020

Printed in the United Kingdom by TJ International, Padstow Cornwall

A catalogue record for this publication is available from the British Library.

Library of Congress Cataloging-in-Publication Data
Names: Dai, J. G. (Jiangang), 1962– author. | Harrison, J. Michael, 1944– author.
Title: Processing networks : fluid models and stability / Jim Dai,
J. Michael Harrison.
Description: Cambridge ; New York, NY : Cambridge University Press, 2020. |
Includes bibliographical references and index.
Identifiers: LCCN 2020012552 | ISBN 9781108488891 (hardback) |
ISBN 9781108772662 (epub)
Subjects: LCSH: Stochastic processes. | Linear programming. | Queuing
networks (Data transmission) | Fluid dynamics. | Stability.
Classification: LCC QA274.A1 D35 2017 | DDC 519.2/3–dc23
LC record available at https://lccn.loc.gov/2020012552

ISBN 978-1-108-48889-1 Hardback

To Liqin and Kevin, Elena and Sasha

Contents

Website

http://www.spnbook.org

This book is accompanied by the above website. The website provides corrections of mistakes and other resources that should be useful to both readers and instructors.

Preface

This book has two purposes. First, it describes a broad class of mathematical system models, called stochastic processing networks (SPNs), that are useful as representations of service systems, industrial processes, and digital systems for computing and communication. The SPN models to be considered include such features as simultaneous resource possession, multi-input operations, and alternative processing modes. No comparably general treatment of network models has appeared previously in book format.

Second, it develops a fluid model methodology for proving SPN stability, by which we mean proving positive recurrence of the Markov chain describing the SPN. Specifically, we develop a theorem that can be informally paraphrased as follows: if the fluid model derived from an SPN is stable (as that phrase is defined later in this preface), then the SPN itself is stable. The significance of that result lies in the relative tractability of fluid models: proving fluid model stability is invariably easier than proving positive recurrence of the Markov chain for which it serves as a surrogate.

As multiple examples will show, proving fluid model stability for a complex SPN can still be challenging, requiring the construction of a suitable Lyapunov function. A large part of the book is aimed at demonstrating how the theorem has been used and can be used to analyze systems of contemporary interest, especially computing and communication networks.

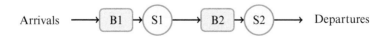

Figure 1 Tandem processing system.

Tandem example. To understand the content of the theorem paraphrased above, it is useful to consider a concrete example. Figure 1
depicts a tandem processing system with two servers, labeled S1 and
S2. Jobs arrive from the outside world one at a time. Each job is
processed first by S1, then by S2, and then exits. If a job arriving
at either server finds that server idle, then the processing of that job
begins immediately. On the other hand, if a job arriving at either server
finds that server busy, then the job waits in a corresponding buffer
(B1 or B2). Each server processes jobs from its associated buffer on a
first-in-first-out (FIFO) basis, and continues working at full capacity
so long as there is any job available for it to process. As a matter of
convention, when reference is made later to "jobs currently occupying
B1," that is understood to include not only waiting jobs but also the
job being processed by S1, if there is one, and similarly for B2. For
concreteness, let us assume that the external arrival process is Poisson
with arrival rate λ_1; that S1 processing times, also called service times,
are independent and identically distributed (i.i.d.) with some phase-
type distribution (see Appendix D, Section D.8, for the meaning of
that term) having mean m_1; and that S2 service times are i.i.d. and
exponentially distributed with mean m_2.

Tandem fluid model. Associated with this discrete-flow network is a
continuous-flow model, or fluid model, that consists of the following
equations: for $t \geq 0$,

(1) $\quad Z_1(t) = Z_1(0) + \lambda_1 t - \mu_1 T_1(t) \geq 0,$

(2) $\quad Z_2(t) = Z_2(0) + \mu_1 T_1(t) - \mu_2 T_2(t) \geq 0,$

(3) $\quad T_i(0) = 0, \quad 0 \leq T_i(t) - T_i(s) \leq t - s \quad$ for $0 \leq s \leq t, \quad i = 1, 2,$

(4) $\quad Z_i(u) > 0$ for all $u \in [s, t]$ implies that $T_i(t) - T_i(s) = t - s$
$\qquad$ for $0 \leq s \leq t$ and $i = 1, 2.$

Here $Z_i(t)$ is interpreted as the fluid content in buffer i at time t, and
$T_i(t)$ is the cumulative amount of time that server i is busy up to time t
$(i = 1, 2)$. The parameter λ_1 is the arrival rate of fluid from the outside,
and $\mu_i := m_i^{-1}$ is the processing speed of server i $(i = 1, 2)$. Equations
(1) and (2) are the flow balance equations, while (3) expresses service
capacity constraints, and (4) dictates that each server operate at full
capacity whenever its buffer is nonempty.

    The fluid model is said to be stable if, for each solution Z of equations
(1) through (4), there exists a time $\delta > 0$ such that $Z_1(t) = Z_2(t) = 0$ for

$t \geq \delta$. One can prove that the fluid model is stable if and only if the following conditions are satisfied:

$$(5) \qquad \lambda_1 m_1 < 1 \quad \text{and} \quad \lambda_1 m_2 < 1.$$

Sequential decomposition of the tandem fluid model. There are many ways to prove that statement. Perhaps the simplest is to analyze the tandem fluid model sequentially, first for fluid in buffer 1, then for fluid in buffer 2. Given that $\lambda_1 < \mu_1$, it is easy to show that there exists a time $\delta_1 \geq 0$ such that $Z_1(t) = 0$ for $t \geq \delta_1$. After time δ_1, the fluid flowing into B1 instantaneously passes into B2, and thus the arrival rate to B2 is λ_1. One can again analyze buffer 2 in isolation, showing there exists $\delta_2 \geq \delta_1$ such that $Z_2(t) = 0$ for $t \geq \delta_2$.

Direct analysis of the discrete-flow model. Under the distributional assumptions stated earlier, we can model the tandem processing system as a continuous-time Markov chain $\{X(t) = (Z_1(t), \eta(t), Z_2(t)), t \geq 0\}$, where $Z_i(t)$ is the number of jobs occupying buffer i at time t, and $\eta(t)$ is a finite-valued phase indicator (see Section D.8) for the S1 service currently under way, if any. (When B1 is empty, $\eta(t) = 0$.) How does one prove that the Markov chain X is positive recurrent under condition (5)?

The first thing to say is that there exists no analog of the sequential decomposition approach we have described. That is, in the discrete-flow setting, stability analysis is *not* decomposable, despite the feedforward structure (that is, unidirectional flow) that is the salient feature of our example. Rather, with rare exceptions, the approach adopted by researchers is to apply the Foster–Lyapunov criterion described in Appendix D, Section D.7. In this approach, the analyst must identify a test function V, hereafter called a Lyapunov function, that satisfies the Foster Lyapunov drift condition (D.36). For our tandem processing example, one can construct a Lyapunov function of the quadratic form

$$V(Z_1, \eta, Z_2) = Z_1^2 + a(Z_1 + Z_2)^2,$$

where a is any constant satisfying $0 < a < \mu_1/\lambda_1 - 1$. An analysis undertaken in Section 8.2, culminating in the inequality (8.19), will show that this function V satisfies the Foster–Lyapunov drift condition, thus proving the positive recurrence of X.

Lyapunov functions for fluid models. The sequential decomposition described earlier to prove stability for our tandem fluid model extends in a direct way to *any* feedforward fluid model. For a general fluid model, however, one proves stability in very much the same way as for a general Markov chain, namely, by identifying a Lyapunov function that satisfies an appropriate drift condition. That general theory will be developed in Chapter 8.

As an example, the simple linear function $V(Z) = Z_1 + Z_2$ satisfies the drift condition for our tandem fluid model, given that (5) holds. However, for reasons explained in Section 8.2, it does *not* satisfy the drift condition for the discrete-flow tandem model. Thus the simplest known Lyapunov function for the discrete-flow tandem model is quadratic, while that for its fluid analog is linear. This illustrates a phenomenon that has often been observed in the analysis of particular model structures: in cases where a Lyapunov function is known both for a discrete-flow SPN model and for the fluid model derived from it, the latter function is substantially simpler.

Control policies and stability conditions. In the preceding paragraphs, we have discussed the stability problem for processing networks as if it were simply one of analysis, that is, as if the central problem were to rigorously prove stability under a given control policy. In general, however, a system designer or system manager first confronts a problem of *synthesis*, namely, he or she must first devise a dynamic control policy, which may be called a *network protocol* or *network algorithm* in a digital system context.

In our tandem example, we have specified FIFO processing by both servers, but the same fluid equations are valid for other *nonidling* policies as well. (Here the term "nonidling" means that each server works at full capacity whenever there is accessible work for that server to do.) Specifically, the fluid model equations (1) and (2) remain valid under any policy such that the number of partially completed jobs at any given time is bounded by some constant L, (3) is valid under any policy, and (4) is valid under any nonidling policy, as will be shown in Section 7.1.

In this book, virtually all effort and attention is directed to the analytical problem of proving stability for a given policy. As a preliminary, we develop in Chapter 5 a general stability condition analogous to (5), based on what is called a *static planning problem*. That condition involves only first-order system data (average arrival rates, average processing times, and routing probabilities or average output quantities), and it is shown to be *necessary* for stability under any control policy.

One tends to feel intuitively that the condition is also sufficient for existence of a policy that achieves stability, but examples will show that expectation is not always correct.

Dominance of the fluid approach. Over the last 25 years, fluid model methodology has come to dominate in studies of network stability, allowing successful treatment of model families that have defied direct analysis. Notable in this regard are the feedforward networks referred to earlier. The method of sequential decomposition makes fluid model stability proofs almost trivial for such networks (see Section 8.3 for elaboration), whereas the feedforward structure may be of little or no help in direct analysis. This contrast is illustrated well by the recent work of Massoulié and Xu (2018) on information processing systems.

There are also important families of non-feedforward networks for which fluid models have been analyzed successfully to prove stability, but no method is known for direct analysis. Another way of saying this is that, for some important families of non-feedforward networks, Lyapunov functions have been successfully devised for their fluid model analogs, but not for the discrete-flow models themselves. This is true, for example, of the FIFO Kelly networks analyzed by Bramson (1996a), also treated in section 5.3 of the monograph by Bramson (2008). Another example is a packet switched communication network with what Walton (2015) called *random proportional scheduling*, to be treated in Chapter 12 of this book. In both those cases, an entropy-type Lyapunov function (see Section 10.5) provides the key to fluid analysis, and there is no known analog for the discrete-flow model itself.

Acknowledgments In writing this book, we have benefited from consultations with many colleagues, including Søren Asmussen, Maury Bramson, Anton Braverman, Doug Down, Sergey Foss, Peter Glynn, Mor Harchol-Balter, Frank Kelly, Sunil Kumar, Yi Lu, Andre Milzarek, Balaji Prabhakar, Devavrat Shah, R. Srikant, Jean Walrand, Neil Walton, Ruth Williams, Qiaomin Xie, Kuang Xu, Lei Ying, Elena Yudovina, Yuan Zhong, and Bert Zwart. Shuangchi He helped with some figures in the book, and Cornell PhD students Chang Cao and Mark Gluzman helped with finalizing the format of references.

Guide to Notation and Terminology

We use $\mathbb{R}$ to denote the set of real numbers, $\mathbb{R}_+$ the set of nonnegative real numbers, $\mathbb{Z}_+$ the set of nonnegative integers, and $\mathbb{N}$ the positive integers. A prime is used to denote the transpose of a vector or matrix, and vectors should be envisioned as column vectors unless something is said to the contrary. For an integer $d > 0$ and a vector $x = (x_1, \ldots, x_d) \in \mathbb{R}^d$, we define the norm $|x| = \sum_{i=1}^{d} |x_i|$, and for two vectors $x, y \in \mathbb{R}^d$, we define the inner product $x \cdot y = \sum_{i=1}^{d} x_i y_i$. For two vectors $x \in \mathbb{R}^d$ and $y \in \mathbb{R}^d$, we write $x \leq y$ to mean that $x_i \leq y_i$ for each $i = 1, \ldots, d$.

The relationship $A := B$ means that A equals B by definition. The letter e is occasionally used to denote the vector $(1, \ldots, 1)$, and we denote by e^j a vector with a 1 as its jth component and all other components equal to zero; in each case, the dimension of the vector should be clear from context. For $x, y \in \mathbb{R}$, we use $x \vee y$ to denote $\max(x, y)$, and $x \wedge y$ to denote $\min(x, y)$.

A square, nonnegative matrix is said to be *substochastic* if each of its row sums is ≤ 1, and to be *stochastic* if each of its row sums is $= 1$. The *spectral radius* of a $d \times d$ substochastic matrix P is $\max_{1 \leq i \leq d} |\lambda_i|$, where $\lambda_1, \ldots, \lambda_d$ are the eigenvalues of P. A substochastic matrix P is said to be *transient* if its spectral radius is < 1, or equivalently, if $P^n \to 0$ as $n \to \infty$.

Throughout the book, we denote by $\mathbb{P}(\cdot)$ the probability measure underlying a model, and by $\mathbb{E}(\cdot)$ the corresponding expectation operator. That is, $\mathbb{P}(A)$ denotes the probability of an event A, and $\mathbb{E}(X)$ is the expected value of a random variable X. For an event A and random variable X, we define the partial expectation $\mathbb{E}(X; A) = \int_A X d\mathbb{P}$.

Phase-type distributions are defined and discussed in Appendix D, Section D.8, where we introduce the following notation for three specific families of nonnegative, univariate distributions: $\exp(r)$ denotes an *exponential* distribution with rate parameter $r > 0$; $\text{Erlang}(2, r)$ denotes an *Erlang* distribution with shape parameter 2 and rate parameter $r > 0$; and $H_d(p,\gamma)$ denotes a *hyperexponential* distribution (see Section D.8 for details).

Stochastic processing networks, also referred to frequently as SPN models, are formally defined in Chapter 2, Sections 2.1 through 2.4, where we introduce notation for model data and model-related processes that continues throughout the entire book. In particular, the uppercase Roman letters A, B, D, E, F, I, J, K, N, S, T, and Z are given more or less permanent meanings in those sections, but such symbols may be reused with new meanings in the appendices. Sets are most often denoted by uppercase script letters; three that appear frequently are $\mathscr{I} = \{1,\dots,I\}$, $\mathscr{J} = \{1,\dots,J\}$, and $\mathscr{K} = \{1,\dots,K\}$.

For a function $f\colon \mathbb{R}_+ \to \mathbb{R}^d$, we use $\dot{f}(t)$ to denote the derivative of f at t. A point $t > 0$ is said to be a *regular point* for f if f is differentiable at t. When the function f is clear from the context, we sometimes call $t > 0$ a regular point without further qualification. Whenever the symbol $\dot{f}(t)$ is used, it is assumed that t is a regular point of f. Occasionally we also use $\frac{d}{dt}f(t)$ to denote $\dot{f}(t)$.

1

Introduction

This book considers a broad class of stochastic system models, focusing on questions and methods related to long-run "stability." To be more precise, we consider stochastic models of multi-resource processing systems, assuming throughout that average input rates and average processing rates are time-invariant, and we focus on the following questions: Do the processing resources have enough capacity to handle the given or hypothesized load, and if so, how can the resources be deployed dynamically to ensure that statistical equilibrium is achieved over the long run?

The first four sections of this introductory chapter strive to explain the breadth and variety of models considered, beginning with the question of how to name them. Sections 1.5 and 1.6 elaborate on the notion of system stability. Section 1.7 describes the structure of the book and its intended audience, and Section 1.8 contains brief comments about sources and literature.

1.1 About the Title of This Book

Our focus is on models of man-made systems in which servers (that is, capacity-constrained processing resources) undertake various activities to satisfy the needs of externally generated jobs. Emphasizing the purpose for which such systems exist, we call these models *processing networks*. The "jobs" to be processed may be digital files requiring transfers, telephone calls from customers seeking information, manufacturing lots that require a particular factory operation, or other possibilities, depending on the application domain.

There is a large literature in applied mathematics concerned with the performance degradation caused by stochastic variability in the functioning of such systems. In that literature, mathematical models are used to explicate the perplexing phenomenon whereby some arriving

1

jobs experience lengthy delays before their processing is completed, even though naive calculations based on average arrival rates and average processing rates indicate that resource capacities are adequate to handle the offered load. Such models are described throughout most of this book as *stochastic* processing networks (SPNs), but for brevity the modifier has been dropped in the book's title.

Queueing theory is a branch of applied mathematics that deals with performance degradation due to stochastic variability, and virtually all of the literature to which this book refers can be placed beneath that very broad umbrella. In fact, "queueing network" might seem to be a reasonable substitute for "stochastic processing network," but the former term has an established meaning in the literature of applied probability (see Sections 1.3, 1.4, and 2.6 for elaboration), and it is narrower than what we mean by the latter term. In particular, the term "queueing network" is generally understood to *not* allow simultaneous resource possession (as occurs in models of bandwidth sharing), nor to allow simultaneous input requirements (as occurs in models of assembly and matching), but those two features *are* allowed in our conception of a stochastic processing network.

What does the word "network" mean in our context? The non-mathematical definition of that word involves multiple distinct entities, such as power plants or supermarkets, that are connected in some way. Adapting that definition to our setting, one may safely say that all the models encountered in this book involve two or more distinct processing resources (or servers, or server pools) that are connected by a workflow. That is, all our models involve a network of capacity-constrained resources that differ in their capabilities and are connected by the need to collaborate, either sequentially or simultaneously, in processing jobs of different types. That high-level characterization is rather vague, and applied mathematicians often attach to the word "network" a narrower meaning that involves *arcs* and *nodes* in a graph structure. Indeed, many important stochastic network models, notably models of communication networks, do involve an underlying graph structure. Such models are somewhat special, however, and in other application domains, the arcs-and-nodes framework of classical network theory is overly restrictive.

1.2 Activity Analysis

Our conception of a stochastic processing network is derived from, and seeks to extend, the deterministic linear model of an operation or enterprise that was developed in the mid-twentieth century by mathematicians and economists working in the area that Koopmans (1951) called "activity analysis." Today that term is viewed as an old-fashioned synonym for "linear programming" (LP) or "linear optimization," but the emphasis here is not on optimization per se. Rather, we seek to develop stochastic analogs of the linear input–output models that define the "feasible region" in various classical LP problem formulations.

In the version of activity analysis that we shall adopt, an economic system or subsystem is characterized in terms of three basic elements: processing *resources* with fixed capacities per time unit; processing *activities* that may be undertaken using those resources; and units of flow, initially referred to as *materials*, that are created, destroyed, or modified by processing activities. In the standard treatment, activity levels and material quantities are treated as continuous variables, but the time parameter (in the case of dynamic models) is discrete. Thus one might say that classical activity analysis uses deterministic fluid models with a discrete time parameter.

The term "server" is commonly used as a synonym for "resource," and we shall follow that practice. However, in many applications a "resource" consists of multiple distinct but functionally interchangeable entities, such as electronic testers, vertical turret lathes, or customer service representatives. Thus we shall formulate and analyze models involving multi-server "pools," with the understanding that servers belonging to the same pool are identical.

Unfortunately, there seems to be no single term that is appropriate in all applications for what were called "materials" in the model description above. For example, one might want to model and analyze activities in which the "materials" processed are telephone calls from customers seeking different kinds of airline reservations, or to model and analyze activities in which the "materials" produced as outputs are electronic components of various types.

Because the word "material" is generally interpreted to have a meaning more narrow than required in these examples, we shall more often use the abstract term "units of flow," or else use concrete terms like "customer," "job," or "packet" that relate to particular application domains. The important point here is that *units of flow will be called by different names at different points in this book.* We are confident that readers, having been alerted to this phenomenon, will find that it causes no confusion.

In his discussion of model formulation in linear programming, Dantzig (1998) described an "activity" as a black box, into which flow inputs and out of which flow outputs, defining "black box" as "any system whose detailed internal structure one willfully ignores" (page 32). In the introductory chapter of that same text, it is said that "the first step [in model formulation] consists in regarding a system under design as composed of a number of elementary functions that are called 'activities' ... The different activities in which a system can engage constitute its technology" (page 6).

Obviously, there is modeling discretion involved in the definition of resources, activities, and units of flow. Different levels of aggregation are possible, and the right choice depends on what decisions are to be supported or informed by the analysis.

1.3 Two Examples of Queueing Networks

Most past research on stochastic processing networks has focused on a narrow class of models called queueing networks, the general definition of which is postponed to Section 2.6. Here we consider two relatively simple examples, using the language of activity analysis to establish parallels with the classical input–output model of linear programming. This discussion provides a stepping stone to the general notion of a stochastic processing network, and allows us to introduce in a concrete setting some concepts and terminology that will be used in more general contexts later.

We begin with the same example used in the preface of this book, namely, the arrangement of two servers in tandem that is pictured in Figure 1.1. There are two servers, labeled S1 and S2 in the figure, that play the role of processing resources. Following standard usage in queueing theory, we call the units of flow *customers*, and assume that new customers arrive from outside the system according to a Poisson process with arrival rate $\lambda_1 > 0$. (The reason for the subscript 1 will become apparent shortly.) There are two storage *buffers* (or waiting rooms) in Figure 1.1, labeled B1 and B2, where customers wait for

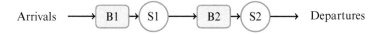

Figure 1.1 Tandem queueing network.

service if they cannot be processed immediately, and where they continue to reside as they are being processed. (The verbs "to serve" and "to process" will be used interchangeably throughout this book, as will the nouns "service time" and "processing time.")

New arrivals enter buffer 1, are processed by server 1 on a first-in-first-out (FIFO) basis, then proceed to buffer 2, are processed by server 2 on a FIFO basis, and then depart from the system. Customers who have not yet completed their first service are designated as *class 1* customers, and those who have completed their first service but not the second one are designated as *class 2*. Thus it is class 1 customers who occupy buffer 1, and class 2 who occupy buffer 2. We assume that the service times or processing times for class i customers are independent and identically distributed (i.i.d.) random variables with mean $m_i > 0$ ($i = 1, 2$), and that the arrival process and the two service time sequences are mutually independent.

There are only two processing activities in this system, namely, the processing of class 1 by server 1 and the processing of class 2 by server 2. A crucial notion for stochastic processing networks generally is that of a *control policy*, by which we mean a rule or set of rules that determine which activities will be undertaken when, based on dynamic observations of system status. (Hereafter the term "control policy" will be routinely shortened to just "policy." It is common in the literature of applied probability to distinguish among various categories of dynamic control, such as routing, sequencing, and input control. Here the term "control," occasionally expanded to "dynamic control" for emphasis, is understood to include all such categories.) For the tandem queueing example pictured in Figure 1.1, there are no meaningful control decisions to be made, given the restriction to FIFO processing stated above: under most commonly used performance criteria, one wants each server to devote its full capacity to processing the first-arriving customer in its associated buffer, as opposed to idling the server or having it work at less than full capacity, which would simply delay completion of the customer's service.

Can we expect the system to achieve long-run stability (see Section 1.5 for an explanation of that term) with FIFO processing? The first thing to consider in that regard is the adequacy of server capacity. Because arrivals occur at an average rate of λ_1 jobs per time unit, the expected time required from server i to complete the processing of jobs that arrive within one time unit is $\lambda_1 m_i$ ($i = 1, 2$). That product, hereafter called the *load factor* for server i, expresses the load imposed on the

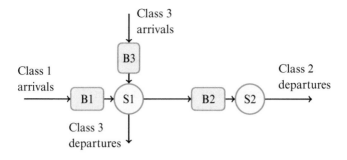

Figure 1.2 The criss-cross network.

server as a fraction of capacity, and it will later be shown that stability is achieved if and only if

$$(1.1) \qquad \lambda_1 m_1 < 1 \quad \text{and} \quad \lambda_1 m_2 < 1.$$

Figure 1.2 pictures a more complicated model, commonly called *the criss-cross network*, where there *are* meaningful control decisions to be made. As in the tandem queueing example, we have two servers and a stream of customers who require one processing operation from each server, but there is also a second stream of customers, and they require just a single processing operation from server 1 before they depart. Customers in the latter stream, whether waiting or being processed, are referred to as *class 3*, and we imagine them as being stored in their own dedicated buffer, as pictured in Figure 1.2. In addition to the assumptions stated earlier for class 1 and class 2 customers, we assume that class 3 customers arrive according to a Poisson process at rate $\lambda_3 > 0$, that their service times are i.i.d. with mean $m_3 > 0$, and that the class 3 arrival process and service time sequence are independent both of one another and of the class 1 input process and the service time sequences for classes 1 and 2.

The criss-cross network is an example of a *multiclass* queueing network, which means that there is at least one server (in this case, server 1) that has responsibility for processing two or more distinct customer classes. If we require that customers within each class be processed on a FIFO basis, which is a reasonable restriction in many settings, then the choice of a control policy amounts to specifying whether class 1 or class 3 customers are to be processed first by server 1, based on the numbers of customers currently occupying the three buffers, and possibly also on other aspects of the system's history or current status. There is no uniquely "best" control policy, because generally speaking, choices that reduce the waiting times of class 3 customers will increase waiting times for customers who require processing from both servers, and vice versa.

With regard to the potential for long-run stability, we first observe that on average, server 1 must work for $\lambda_3 m_3$ time units to process the class 3 jobs that arrive in one time unit, so the total load factor for server 1 is $\lambda_1 m_1 + \lambda_3 m_3$, while the load factor for server 2 is the same as in our tandem queueing example. Thus we have the following analog of the stability condition (1.1):

$$(1.2) \qquad \lambda_1 m_1 + \lambda_3 m_3 < 1 \quad \text{and} \quad \lambda_1 m_2 < 1.$$

The criss-cross network is a *feedforward* queueing network, which means that servers can be numbered in such a way that customers never move from higher numbered servers to lower numbered ones. As a consequence, any non-idling dynamic control policy (that is, any policy that requires each server to work at full capacity except when all its associated buffers are empty) will achieve long-run stability if both inequalities in (1.2) hold. This will be proved in Chapter 8 under mild additional assumptions. In contrast, Section 1.6 provides examples of *non*-feedforward networks where long-run stability is *not* achieved by a particular nonidling policy, even though stability *is* achievable using other, more intelligent policies. Examples of this kind, which first came to light in the early 1990s, provided the motivation for much of the research recounted in this book.

1.4 SPN Examples with Additional Features

As we shall see in future chapters, multiclass queueing networks can be much more complicated than the examples discussed above, but our conception of a stochastic processing network (SPN) further allows various interesting and realistic structural features that are *not* allowed in the queueing network framework. Some of those features are illustrated by the four examples discussed in this section, each of which falls within some model family treated later in the book.

Alternate routing example. The system pictured in Figure 1.3 differs from the criss-cross network (Figure 1.2) in that either server 1 or

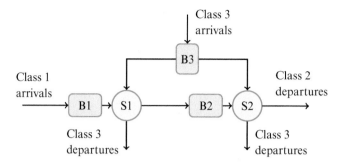

Figure 1.3 A system with alternate routing.

server 2 can provide the single processing operation needed by class 3 customers; the processing time distribution may or may not be different depending on which server is chosen. (We are assuming here that the decision about which server will process a class 3 arrival need not be made until the service is about to begin, so class 3 arrivals are held in a single buffer while awaiting service. Alternatively, one might assume that a commitment must be made at the moment of arrival; see Section 4.2 for further discussion.) In that case, there are four distinct activities: the processing of class 1 by server 1, of class 2 by server 2, of class 3 by server 1, and of class 3 by server 2. Thus a control policy must allocate the capacity of each server to two potential activities over time, based on observed system status.

In the simple example of alternate routing just described, class 3 customers depart from the system after being processed, regardless of which server does the processing. But more generally, when a customer class can be processed by any of several different servers, one may allow the future routing of such customers to depend (in the probabilistic sense) on which server is chosen. Obviously, alternate routing capabilities create a rich and complex environment for dynamic system control.

Bandwidth sharing example.   Figure 1.4 pictures a network with two servers, labeled S1 and S2, that process *jobs* of three classes. Servers 1 and 2 have *capacities* b_1 and b_2, respectively, the significance of which will be explained shortly. The long-run average arrival rate for class i is λ_i, and each class i arrival has a *size* that is drawn from a class-specific distribution with mean $m_i > 0$ ($i = 1, 2, 3$). Job sizes for the three classes

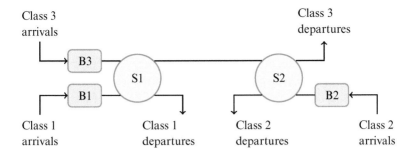

Figure 1.4 A two-link bandwidth sharing network.

form mutually independent sequences of i.i.d. random variables that are also independent of the arrival processes. For concreteness, one may assume that arrivals of the three classes occur according to independent Poisson processes.

The processing of a job is accomplished by allocating a *flow rate* to it over time: newly arriving jobs are stored in a class-specific buffer, and a job departs from the system when the integral of its allocated flow rate equals its size. Processing a class 1 job consumes the capacity of only server 1, and processing a class 2 job consumes the capacity of only server 2, but processing a class 3 job consumes the capacities of servers 1 and 2 simultaneously and at equal rates. (Expressing this last feature in generic language, it may be said that there are three processing activities in the model under discussion, one of which involves *simultaneous resource possession*.) To be more precise, if we denote by x_i the flow rate allocated to class i at a given point in time, those allocations must satisfy the following capacity constraints:

$$(1.3) \qquad x_1 + x_3 \le b_1 \quad \text{and} \quad x_2 + x_3 \le b_2.$$

Finally, a crucial model assumption is that the flow rate x_i at any given time is *divided equally among all the class i jobs then present in the system*. The flow rate received by any one job may vary over time, depending on the contents of the three buffers and on the resource allocation policy followed, and jobs of a given class will not necessarily finish processing in the order of their arrival.

Interpreting this example as a data communication model, one may equate servers to links of a communication network, and jobs to files that require transmission over different routes: jobs of class 1 traverse only link 1, those of class 2 traverse only link 2, and those of class 3 traverse both links without intermediate storage. Job sizes might then be expressed in bits, with flow rates x_i and link capacities (or *bandwidths*) b_i expressed in bits per second. Our "equal sharing" assumption, whereby all jobs seeking transmission over a given route at a given time share equally in the flow rate allocated to that route, is in fact a common feature of communication networks, motivated by considerations of fairness. Section 4.5 describes and analyzes a class of bandwidth sharing models that generalize the example portrayed in Figure 1.4.

Data switch example. The 2×2 data switch pictured in Figure 1.5 consists of two input ports (also called *ingresses*) labeled a and b, two output ports (*egresses*) labeled c and d, and a switching fabric that connects them. (More generally, an $m \times n$ switch has m input ports and n output ports.) This system operates in discrete time, the units of time being called *timeslots*, and its units of flow are data *packets* of uniform size.

During each timeslot, a random number of packets arrive at each ingress (that random number may be zero), and packets arriving at each ingress are logically separated into two virtual buffers according to the egress through which they must exit. Thus there are four virtual buffers on the input side of the switch, one for each combination of ingress and egress. For reasons we need not go into, this system may

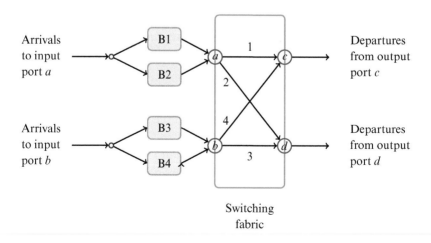

Figure 1.5 A 2×2 data switch.

be referred to as an *input-queued switch*, an *input-buffered switch*, or a *crossbar switch*. The ingress–egress combinations, which may also be called *paths through the switch*, are numbered 1 through 4 in Figure 1.5. We denote by B1 the virtual buffer (also called a virtual output queue, abbreviated VOQ) containing packets that follow path 1, and similarly for the other VOQs.

One timeslot is required to transfer a packet from its associated ingress to its associated egress, and such transfers are subject to the following capacity constraints: at most one packet can be transferred from each ingress per timeslot, and at most one packet can be transferred to each egress per timeslot. Thus there are effectively just two modes or *matchings* that characterize the switch's function during any given timeslot: ingress a can be matched with egress c (that is, the first-arriving packet in B1 will be served during that timeslot), while ingress b is simultaneously matched with egress d; or else ingress a is matched with egress d, and ingress b with egress c.

During each timeslot, the system manager may choose between these two matchings, taking into consideration the backlogs accumulated in the four virtual buffers. Of course, it will not always be possible to fully exploit the capabilities created by the choice of a particular matching. Suppose, for example, that the first matching described is chosen, and that B1 is empty; then no packet will be transferred from ingress a or to egress c during the timeslot. Similarly, if both B1 and B3 are empty, then at most one packet can be transferred during the timeslot, regardless of which matching is chosen.

One way to describe a 2×2 data switch in the language of activity analysis is to identify each ingress and each egress as a different "server," and to identify four different activities, each of which consists of serving a packet from a particular virtual buffer (that is, transferring a packet from a particular ingress to a particular egress). Taking that view, each activity operates on a single input packet of a particular class and requires the simultaneous use of two particular servers, giving us another example of simultaneous resource possession.

Alternatively, one may view the entire switch as a single "server," and define a different activity for each set of virtual buffers that can be served simultaneously during a timeslot, such as $\{1,3\}$, $\{2,4\}$, $\{1\}$, and $\{3\}$. Taking that view, all processing activities involve a single server, but some of them operate on multiple input packets simultaneously. Chapter 12 is devoted to slotted-time models (that is, discrete-time models) that generalize this example in various ways.

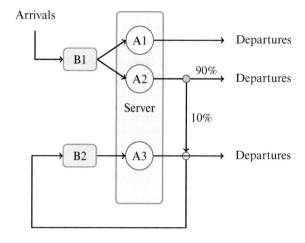

Arrivals

Figure 1.6 System with alternative processing modes.

Example with alternative processing modes. Figure 1.6 pictures a single-server system with a single external input process. Newly arrived jobs are designated as class 1, and each of them can be processed in one of two ways: the server can employ a slow but steady processing mode (this is designated as activity 1, denoted A1 in the diagram), or can employ a fast but error-prone processing mode (activity 2, denoted A2). Services of the former kind have a longer mean duration and are always successful, whereas those of the latter kind have a 10% chance of failing, after which the job is designated as class 2 and must return for a later "rework" operation (activity 3). Assume for concreteness that the external arrival process is Poisson and all three service time distributions are exponential.

Let us denote by T_1 the service time of a job that is processed via A1, and by T_2 the *total* service time required by one that is processed via A2, including the rework operation if it is needed. A plausible scenario is that where T_2 has a smaller mean but larger variance than T_1. Let us suppose that is the case. If the system manager wants to minimize the long-run average number of jobs in the system, is it true that he or she should either use A1 to process all new arrivals or else use A2 to process all new arrivals, irrespective of system status? Perhaps surprisingly, the answer is negative: there exist parameter values for which an optimal control policy uses both processing modes, with the choice depending on how many jobs are waiting to be processed.

1.5 Stability

In this book, we focus exclusively on *time-invariant* operating environments, which means that resource capacities and first-order system data (arrival rates, mean processing times, and routing probabilities) remain constant over time. In such an environment, the most basic question is that of *stability*: does the system achieve statistical equilibrium over the long run? In choosing a dynamic control policy (that is, a policy that routes new arrivals and schedules processing activities dynamically, based on current system status), one wants to first ensure stability, after which attention can be directed to more refined aspects of performance, such as the long-run average number of jobs in the system. We now consider several alternative notions of "stability" for stochastic systems.

Positive recurrence. Consider a single-server processing system in which arrivals occur according to a renewal process, service times are i.i.d. and independent of the arrival process, and arriving jobs are served in the order of their arrival. This is commonly called a $G/G/1$ queueing system, or $G/G/1$ queue; the first G in that designation means that a general distribution of interarrival times is allowed, and the second one means that a general distribution of service times is allowed. A special case of interest is a so-called $M/M/1$ system, in which the arrival process is assumed to be Poisson (that is, interarrival times are exponentially distributed), and the service time distribution is also exponential. (The letter M in that designation may be viewed as mnemonic for either "memoryless" or "Markov.")

Let us denote by $Z(t)$ the number of jobs present in an $M/M/1$ system at time t, assuming for concreteness that $Z(0) = 0$. It follows from the memoryless property of the exponential distribution that the job count process $Z = \{Z(t), t \geq 0\}$ is a continuous-time Markov chain (CTMC), and a natural notion of stability for such a model is *positive recurrence*; that is, we say that the $M/M/1$ system is stable if Z is positive recurrent.

For an $M/M/1$ queueing system, and for a broader but still very restrictive class of Markovian queueing networks, one can prove stability (that is, prove positive recurrence) by explicitly solving for the stationary distribution of the job count process. For more complex Markovian models, however, one must establish stability by indirect means, and the Foster–Lyapunov method is the overwhelmingly dominant tool for that purpose. Using that method effectively is an art that often involves creativity. To be more specific, it requires the construction of a Lyapunov function that is suitable for the case at hand.

Positive Harris recurrence. For the $G/G/1$ queueing system just described, the job count process Z is not in general a Markov process. A Markov representation of the system can be achieved by adding to Z certain "supplementary variables," but the resulting Markov process $X = \{X(t), t \geq 0\}$ does not have a discrete (countable) state space. Such a construction will be spelled out in Section 3.8, where the result is a continuous-time Markov process X that lives in the uncountable space $\mathbb{Z}_+ \times \mathbb{R}_+ \times \mathbb{R}_+$. In this case, one can define system stability as *positive Harris recurrence* of X. There is a Foster–Lyapunov criterion for proving positive Harris recurrence, but the success of that approach with nondiscrete state spaces has been limited. Even in the case of a so-called generalized Jackson network, which has a simple structure but general interarrival and service time distributions, it took many years before a Foster–Lyapunov proof of stability was devised by Meyn and Down (1994), and that proof is by no means simple.

Phase-type distributions. To circumvent the technical complexities that arise with general state spaces, we shall restrict attention in this book to stochastic structures and distributional families that allow a *discrete* (but possibly complicated) Markov representation of the system under study. In particular, we assume that service times have phase-type distributions. This modeling choice allows an elementary and self-contained mathematical exposition with very little sacrifice in terms of practical relevance, since phase-type distributions are known to be dense in the space of nonnegative distributions (see Appendix D). This feature should be attractive to students of engineering and operations research. Positive Harris recurrence will be discussed briefly in Section 3.8, mainly to emphasize that the added generality does not provide much additional insight.

Rate stability. One may generalize the $G/G/1$ queueing model by relaxing the requirement that either the interarrival times or the service times be i.i.d. In the absence of such structure, Markov process theory is not a useful framework for studying stability, but it may still be possible

to prove *rate stability* of the single-server processing system. The system
is said to be rate stable if, with probability one,

$$(1.4) \qquad \lim_{t \to \infty} \frac{Z(t)}{t} = 0$$

for any initial configuration. In words, rate stability means that the
backlog of unprocessed jobs does not accumulate at a positive rate.

To establish rate stability of a stochastic processing system, one
typically assumes that the model's primitive stochastic elements (arrival
processes, service time sequences, and routing or output sequences)
each satisfy a strong law of large numbers (SLLN). In the case of a
single-server processing system, those strong laws establish the existence
of a long-run arrival rate λ and a long-run service rate μ. If $\lambda \leq \mu$, it
can be shown that (1.4) holds; that result can be expressed verbally by
saying that the server's long-run average output rate equals the long-run
average input rate λ. In general, rate stability can be studied under the
minimal assumption that all relevant long-run averages are well defined.

Fluid model stability. For a single-server processing system, let us
assume that the long-run arrival rate λ and long-run service rate μ
are well defined and strictly positive. The corresponding *fluid model*
is a deterministic, continuous-time system defined by the following
equations and conditions, in which $T(t)$ represents the amount of time
that the server spends working (rather than idle) over the interval $[0, t]$:

$$Z(t) = Z(0) + \lambda t - \mu T(t) \quad \text{for } t \geq 0,$$
$$Z(t) \geq 0 \quad \text{for } t \geq 0,$$
$$T(0) = 0 \text{ and } 0 \leq T(t) - T(s) \leq t - s \quad \text{for } t > s \geq 0,$$
$$Z(u) > 0 \text{ for all } u \in [s, t] \text{ implies } T(t) - T(s) = t - s \quad \text{for } t > s \geq 0.$$

These relationships are valid under most control policies that are
nonidling, by which we mean policies that keep the server working so
long as there are unfinished jobs in the system. They have a unique
solution for a single-server system, but fluid model solutions are *not*
unique for most systems studied in this book. *A fluid model is said to
be stable* if there exists a constant $\gamma > 0$ such that, for any fluid model
solution $Z = \{Z(t), t \geq 0\}$, one has $Z(t) = 0$ for all $t \geq \gamma |\hat{Z}(0)|$.

There are two primary motivations for studying fluid models
and their stability. (a) Fluid model stability is sufficient for positive

recurrence, positive Harris recurrence, and rate stability, provided that the system primitives satisfy certain natural conditions; proving and exploiting that linkage is the main focus of this book. Moreover, fluid model stability is often necessary for stochastic model stability, or very nearly so. Consider, for example, a single-server processing system. The condition

$$(1.5) \qquad\qquad \lambda < \mu$$

is necessary and sufficient for positive recurrence in the $M/M/1$ case, necessary and sufficient for positive Harris recurrence in the $G/G/1$ case, and necessary and sufficient for fluid model stability. It is also a sufficient condition for rate stability. In the "critical" case where $\lambda = \mu$, the system is not positive recurrent or positive Harris recurrent, but it is still rate stable, and the fluid model is *weakly stable*, meaning that if $Z(0) = 0$, then $Z(t)$ remains at zero for all $t \geq 0$. (b) In addition to providing a unified framework for the study of system stability, a fluid model is more easily analyzed than its discrete counterpart, thanks to the elegant and powerful fluid calculus that will be developed in Chapter 8. Constructing an appropriate Lyapunov function in a fluid model context typically requires some ingenuity, and sometimes requires a great deal of it, but doing so in a discrete stochastic model context is invariably more difficult, often much more difficult. When Lyapunov functions can be found for both the fluid and discrete models, the Lyapunov function for the former is often simpler and more elegant. For some stochastic processing networks, Lyapunov functions have been found for the corresponding fluid model but not for the discrete stochastic model itself.

Necessary conditions for stability.   Throughout the remainder of this section, we use the term "stability" to mean positive recurrence, restricting attention to system models that can be represented as CTMCs. (That restriction involves both the primitive stochastic elements of the model, including arrival processes and service time distributions, and the control policy that is assumed. See Chapter 3 for elaboration.) Of course, stability is not always achievable. For multiclass queueing networks, such as the two examples discussed earlier in Section 1.3, an obvious necessary condition is that the long-run average load imposed on each resource by the external arrival processes be less than the resource's capacity. Conditions (1.1) and (1.2) express that requirement for our two queueing network examples. For a general stochastic processing network, we develop an analogous condition in Chapter 5, expressing it solely in terms of the first-order data referred to in the first sentence of this section.

Maximally stable control policies. Assuming that the necessary condition is satisfied, how does one synthesize a control policy that achieves long-run stability? The examples provided in the next section show that reasonable looking policies may fail to do so. In Chapter 9, however, we consider the *back-pressure policy* propounded by Tassiulas and Ephremides (1992), and show the following: if the system's first-order data satisfy the necessary condition referred to in the previous paragraph, and if certain additional restrictions are met, then the SPN is stable under the back-pressure policy. Because the class of models embraced by that result is so extremely broad, the control policy is necessarily expressed in rather abstract terms: it chooses at each point in time a vector of activity levels that solves a linear optimization problem, the objective coefficients of which involve current buffer contents.

The result described in the previous paragraph is one that establishes, under certain restrictions, the *maximal stability* of the back-pressure policy: that is, the back-pressure policy achieves stability if *any* policy can do so. In addition to that general result, we prove the maximal stability of various other control policies that are tailored to particular network structures. For example, it is shown in Section 8.3 that any nonidling control policy is maximally stable in a feedforward queueing network, and in Chapter 10 that *proportionally fair* capacity allocation is maximally stable in a bandwidth sharing network.

1.6 Illuminating Examples of Instability

In this section, we present three SPN examples, each of which has the following characteristics: (a) analysis of first-order data indicates that resource capacities are adequate to handle the load imposed by external arrivals; and (b) a control policy is specified that looks reasonable, even attractive; but (c) simulation results indicate that the system is nonetheless unstable under the specified policy. These examples show that stability questions can be subtle, thus motivating material to come in future chapters.

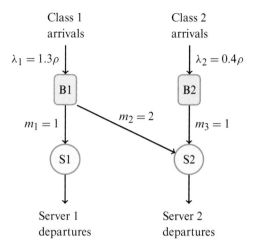

Class 1   Class 2
arrivals   arrivals

$\lambda_1 = 1.3\rho$ $\lambda_2 = 0.4\rho$

B1   B2

$m_1 = 1$ $m_2 = 2$ $m_3 = 1$

S1   S2

Server 1   Server 2
departures   departures

Figure 1.7 N-model example.

***N*-model example.** The stochastic processing network portrayed schematically in Figure 1.7 has two servers and two job classes, with members of each class residing in their own dedicated buffer. Jobs of class i arrive according to a Poisson process at an average rate of λ_i per minute $(i = 1, 2)$, and the two input streams are assumed to be independent. As indicated in Figure 1.7, we take the average arrival rates to be $\lambda_1 = 1.3\rho$ and $\lambda_2 = 0.4\rho$, where ρ is a parameter to be specified. Each job requires a single service before it departs, and class 1 can be processed by either server 1 or server 2, whereas class 2 can be processed only by server 2. (This is called an *N*-model, because of the pattern formed by the arrows that indicate routing options in Figure 1.7.) It is not necessary to decide which server will process a class 1 job until the service actually begins. For concreteness, we assume that a service must be completed without interruption once it has begun. We identify three different processing activities, as follows:

Activity 1 = processing of class 1 jobs by server 1
Activity 2 = processing of class 1 jobs by server 2
Activity 3 = processing of class 2 jobs by server 2

To be concrete, let us assume that service times for activity j are exponentially distributed with mean m_j minutes, where $m_1 = m_3 = 1$ and $m_2 = 2$, as shown in Figure 1.7. That is, it takes an average of one minute for server 1 to process a job of class 1 or for server 2 to process a job of class 2, but it takes an average of two minutes for server 2 to process a job of class 1. To assess the adequacy of available capacity, let us consider the case $\rho = 1$. Devoting all its time to class 1, server 1 can handle only one of the 1.3 class 1 jobs that arrive per minute, and server 2 must spend 60% of its time processing the class 1 jobs that are left over (0.3 jobs per minute $\times$ 2 minutes per job = 0.6). Server 2 has 40% of its time left for class 2 jobs, which is just adequate to handle the arrival rate of $\lambda_2 = 0.4$. Thus it appears that stability is achievable (that is, capacity is more than adequate to handle the offered load) if ρ is strictly less than 1. To complete the model specification, let us assume that holding costs are continuously incurred at a rate of h_i dollars per minute for each class i job that remains within the system, with the specific numerical values

$$(1.6) \qquad\qquad h_1 = 3 \quad \text{and} \quad h_2 = 1.$$

Each time a service is completed, the system manager must decide which job class the newly freed server will process next, if there is in fact a choice, or the manager can choose to keep the server idle if he or she chooses; a similar control decision must be made each time a new job arrives to find one or both servers idle. In making these control decisions, the manager strives to minimize holding costs. Let us denote by $\mu_j = 1/m_j$ the average service rate for activity j, and further define

$$(1.7) \qquad c_1 = h_1 = 3, \quad c_2 = h_1 = 3, \quad \text{and} \quad c_3 = h_2 = 1.$$

Thus c_j is the holding cost rate for the job class that is served in activity j. When the system contains z_1 jobs of class 1 and z_2 jobs of class 2, holding costs are continuously incurred at a total rate of $h_1 z_1 + h_2 z_2$, and activity j decreases this total at an average rate of $c_j \mu_j$ per minute spent in the activity.

The following is a "greedy" control policy that makes each decision so as to drive down as fast as possible the total holding cost rate. First, server 1 should spend as much time as it can processing class 1, going idle only when buffer 1 is empty. Second, when server 2 completes the processing of a job and finds new jobs waiting in both buffer 1 and buffer 2, it should choose between activities 2 and 3 so that the chosen activity j maximizes $c_j \mu_j$. (If there are waiting jobs in only one buffer, the server should take one of them, and if both buffers are empty it

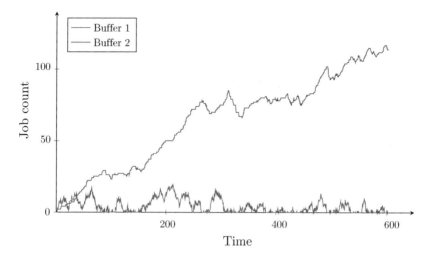

Figure 1.8 Job counts in the N-model with greedy control ($\rho = 0.95$).

must go idle.) Since $c_2\mu_2 = 3/2$ and $c_3\mu_3 = 1$, this means that server 2 will always choose to serve class 1 when confronted with a choice.

System behavior under the greedy control policy has been simulated for the case $\rho = 0.95$, and the results are presented in Figure 1.8, which shows the simulated job counts for class 1 and class 2 (including any jobs that may be in service) as a function of time. One sees from this plot that the greedy rule is disastrously ineffective. In fact, the system fails to achieve a statistical equilibrium: the number of class 2 jobs grows in roughly linear fashion as a function of time.

Upon reflection, the source of this instability becomes obvious. When server 2 gives priority to class 1 jobs, it has too little leftover capacity to handle the flow of class 2 jobs, and it also leaves server 1 with no work to do some of the time. (In 5,000 minutes of simulated operation, server 2 spent 71% of its time on class 1 jobs, while server 1 experienced 13% idleness.) To achieve stability, one must balance the desire for immediate cost reduction with a concern for efficient use of available capacity.

Rybko–Stolyar example. Figure 1.9 pictures a multiclass queueing network made famous by Rybko and Stolyar (1992). It has two servers, which will occasionally be referred to as single-server processing *stations*, and four job classes. Jobs of class 1 arrive according to a Poisson processes at rate $\lambda_1 = 1$ and are processed by server 1 (hereafter

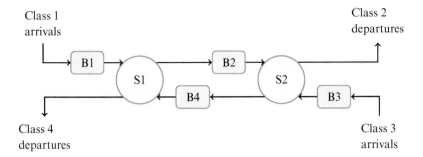

Class 1
arrivals

Class 2
departures

Class 4
departures

Class 3
arrivals

Figure 1.9 The Rybko–Stolyar network.

abbreviated S1). When that service is complete, they become class 2 jobs and are processed by server 2 (similarly abbreviated S2), after which they leave the system. In similar fashion, class 3 jobs arrive according to a Poisson process at rate $\lambda_3 = 1$, are processed by S2, then transition to class 4 and are processed by S1, after which they leave the system. Service times for class i jobs are exponentially distributed with mean $m_i > 0$ ($i = 1, \dots, 4$). Each server works on jobs one at a time, and the service of a job must be completed without interruption once it has begun. Given that $\lambda_1 = \lambda_3 = 1$, the load factor for S1 (that is, the expected amount of time required from S1 to complete the processing of all jobs that arrive within one time unit) is $m_1 + m_4$, and the corresponding figure for S2 is $m_2 + m_3$, so the obvious analog of the stability conditions (1.1) and (1.2) is the following:

(1.8) $$ m_1 + m_4 < 1 \quad \text{and} \quad m_2 + m_3 < 1. $$

Suppose that the system manager, seeking to minimize the total number of jobs in the system, adopts a greedy strategy in pursuit of that objective, meaning that each control decision is made so as to drive down as quickly as possible the total system job count. Thus S1 gives priority to class 4 over class 1, because jobs of the former class leave the system immediately upon completion of their impending service, whereas those of the latter class must complete another service before exiting; in similar fashion, the greedy policy dictates that S2 give priority to class 2 over class 3.

Figure 1.10 presents simulation results using the greedy control policy and the following mean service times: $m_1 = 0.3$, $m_2 = 0.6$, $m_3 = 0.3$, and $m_4 = 0.6$. Thus the simulation focuses on a symmetric case where

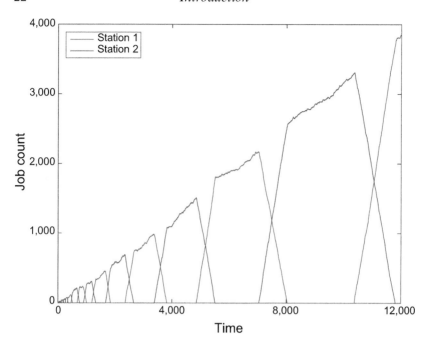

Figure 1.10 Job counts at the two stations of the Rybko–Stolyar network.

newly arrived jobs have relatively short service times and those about to exit have relatively long service times. Condition (1.8) is satisfied, but Figure 1.10 shows that the system is nonetheless unstable: job counts at each station build and dissipate in a cyclic fashion, with the cycles getting longer and the build-ups getting larger as time passes. This result is even more surprising than the instability of our N-network example under greedy control, because the current example involves so little control discretion: there are no decisions to be made about which server will do which kinds of work, only decisions about the *order* in which jobs will be processed from the two buffers for which a server is responsible. The greedy policy is nonidling (that is, it never holds a server idle when there is work available for that server to do), and (1.8) suggests that each server has enough capacity to handle its responsibilities, and yet the system spirals out of control.

The instability evident in Figure 1.10 derives from a phenomenon called *mutual blocking*, which is most easily seen in the following modified version of the Rybko–Stolyar model. Let us suppose that servers 1 and 2 give *preemptive-resume* priority to classes 4 and 2, respectively. This means, for example, that if S1 is processing a class 1 job at the moment of a class 4 arrival (due to a class 3 service completion by S2), then S1 will put aside the class 1 job, proceed with service of the newly arrived class 4 job, and return to the interrupted class 1 service only when it has emptied buffer 4 (hereafter abbreviated B4, and similarly for other buffers). We assume in this variant of the model that the class 1 service is resumed at precisely the point where it was interrupted, with no efficiency loss. That is, the total service time required by a class 1 customer is the same, no matter how many times or for how long its service may be interrupted.

In the preemptive-resume version of the Rybko–Stolyar model, let us suppose that both B2 and B4 are initially empty. We denote by $\tau_1 > 0$ the first time at which a job of either class 2 or class 4 is created, those being the two "exit classes" that are given priority by the greedy policy. For concreteness, suppose that a class 2 job is created first, so it begins service at time τ_1. (There is a precisely similar story in the opposite case.) B4 is thus empty at time τ_1, and it will remain so until the first time $\tau_2 > \tau_1$ at which B2 is emptied, because S2 is continuously occupied with serving class 2 jobs over the interval $[\tau_1, \tau_2]$, which means that no class 3 services can be completed over that interval, and it is class 3 service completions that create class 4 arrivals.

To repeat, B4 remains empty over the interval $[\tau_1, \tau_2]$, during which S2 is continuously occupied with class 2 services. Both B2 and B4 will be empty at time τ_2, and then the whole story will repeat: there will be an interval (τ_2, τ_3) during which both B2 and B4 remain empty, followed by an interval $[\tau_3, \tau_4]$ during which exactly one of them will be empty, followed by an interval (τ_4, τ_5) during which both of them will be empty, and so on *ad infinitum*. Thus we see that *class 2 jobs and class 4 jobs can never be served simultaneously*. Given that $\lambda_1 = \lambda_3 = 1$, and that each class 1 arrival eventually necessitates one class 2 service, and that each class 3 arrival eventually necessitates one class 4 service, we conclude that the following must be added to (1.8) as a necessary condition for stability:

$$(1.9) \qquad\qquad m_2 + m_4 < 1.$$

The numerical values assumed for our simulation study included $m_2 = m_4 = 0.6$, so (1.9) is *not* satisfied: one simply cannot complete all class 2 and class 4 services using the alternating cycles previously

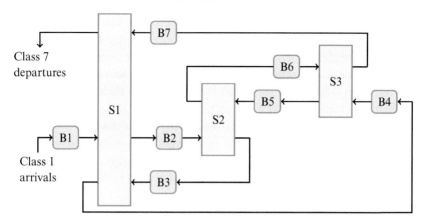

Figure 1.11 A FIFO reentrant line.

described, which leads to the increasingly long cycles and increasingly large build-ups that we see in Figure 1.10. Of course, the stability condition (1.9) was derived for the preemptive-resume version of the Rybko–Stolyar model, whereas Figure 1.10 shows simulation results for the original (nonpreemptive) version, but the large-scale behavior of the two models is the same.

A FIFO reentrant line. Figure 1.11 pictures a multiclass queueing network with three single-server processing stations and a single external arrival process. Each arriving customer follows the same deterministic route, receiving a total of seven services at various stations, as indicated in the figure, and then leaving the system. Models like this, with a single input flow, deterministic routing, and jobs revisiting some servers as they progress through their routes, were called *reentrant lines* by Kumar (1993), whose study was motivated primarily by semiconductor manufacturing applications. Customers seeking the ith service on their route are designated as class i. As usual, we denote by $\lambda_1 > 0$ the external arrival rate into buffer 1, and by $m_i > 0$ the mean service time for class i jobs. We shall assume the following numerical values: $m_1 = 0.1, m_2 = 0.5, m_3 = m_6 = 0.05, m_4 = 0.9, m_5 = 0.45, m_7 = 0.8$, and $\lambda_1 = 1$.

For simplicity, we assume the external arrival process to be Poisson and all seven service time distributions to be exponential. We also

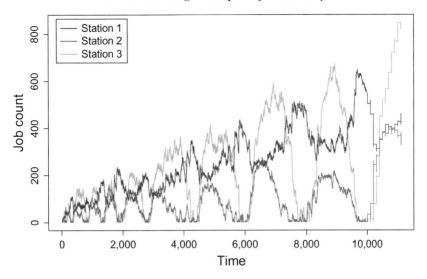

Figure 1.12 Job counts for the FIFO reentrant line.

assume FIFO control, which means that jobs at each station are pro-cessed in the order of their arrival *to their current buffer* (as opposed to the order of their arrival from the outside world, for example). To ensure that each server has adequate capacity, we restrict attention to parameter values satisfying the following conditions:

$$(1.10) \quad \lambda_1(m_1 + m_3 + m_7) < 1, \quad \lambda_1(m_2 + m_5) < 1,$$
$$\text{and} \quad \lambda_1(m_4 + m_6) < 1.$$

The quantities on the left sides of these three inequalities are the load factors for server 1, server 2, and server 3, respectively. With the numerical values assumed previously, the left side of each inequality is 0.95, so we expect stability.

The simulation results presented in Figure 1.12 show that is not the case. Rather, one sees there a cyclic pattern with increasingly long cycles and increasingly large build-ups, as in our earlier analysis of the Rybko–Stolyar example. This suggests that FIFO control somehow produces a "mutual blocking" phenomenon in the reentrant line, similar to what was explained earlier for the Rybko–Stolyar model with preemptive-resume priorities, but the interactions in the current example are more complex and subtle.

1.7 Structure of the Book and Intended Audience

Chapter 2 lays out our general formulation of an SPN and describes special cases of interest. There we also define what are called "simply structured" control policies, on which attention will be focused in later chapters. In Chapter 3, we impose additional assumptions that allow our study of SPN stability to be undertaken within the framework of CTMC theory. Chapter 4 describes several important extensions of our basic theoretical framework, which are suppressed in Chapters 2 and 3 to simplify the exposition, and elaborates on that framework in various ways.

Chapter 5 concerns the necessary condition for stability that was mentioned in Section 1.5, along with several related topics. Chapters 6 through 8 develop the basic fluid model methodology, and demonstrate its relevance for the study of SPN stability. Chapter 9 focuses on max-weight and back-pressure control polices, while Chapter 10 treats policies based on the notion of "proportional fairness," or "proportionally fair" resource allocation; in each case, it is shown that policies of the kind being studied are maximally stable given certain restrictions on network structure.

Our last two chapters have a somewhat different character, each being focused on an application domain that has risen to prominence in the digital age. Chapter 11 is concerned with workflow management in server farms, or data centers, with particular emphasis on the issue of data locality, and Chapter 12 is devoted to specially structured models of digital communications, exemplified by the 2×2 data switch discussed earlier in Section 1.4. In addition to those application-specific final chapters, there is a substantial treatment of another digital-age topic, commonly referred to as *bandwidth sharing*, embedded in earlier chapters (specifically, see Sections 4.5 and 8.2).

Target audience. This book is aimed at researchers and graduate students in engineering and applied mathematics, especially electrical engineering and computer science, where stochastic network models generally, and stability analysis specifically, have become mainstream topics over the last 20 years. As mathematical background, we assume that readers know the basics of real analysis, stochastic process theory, and linear optimization. Our continuous-time models are based on phase-type distributions and Poisson arrival processes (later generalized to Markovian arrival processes), so all stochastic processes encountered in the book are Markov chains with either a discrete or continuous time parameter.

It should be acknowledged at the outset that parts of this book are mathematically demanding. To be specific, the following three segments are likely to test the patience of readers who are unaccustomed to mathematical formalism.

(i) To achieve a suitable level of generality, the development of our SPN framework in Sections 2.1 through 2.4 is necessarily abstract. In particular, our initial definition of a "control policy" is expressed in terms far removed from practical engineering language. In later chapters, however, attention is focused on specially structured policies described in operationally meaningful terms.

(ii) An elaborate mathematical setup is required to define fluid limits in Section 6.4, and the associated notation in Sections 6.3 and 6.4 is daunting. In particular, fluid limits are defined in the almost sure sense, and a full understanding requires a constant awareness of the probability space underlying the model under study. Here again, the reader's tolerance for mathematical abstraction will be tested.

(iii) The most difficult proofs of fluid model stability involve lengthy forays into real analysis, with a profusion of deltas and epsilons seldom seen in applied work. In this book, the apotheosis of that phenomenon occurs in the proof of Theorem 10.5 (Section 10.5), where we study proportionally fair resource allocation using an entropy Lyapunov function. Some readers will want to skip that lengthy and elaborate argument, but those seeking mastery of fluid-based methods will find value in its study.

For ease of reference, we have included appendices that recapitulate essential results from real analysis, probability, and Markov chain theory, plus the basic theory of phase-type distributions and Markovian arrival processes. Most chapters end with a section on sources and literature, in which we identify the original research on which our treatment is based, and in some cases give pointers to books and articles that provide useful background or develop the topic more fully.

1.8 Sources and Literature

The general notion of a stochastic processing network propounded in this book is a somewhat restricted version of the framework developed in a sequence of papers by Harrison (1988, 2000, 2003); the restrictions will be explained in the final paragraph of Section 2.1. Section 1.2 is adapted from a related expository article by Harrison (2002).

The crisscross network (Figure 1.2) first appeared in a paper by Harrison and Wein (1989), and the alternate routing model in Figure 1.3 was introduced by Wein (1991). The three examples of unstable networks that we discuss in Section 1.6 are drawn from the work of Harrison (1998), Rybko and Stolyar (1992), and Dai and Wang (1993), respectively. Lu and Kumar (1991), Rybko and Stolyar (1992), and Bramson (1994) are among the pioneering papers that stimulated a wave of stability research in the 1990s, of which this book is an outgrowth.

2

Stochastic Processing Networks

This chapter develops a general model of a stochastic processing network (SPN), filling out the sketchy picture provided in Chapter 1. Actually, we develop two such models, called the *basic* and *relaxed* formulations. They will be described in Sections 2.3 and 2.4, respectively, after elements common to the two models have been laid out in Sections 2.1 and 2.2. In each of our two model contexts, attention will be focused on what we call *simply structured* control policies, but the analytical framework developed in this chapter and the next one can accommodate policies of a more general form; a policy that is *not* simply structured will be analyzed later in Section 7.3.

Units of flow are initially called "items" in this chapter, but are later called "jobs" in contexts where that term seems more natural. Also, as a matter of linguistic convention, we imagine each item class as having its own dedicated storage buffer; the terms "class" and "buffer" are used essentially as synonyms throughout this book. As in Chapter 1, an item of any given class is viewed as occupying that class's buffer regardless of whether it is waiting or being processed. What we call a "buffer" might better be described as a "virtual buffer," because what matters is that items of different classes are logically distinguished from one another, regardless of whether they are physically separated. See Section 2.7 for elaboration on the important modeling concept of an item class, or job class.

2.1 Common Elements of the Two Model Formulations

Each of our models has I buffers (or item classes), J activities (also called *service types*), and K resources. For notational convenience, let $\mathscr{I} = \{1, \ldots, I\}$, $\mathscr{J} = \{1, \ldots, J\}$, and $\mathscr{K} = \{1, \ldots, K\}$.

Capacity vector b. There is given a strictly positive K-vector $b = (b_k)$, elements of which have different interpretations in our basic and relaxed model formulations. In the basic SPN model (Section 2.3), components of b are positive integers, and by assumption, resource k is a pool of b_k identical, interchangeable servers ($k \in \mathcal{K}$). In the relaxed SPN model (Section 2.4), resource k is by assumption a single server (or perhaps several servers with fungible capacity, which are therefore functionally equivalent to a single server), and b_k is the capacity of server k (a positive number, not necessarily integer), expressed in units like megabytes per second or gallons per hour ($k \in \mathcal{K}$).

Services. We shall speak in terms of a *system manager* who undertakes processing activities in a centralized fashion, but that is just a convenient expository device: many of the control policies to be considered in this book are in fact amenable to decentralized implementation, and that consideration will be explicitly discussed at various points. The only role of the system manager in our basic model (Section 2.3) is to initiate *services* of types $1, \ldots, J$, each of which requires certain inputs and certain resources, as specified in this chapter. The system manager also initiates services in our relaxed formulation, but then has decisions to make about allocation of service effort; see Section 2.4 for elaboration.

A service is said to be *open* at a given time if (a) it was initiated at or before that time, but (b) it has not been completed by that time. In our basic model, a synonym for "open" might be "ongoing," but the latter term is potentially misleading in our relaxed formulation, where partially completed services can be suspended.

Initial system state. We take as given an I-vector $Z(0)$ and a J-vector $N(0)$, both deterministic, that have nonnegative integer components $Z_1(0), \ldots, Z_I(0)$ and $N_1(0), \ldots, N_J(0)$, respectively; one interprets $Z_i(0)$ as the number of class i items that reside in buffer i at time zero, and $N_j(0)$ as the number of type j services that are open at time zero. In addition to $Z(0)$ and $N(0)$, the initial state of our model may include remaining service times and corresponding output vectors for services that are open at time zero; see elaboration later in this chapter.

It is important for our purposes to allow a general initial state (see Chapter 6, especially Section 6.3), but to grasp the main ideas with a minimum of distractions, readers may focus initially on the special case where the initial state is given by $Z(0)$ and $N(0)$ only.

Core stochastic elements. There is given a right-continuous, I-dimensional *external arrival process* $E = \{E(t), t \geq 0\}$ with components E_i that are nondecreasing and integer-valued with $E_i(0) = 0$, $i \in \mathscr{I}$. We allow batch arrivals in our general formulation, including batches that contain items of several different classes, but assume that arrival batches all come from some finite space Δ_0. That is, the I-dimensional arrival process E is a pure jump process whose jumps all belong to a finite set Δ_0. The standard case in queueing theory is that where all arrivals are singletons, so Δ_0 is a subset of $\{e^1, \ldots, e^I\}$. (Here e^i denotes the I-vector with a 1 as its ith component and all other components equal to zero.)

There are also given, for each $j \in \mathscr{J}$, strictly positive *service times* $\{v_j(\ell), \ell = 1, 2, \ldots\}$ and I-dimensional *output vectors* $\{\varphi^j(\ell), \ell = 1, 2, \ldots\}$ that are nonnegative and integer-valued, taking values in some finite set Δ_j. Service times and output vectors will be referred to collectively as *processing variables*. The *core stochastic elements* of an SPN model are its external arrival process E and the processing variable sequences introduced in this paragraph.

For the basic SPN formulation developed in Section 2.3, one can interpret $v_j(\ell)$ simply as the time required for completion of the ℓth-type j service initiated after time 0, but for the "relaxed" formulation developed in Section 2.4, it is more accurately described as the *cumulative service effort* required for completion of that service. For example, $v_j(\ell)$ can be the size (in megabytes, say) of the ℓth-type j file that needs to be transferred over a communication link. The time to complete the transfer depends on the bandwidth (a portion of the link capacity) that is allocated dynamically to the file.

Components of the random I-vector $\varphi^j(\ell)$ are denoted $\varphi_1^j(\ell), \ldots, \varphi_I^j(\ell)$, and we interpret $\varphi_i^j(\ell)$ as the number of class i items that are created upon completion of the ℓth type j service. For concreteness, let us agree that the numbering of services of any given type corresponds to the order in which those services are *begun*. As we shall see, two or more services of the same type may be conducted simultaneously, so services of a given type need not finish in the same order they were begun.

By assumption, the core stochastic elements of our SPN model are exogenously specified (that is, unaffected by the system manager's decision making), but up to now nothing has been said about their joint distribution. In the next section, we shall impose strong distributional assumptions that remain in force unless something is said to the contrary, but those "baseline stochastic assumptions" are motivated in large part by expositional considerations, and extensions that allow more general stochastic structures will be discussed in Section 4.1.

Initial processing variables.   The processing variables introduced earlier in this section, which are indexed by $\ell = 1, 2, \ldots$, correspond to services that are initiated *after time zero*. In addition, there are $N_j(0)$ type j services open at time zero, and those will be indexed by $\ell = -N_j(0) + 1, \ldots, -1, 0$. (This notation is admittedly ungainly, but every alternative seems to be worse.) To simplify typography now and later, we define

$$(2.1) \qquad \mathscr{L}_j^0 := \{-N_j(0) + 1, \ldots, -1, 0\},$$

(thus $\mathscr{L}_j^0$ is the empty set if $N_j(0) = 0$), and then set

$$(2.2) \qquad \mathscr{L}_j := \mathscr{L}_j^0 \cup \{1, 2 \ldots\}.$$

If $N_j(0) > 0$, then there are given strictly positive random variables $v_j(\ell)$ and non-negative, integer-valued random I-vectors $\varphi^j(\ell)$ for $\ell \in \mathscr{L}_j^0$; in the obvious way, $v_j(\ell)$ is interpreted as a *remaining* service time, or residual service time, for nonpositive integers ℓ, and $\varphi^j(\ell)$ is the corresponding output vector. Remaining service times for services ongoing at time zero, together with their associated output vectors, will be referred to as the *initial processing variables* of the SPN model. For later purposes (see specifically Section 6.3), we establish the following notation for the complete set of initial processing variables:

$$(2.3) \qquad \Psi := \{(v_j(\ell), \varphi^j(\ell)), j \in \mathscr{J}, \ell \in \mathscr{L}_j^0\}.$$

Hereafter, Ψ will be called the model's *IP set*, and the pairs of initial processing variables (v, φ) that belong to Ψ will be called *IP pairs*. We assume that the IP pairs are mutually independent, although dependence is allowed between the components v and φ of any one such pair, and are also independent of the core stochastic elements introduced previously.

Resources and inputs required for processing activities.   There are given a $K \times J$ *server requirements matrix* $A = (A_{kj})$ and an $I \times J$ *material requirements matrix* $B = (B_{ij})$, both of which are binary (that is, each element is either 0 or 1). In each case, the value $B_{ij} = 1$ is interpreted to mean that an item from buffer i is required for execution of a type j service. If a column of B contains more than a single 1, it means that the corresponding service type requires input items from two or more buffers, as occurs in the modeling of assembly and matching operations.

In our basic SPN model, the value $A_{kj} = 1$ is interpreted to mean that a server from pool k is required for execution of a type j service. In the relaxed SPN model, where we have only single-server pools, the value $A_{kj} = 1$ similarly means that execution of a type j service involves server k. However, in the relaxed setting a type j service can be conducted at a variable rate $\beta_j \geq 0$, and the amount of server k capacity consumed varies in proportion to that service rate.

In either the basic or relaxed model context, an activity is said to involve *simultaneous resource possession* if the corresponding column of A contains two or more positive elements. To avoid trivialities, we assume that no column of either A or B consists entirely of zeros.

In our basic SPN model (Section 2.3), the associated servers are committed throughout the service time, being released only when the service is complete, but in our relaxed formulation the situation is more complex; see Section 2.4. In both of our formulations, the input items required for a service are committed from the moment of its initiation to the moment of its completion, and they disappear from their respective buffers upon its completion; also, at the moment when the ℓth-type j service is completed, $\varphi_i^j(\ell)$ "new" items appear in buffer i ($i \in \mathscr{I}, j \in \mathscr{J}, \ell = 1, 2, \ldots$). (In most application contexts, a service completion does not really create any "new" items; rather, an "old" item moves from one buffer to another, receiving a new class designation in the process.) The completion of a service may either decrease the total number of items in the system, as when the associated output vector has all components zero, or increase the total number if there are multiple output items.

General notation and system equations. The two model formulations described in Sections 2.3 and 2.4 differ with regard to the type of control exerted by the system manager. In both cases, however, the system manager directly determines a J-dimensional *cumulative service starts* process $S = \{S(t), t \geq 0\}$ whose jth component $S_j(t)$ is the number of type j services initiated over the time interval $(0, t]$. This process S must be right-continuous, nondecreasing, and integer-valued with $S(0) = 0$. The last of those restrictions reflects the view, adopted throughout this book, that the system state at $t = 0$ already incorporates decisions made at time zero, if any. More will be said later about the logic used to determine service starts.

For $j \in \mathcal{J}$ and $\ell = 1, 2, \ldots$ we denote by $\tau_j(\ell)$ the time at which the ℓth type j service is initiated. These *service initiation times* satisfy

$$(2.4) \qquad 0 < \tau_j(1) \leq \tau_j(2) \leq \cdots \quad \text{for each } j \in \mathcal{J},$$

and for completeness we set

$$(2.5) \qquad \tau_j(\ell) := 0 \quad \text{for } j \in \mathcal{J} \text{ and } \ell \in \mathcal{L}_j^0.$$

Service initiation times can be determined from cumulative service starts via the obvious relationship

$$(2.6) \qquad \tau_j(\ell) = \inf\{t \geq 0 : S_j(t) \geq \ell\} \quad \text{for } j \in \mathcal{J} \text{ and } \ell \in \mathcal{L}_j.$$

We denote by $F = \{F(t), t \geq 0\}$ the right-continuous, J-dimensional processes whose jth component $F_j(t)$ is the number of type j services completed (or *finished*) over $(0, t]$. In our basic SPN model (Section 2.3), where the system manager's only means of control is choosing when to initiate services, F is completely determined by S and the service time random variables that are taken as primitive. In our relaxed formulation, however, the timing of service completions is further influenced by allocation-of-effort decisions. Those relationships will be explicated in Section 2.4.

Let $N = \{N(t), t \geq 0\}$ be the J-dimensional process whose jth component $N_j(t)$ is the number of type j services that are open at time t. We then have

$$(2.7) \qquad N_j(t) = N_j(0) + S_j(t) - F_j(t) \quad \text{for } j \in \mathcal{J} \text{ and } t \geq 0.$$

Our next step is to construct I-dimensional processes $D = \{D(t), t \geq 0\}$ and $Z = \{Z(t), t \geq 0\}$, whose components have the following interpretations: $D_i(t)$ is the number of items *departing* from buffer i over $(0, t]$, and $Z_i(t)$ is the number of items residing in buffer i at time t. (One can think of Z as mnemonic for *zahl*, which means "number" in German.) For that purpose, it will be convenient to define the I-dimensional sums

$$(2.8) \qquad \Phi^j(n) := \varphi^j(1) + \ldots + \varphi^j(n) \quad \text{for } j \in \mathcal{J} \text{ and } n = 1, 2, \ldots,$$

so that $\Phi^j(n)$ represents the (random) I-vector of cumulative output flows created by the first n type j service completions. We denote by $\Phi_i^j(n)$ the ith component of $\Phi^j(n)$. The following two equations define

D and Z in terms of the service completion process F and primitive model elements:

$$(2.9) \qquad D_i(t) := \sum_{j=1}^{J} B_{ij} F_j(t) \quad \text{for } i \in \mathscr{I} \text{ and } t \geq 0,$$

and

(2.10)

$$Z_i(t) := Z_i(0) + E_i(t) + \sum_{j=1}^{J} \Phi_i^j(F_j(t)) - D_i(t) \quad \text{for } i \in \mathscr{I} \text{ and } t \geq 0.$$

The two processes of greatest interest in studies of system performance are N and Z. They will be called the *service count process* and *buffer contents process* (or item count process), respectively. Readers may verify that, because the external arrival process E is right-continuous by assumption, all of the other continuous-time processes we have defined, including N and Z, are right-continuous as well. In our basic SPN model (Section 2.3), constraints will be imposed on the system manager's control decisions to ensure that the following *capacity constraints* and *item availability constraints* are satisfied:

$$(2.11) \qquad\qquad AN(t) \leq b \quad \text{for all } t \geq 0,$$

and

$$(2.12) \qquad\qquad BN(t) \leq Z(t) \quad \text{for all } t \geq 0.$$

As noted earlier, (2.12) also applies in our relaxed formulation, but (2.11) is replaced by other requirements that involve service effort levels.

Limitations of the general framework. The notion of a stochastic processing network advanced in this book is one where exogenous input flows (external arrivals) are processed to completion, perhaps with delays and interruptions along the way, and then they leave. The input flows in our models are not controllable in any sense. Another important limitation of our framework is that it disallows reneging or abandonment: the jobs or items that flow through the network do not become impatient and leave of their own accord. This last feature particularly limits the value of our framework for modeling telephone call centers, where caller abandonment is a ubiquitous phenomenon. Of course, there is still insight to be gained by analyzing a call center model in which abandonment is ignored; one would like to know, for example, whether resource capacities are adequate to achieve stability even in the absence of abandonment, and if so, what policies for call routing and operator allocation are able to achieve such stability.

2.2 Baseline Stochastic Assumptions

The following probabilistic assumptions are imposed throughout the remainder of this chapter, and except as noted to the contrary, through-out the following chapters as well. They are made in the interest of tractability, and more particularly, to ensure a Markov representation of the SPN under study (see Chapter 3). In Section 4.1, it will be shown that these "baseline" assumptions can be weakened to allow a much more general external arrival process. However, doing so complicates both notation and exposition, so we shall first develop the baseline scenario and explain later how to extend it. Readers are reminded (see Section 2.1) that we use $\ell = 1, 2, \ldots$ to index services initiated *after* time zero, so the following does not restrict in any way the distribution of initial processing variables. Also, recall that we use primes to denote transposes of vectors and matrices, so Γ^j is being defined as a column vector in Assumption 2.1.

Assumption 2.1 (baseline stochastic assumptions).   (a) The external arrival processes $E_1, \ldots, E_I$ are independent Poisson processes with arrival rates $\lambda_1, \ldots, \lambda_I$, respectively; when $\lambda_i = 0$, there are no external arrivals into buffer i. (b) For each activity or service type $j \in \mathscr{J}$, the matched pairs of processing variables $\{(v_j(\ell), \varphi^j(\ell)), \ell = 1, 2, \ldots\}$ form an i.i.d. sequence with finite means

$$(2.13) \qquad \mathbb{E}(v_j(\ell)) := m_j > 0 \quad \text{and} \quad \mathbb{E}(\varphi^j(\ell)) := \Gamma^j \geq 0,$$

where $\Gamma^j = (\Gamma_{1j}, \ldots, \Gamma_{Ij})'$. (Here prime denotes transpose as usual, emphasizing that Γ^j is to be viewed as a column vector.) (c) For each $j \in \mathscr{J}$, the matched pairs of processing variables referred to in part (b) have a joint phase-type distribution, as defined in Appendix D.9. (d) The initial processing variables in (2.3), the external arrival process E, and the J different sequences of processing variables referred to in part (b) are mutually independent.

Given that Assumption 2.1 holds, it will be convenient to define the mean service rates $\mu_j := m_j^{-1}$ for $j \in \mathscr{J}$ and to denote by Γ the $I \times J$ matrix whose (i, j)th component is Γ_{ij}. Thus Γ^j is the jth column of Γ.

Assumption 2.1 allows dependence between $v_j(\ell)$ and $\varphi^j(\ell)$. That is, we do *not* require that service times and their corresponding output vectors be independent. Moreover, as noted in Appendix D.9, any joint distribution of service time and output vector can be approximated to an arbitrary degree of accuracy by a joint phase-type distribution. Thus our baseline stochastic assumptions allow very general processing variable sequences, but are restrictive with respect to external arrivals. The more general class of arrival processes referred to in the first paragraph of this section are the Markovian arrival processes defined in Appendix E.1, which can be used to approximate, to any desired degree of accuracy, any stationary marked point process.

Parts (a) and (b) of Assumption 2.1 immediately imply the following strong laws of large numbers (SLLNs) for core stochastic elements of the SPN model. Our fluid model methodology relies on these strong laws in an essential way, whereas the full force of Assumption 2.1 is *not* required in principle. To be more precise, the essential requirements for our methodology are the strong laws (2.14) and (2.15), plus Assumption 3.1 in the following chapter, which ensures that the system under study has a time-homogeneous Markov representation; our baseline stochastic assumptions, together with suitable restrictions on the control policy employed, are sufficient but not necessary for those three requirements to be satisfied.

Proposition 2.2 (SLLN for external arrivals). *For each $i \in \mathscr{I}$, with probability one,*

$$(2.14) \qquad \frac{1}{t} E_i(t) \to \lambda_i \quad \text{as } t \to \infty.$$

Proposition 2.3 (SLLN for processing variables). *For each activity or service type $j \in \mathscr{J}$, with probability one,*

$$(2.15) \qquad \frac{1}{n}\sum_{\ell=1}^{n} v_j(\ell) \to m_j \quad \text{and} \quad \frac{1}{n}\sum_{\ell=1}^{n} \varphi^j(\ell) \to \Gamma^j \quad \text{as } n \to \infty.$$

2.3 Basic SPN Model

In this basic version of our model, services cannot be interrupted once they have begun, servers work either at full capacity or not at all, and there is no concept of server sharing; that is, a processing activity either proceeds at full speed or not at all, and its associated servers and input items are unavailable for other purposes throughout the service. Thus the system manager's only role is to determine the cumulative service starts process S, or equivalently, to determine the service initiation times $\tau_j(\ell)$ defined in Section 2.1. Components of the cumulative service completion process F are then given by the following relationship, where #{·} denotes the number of elements in a finite set:

$$(2.16) \quad F_j(t) = \#\left\{\ell \in \mathscr{L}_j : \tau_j(\ell) + v_j(\ell) \leq t\right\} \quad \text{for } j \in \mathscr{J} \text{ and } t \geq 0.$$

Simply structured control policies. Throughout this book, attention is focused primarily on what we call *simply structured* control policies. For the basic SPN model, such a policy is defined by two features, the first of which is easy to state: new services are initiated only at *decision times* $t > 0$ when either an external arrival occurs or a service completion occurs. Under our baseline stochastic assumptions (Section 2.2), the probability of two service completions occurring simultaneously is zero, as is the probability of an arrival and a service completion occurring simultaneously. But even if such simultaneous events were possible, the proposed definition of a decision time would still be meaningful.

A bit more notation is needed to state the second defining feature of a simply structured policy. For an arbitrary decision time t, let us denote by n and z the service count vector and buffer contents vector, respectively, just *before* time t; that is, $n = N(t-)$ and $z = Z(t-)$. Then let $\hat{n}$ and $\hat{z}$ be the corresponding quantities after we account for whatever arrivals or service completions may occur at t, but before we account for new services that may be initiated at t. If, for example, there is a class i arrival at the decision time and there are no simultaneous service completions, then we have

$$(2.17) \qquad \hat{n} = n \quad \text{and} \quad \hat{z} = z + e^i,$$

where e^i is the I-vector with a 1 as its ith component and all other components equal to zero. Alternatively, suppose that a single service completion occurs at the decision time, the service type being $j \in \mathscr{J}$ and the associated output vector being $\delta \in \Delta_j$, and there is no simultaneous arrival. Then we have

$$(2.18) \qquad \hat{n} = n - e^j \quad \text{and} \quad \hat{z} = z - B^j + \delta,$$

where e^j denotes the J-vector with a 1 as its jth component and all other components equal to zero, and B^j is the jth column of the input requirements matrix B. The corresponding accounting relationships when two or more events occur simultaneously can be resolved sequentially, one event at a time; one can verify that the final pair $(\hat{n}, \hat{z})$ is independent of the order in which simultaneous events are resolved. (Here, an "event" is either a service completion or an external arrival.) Hereafter $\hat{n}$ and $\hat{z}$ will be referred to as the *updated service count vector* and *updated buffer contents vector*, respectively, at decision time t.

For each $j \in \mathcal{J}$, let us denote by u_j the number of type j services to be initiated at the decision time t, and let $u = (u_1, \ldots, u_J)$. To satisfy the capacity constraints and material availability constraints specified in (2.11) and (2.12), our control policy must choose u to lie in the set

$$(2.19) \qquad \mathcal{U}(\hat{n}, \hat{z}) := \left\{ u \in \mathbb{Z}_+^J : A(\hat{n} + u) \leq b, \; B(\hat{n} + u) \leq \hat{z} \right\}.$$

The second defining feature of a simply structured policy is that it chooses u as a function g of $\hat{n}$ and $\hat{z}$ (only) such that

$$(2.20) \qquad u = g(\hat{n}, \hat{z}) \in \mathcal{U}(\hat{n}, \hat{z}).$$

Completing the notational system begun, let us denote by $\tilde{n}$ and $\tilde{z}$ the service count vector and buffer contents vector, respectively, *after* we account for new services that may be initiated at t. Then we have

$$(2.21) \qquad \tilde{n} = \hat{n} + g(\hat{n}, \hat{z}) \quad \text{and} \quad \tilde{z} = \hat{z}.$$

Construction of sample paths. Suppose we are given a function g that defines a simply structured control policy via (2.20), plus initial data $N(0)$ and $Z(0)$, plus realizations of the processing variables $\{v_j(\ell)\}$ and $\{\varphi^j(\ell)\}$ for both services ongoing at time zero and services initiated thereafter, plus the sample path of the arrival process E. It is then a simple matter to construct the corresponding sample paths of all the network processes defined in Section 2.1, such as Z, F, and D. This is a recursive process that begins with determination of the first decision time $\sigma_1 > 0$, followed by determination of the pairs (n, z) and $(\tilde{n}, \tilde{z})$ that describe the system state immediately before and immediately after that decision time, followed by determination of the second decision time $\sigma_2 > \sigma_1$, and so forth. It follows easily from Assumption 2.1 (baseline stochastic assumptions) that $\sigma_n \to \infty$ as $n \to \infty$ with probability one.

In that recursive construction, the policy function g comes into play as part of the logic that determines $(\tilde{n}, \tilde{z})$ from (n, z) at each decision time. Except for brief comments made in passing, this book treats just one example of a control policy that is *not* simply structured, namely, a queueing network where service is by order of arrival at each station, without regard to the class designations of the jobs being served (see Sections 3.3 and 7.3). In that case, readers will see that a similar recursive procedure is obviously available for constructing sample paths.

General control policies. Removing the restriction to simply structured policies, one may object that the general definition of a "control policy" advanced in this section is overly broad for purposes of application. Specifically, readers familiar with the theory of stochastic control will be expecting a requirement that control actions be "non-anticipating" in some appropriate sense, but we have not imposed such a restriction in this initial development. Rather, a requirement of that type will be added in Section 3.1, where we assume that the SPN has a time-homogeneous, discrete-state Markov representation.

The cumulative service effort process T. A central task in future chapters is to determine the fluid limit of a given SPN. For that purpose, it will be useful to introduce a nondecreasing J-dimensional process $T = \{T(t), t \geq 0\}$ whose jth component $T_j(t)$ represents the cumulative amount of *service effort* (expressed in service-hours, for example) devoted to activity j over the time interval $(0, t]$. For the basic SPN model currently under discussion, this is expressed mathematically by the following definition:

$$(2.22) \qquad T_j(t) := \int_0^t N_j(u)\,du \quad \text{for } j \in \mathscr{J} \text{ and } t \geq 0.$$

Analogous definitions for the relaxed formulation will be provided in Section 2.4. Given definition (2.22), the capacity constraint (2.11) can be alternatively expressed in terms of T as follows:

$$(2.23) \qquad A\big(T(t) - T(s)\big) \leq b(t - s) \quad \text{for any } 0 \leq s < t.$$

There are two reasons for introducing the cumulative service effort process T. First, T captures the essence of a control policy for purposes of our fluid model methodology, as will become apparent in Chapter 6. Second, it follows from (2.23) that each sample path of T is Lipschitz continuous (see Definition 8.1), and therefore the set of its fluid-scaled paths is automatically precompact (see Corollary A.9); this fact allows us to easily define a fluid limit in Section 6.4.

2.4 Relaxed SPN Model

In this section, we define a class of control policies that are "relaxed" in two regards. First, we allow services to be conducted at continuously variable rates, with capacity consumption varying in proportional fashion. In particular, the system manager can temporarily suspend a service by reducing its associated service rate to zero (this is commonly called a *service interruption*), and then resume the service later. The second relaxation is to allow *server sharing*, which means that a server may divide its capacity among two or more ongoing services. As noted earlier in Section 2.1, the processing resources in our relaxed SPN model are single-server pools, and components of the K-vector b are interpreted as capacities of the various servers, expressed in units like megabytes per second.

Recall that elements of the $K \times J$ server requirements matrix A are binary, and a type j service can be conducted at a continuously variable rate $\beta_j \geq 0$; by definition, activity j consumes server k capacity at rate β_j if $A_{kj} = 1$, and consumes none of that capacity if $A_{kj} = 0$. (The service rate β_j may also be referred to a *service effort level*.) By assumption, the ℓth-type j service concludes when the integral of its associated service rate first equals the "service time" random variable $v_j(\ell)$, which we take as primitive (see Section 2.1). Readers may prefer to think of $v_j(\ell)$ as a "service size," expressed in units like megabytes or gallons, interpreting the service rate β_j as a rate at which the remaining service size is reduced as a result of service effort.

Relaxed SPN formulations of the kind discussed here are motivated primarily by digital system applications in which either computers or communication links (or possibly both) play the role of servers. A notable application of our relaxed SPN formulation is the bandwidth sharing model discussed in Section 4.5. Actually, as readers will see in that section, the more precise statement is as follows: a bandwidth sharing network is stochastically equivalent to a head-of-line relaxed SPN, which is a particular case of the general model formulated in this section.

Simply structured control policies. The relaxed policies to be considered are all simply structured, which means that they have the following three features. First, for each service type $j \in \mathscr{J}$, *no more than one type j service can be open at any given time*. This is essentially equivalent to requiring that all of the service effort allocated to any given service type at any given time be directed to the oldest service of that type then open. Note that we do not impose any bound on the service rate that can be allocated to any one service, except for bounds that may be implied by the capacities of the servers involved.

Second, as in our basic SPN formulation (Section 2.3), decisions are made only at times $t > 0$ when an external arrival occurs or a service completion occurs. At each decision time t, the system manager observes the updated service count vector $\hat{n}$ and updated buffer contents vector $\hat{z}$ that were defined in the previous section. In the current context, $\hat{n}_j = 1$ if a type j service has been initiated before time t but is not yet completed, and otherwise $\hat{n}_j = 0$. Also, $\hat{z}_i$ is the number of items in buffer i at time t, possibly including newly arrived items, and some or all of them may be committed to currently open services. To be specific, the ith component of the I-vector $B\hat{n}$ is the number of type i items already committed to open services at time t, excluding any new services that may be initiated at t. (Here B is the material requirements matrix introduced in Section 2.1.)

The third defining feature of a simply structured relaxed control policy is that, at each decision time t, the system manager chooses a nonnegative J-vector β of *service rates*, based solely on the updated state descriptors $(\hat{n}, \hat{z})$. That is,

$$(2.24) \qquad\qquad \beta = h(\hat{n}, \hat{z}) \in \mathbb{R}_+^J,$$

where $h(\cdot, \cdot)$ is a policy function satisfying both material availability constraints and capacity constraints (discussed later in this section). One interprets β_j as the service rate applied to the open type j service, if there is one, going forward from t (that is, to be maintained until the next decision time). As in the previous section, let us denote by u_j the number of type j services to be initiated at the decision time t, and let $u = (u_1, \ldots, u_J)$. The choice of a service rate vector β determines u via the following obvious rule:

$$(2.25) \qquad\qquad u_j = \begin{cases} 1 & \text{if } \beta_j > 0 \text{ and } \hat{n}_j = 0, \\ 0 & \text{otherwise.} \end{cases}$$

The material availability and capacity constraints that h must satisfy are the following:

$$(2.26) \qquad\qquad B(\hat{n} + u) \leq \hat{z}$$

and

(2.27) $$A\beta \leq b.$$

Remark 2.4. All of the analysis to follow remains valid if elements of A are allowed to take arbitrary nonnegative values. In that case, A_{kj} is simply a constant of proportionality that one uses to translate the service rate for activity j into a rate of consumption for server k capacity. (Thus a better name for A in the general case would be *capacity consumption matrix*.) In section 5.5 of their paper, Kang et al. (2009) explain how the more general setup can arise in the treatment of bandwidth sharing models with *multi-path routing*. Here we have assumed a binary A matrix for ease of exposition, and more specifically, to minimize differences between our basic and relaxed SPN formulations.

Construction of sample paths. Given the function h that defines a simply structured relaxed control policy, plus initial data $N(0)$ and $Z(0)$, plus realizations of the processing variables $(v_j(\ell), \varphi^{(\ell)})$ for both services that are open at time zero and services initiated thereafter, plus the sample path of the arrival process E, one can construct the corresponding sample paths of all other network processes exactly as in our basic SPN model (Section 2.3).

The cumulative service effort process T. As for the basic SPN model (Section 2.3), we denote by $T = \{T(t), t \geq 0\}$ a nondecreasing J-dimensional process whose jth component $T_j(t)$ represents the cumulative amount of *service effort* devoted to activity j over the time interval $(0, t]$. In the relaxed model context, this is expressed mathematically as

(2.28) $$T_j(t) := \int_0^t \beta_j(u)\,du \quad \text{for } j \in \mathscr{J} \text{ and } t \geq 0,$$

so the capacity constraints (2.27) can again be expressed in terms of T via (2.23), which we rewrite here for ease of reference:

(2.29) $$A\big(T(t) - T(s)\big) \leq b(t - s) \quad \text{for any } 0 \leq s < t.$$

Also, the cumulative service completion process F_j can be represented in terms of T_j as follows ($j \in \mathscr{J}$, $t \geq 0$):

$$(2.30) \quad F_j(t) = \begin{cases} \sup\{\ell \geq 1 : T_j(t) \geq V_j(\ell)\} & \text{if } N_j(0) = 0, \\ \sup\{\ell \geq 1 : T_j(t) \geq V_j(0) + V_j(\ell - 1)\} & \text{if } N_j(0) = 1, \end{cases}$$

where, for an integer $n \geq 0$, $V_j(n)$ is the sum of the first n type j service times, and the supremum in each case equals zero by convention if the indicated set is empty. Equation (2.30) expresses in mathematical terms a model assumption that was stated verbally earlier, namely, that the ℓth-type j service concludes when the integral of its associated service rate first equals $v_j(\ell)$.

2.5 Recap of Essential System Relationships

In Chapter 6, we shall define and characterize fluid model analogs of the multidimensional processes D, F, T, and Z that were defined in Sections 2.1, 2.3, and 2.4. In this section, we recapitulate the key properties of those processes, and key relationships among them, that will be needed in that later analysis. These properties and relationships involve the model data and primitive stochastic elements that were introduced in Section 2.1.

First, the system equations (2.9) and (2.10) will be used directly in their stated form. Second, for our basic SPN formulation, it follows from the definition (2.22) of the cumulative service effort process T that

$$(2.31) \qquad T(\cdot) \text{ is nondecreasing with } T(0) = 0,$$

and the same is true for the relaxed formulation with T redefined via (2.28). Also, we shall use the relationship

$$(2.32) \qquad A\big(T(t) - T(s)\big) \leq b(t - s) \quad \text{for any } 0 \leq s < t,$$

which was displayed as (2.23) for our basic SPN formulation, and again as (2.29) for the relaxed formulation. Finally, for both our basic and relaxed SPN formulations, the relationship between the cumulative service effort process T and cumulative service completion process F can be expressed in the form

$$(2.33) \qquad T_j(t) = V_j(F_j(t)) + \epsilon_j(t) \quad \text{for all } j \in \mathscr{J} \text{ and } t \geq 0,$$

where, for an integer $n \geq 0$, $V_j(n)$ is again the sum of the first n type j service times, and $\epsilon_j(t)$ is a quantity that is negligible for purposes of our fluid model derivation; see (6.51) for a precise statement of that property.

Bounded service counts.   We conclude this section by stating a bound that plays an important role in future analyses. Recall that $N_j(t)$ denotes the number of type j services that are open (that is, initiated but not yet completed) at time $t \geq 0$, and $N(t)$ is a J-vector having those service counts as its components. Under the assumptions imposed in this chapter, one has the following for both our basic SPN formulation (Section 2.3) and the relaxed formulation (Section 2.4):

$$(2.34) \qquad\qquad N(t) \leq \kappa e \quad \text{for all } t \geq 0,$$

where e is the J-vector of ones and $\kappa > 0$ is a constant to be specified. In the relaxed formulation, we have stipulated that no more than one service of each type can be open at any given time, so (2.34) holds with $\kappa = 1$. For the basic formulation, where every open service must proceed at full speed until it is completed, our capacity constraints require $AN(t) \leq b$ for all $t \geq 0$. Thus, defining

$$\mathscr{A} = \{n \in \mathbb{Z}_+^J : An \leq b\},$$

it follows that (2.34) holds with

$$\kappa = \max_{n \in \mathscr{A}} \max_{j \in \mathscr{J}} n_j.$$

It will be shown in (6.51) that the term $\epsilon_j(t)$ in (2.33) is closely related to $N_j(t)$; the bound (2.34) then plays a critical role in proving that $\epsilon_j(t)$ is negligible under the fluid limit scaling to be developed in Section 6.2.

Unbounded service counts with processor sharing.   The validity of (2.34) in the relaxed SPN formulation depends on our assumption, as part of the restriction to "simply structured" control policies, that the following holds: if there is a service of a given type currently open, then no second service of that type can be initiated until the one currently open has been completed. Are there any relaxed control policies of either practical or theoretical interest which violate that restriction, and hence fail to satisfy (2.34)?

At least superficially, the answer is affirmative, as follows. Consider a single-server queueing model with Poisson arrivals and a general phase-type distribution of service times, in which the server, at each time

$t \geq 0$, divides its capacity equally among all jobs present in the system. This is the classical *processor sharing* (PS) control policy, introduced by Kleinrock (1967) as an idealized model of round-robin processing in a time-shared computer system.

Under this policy, a new service is opened every time there is a new arrival, and one has $N(t) = Z(t)$ at every time t (that is, the number of open services equals the number of jobs in the system), so (2.34) is eventually violated for any choice of κ. However, in Section 4.4 it will be shown (in a more general model context, but still under our baseline stochastic assumptions) that classical PS is equivalent, in a distributional sense, to a policy called head-of-line proportional processor sharing (HLPPS), which *does* satisfy (2.34) and will be analyzed later in the book (see Sections 4.6 and 8.2). A similar distributional equivalence for bandwidth sharing models is discussed in Section 4.5.

2.6 Unitary Networks and Queueing Networks

In this section, we first define the family of models that are called *unitary networks* in this book. (Unfortunately, that terminology is not standard.) As we shall explain, such models have a special structure that allows one to compute a unique "load factor" for each server pool, and to define a unique "route" for each job or item passing through the system, both of which substantially simplify analysis.

A still narrower family of models, also defined in this section, are the queueing network models, or multiclass queueing networks, that dominate the scholarly literature on stochastic processing networks. A second important subclass of unitary networks are bandwidth sharing networks, which were discussed briefly in Section 1.4 and will be treated systematically in Section 4.5. One particular control policy for unitary networks, called proportionally fair capacity allocation, will be discussed in Chapter 10 (see in particular Section 10.4), where we also give an example (see Figure 10.1) of a unitary network that is neither a queueing network nor a bandwidth sharing network, as those terms are used in this book.

Unitary networks. For purposes of this book, an SPN will be called *unitary* if it satisfies four assumptions, numbered (i) through (iv) in the paragraphs that follow.

(i) The first assumption is that $J = I$, which means we have a one-to-one relationship between processing activities on the one hand, and job classes (or buffers) on the other.

(ii) Second, the input requirements matrix B is simply the $I \times I$ identity matrix, which means that each activity $i \in \mathscr{I}$ requires as its input a single item or job from buffer i. Thus each activity consists of one or more servers from specified pools processing a single job from a specified class, and there is only one way to process jobs of any given class.

(iii) The third assumption defining a unitary network is the following: with probability one, each of the output vectors $\varphi^j(\ell)$ consists either of a single one and the rest zeros, or else all zeros. This means that when the processing of an item or job of any class is complete, the item or job either proceeds to a randomly selected output buffer to await further processing, or else it leaves the system. Thus the assumptions defining a unitary network forbid activities that require multiple inputs or produce multiple outputs.

When we speak of the *route* followed by a job in a unitary network, this means the ordered sequence of buffers visited by the job, starting with the buffer into which it arrives from the outside world; the system manager has no control over the routes followed by jobs in a unitary network, and routes may be stochastic. In networks where processing activities can have multiple inputs or multiple outputs, the notion of a route is not well defined.

Under the assumptions stated thus far, there is no guarantee that the routes followed by jobs in a unitary network will all be finite. That is, given only the assumptions stated so far, there may be jobs that remain in the system forever under some control policies, or even under *all* control policies, but such phenomena will be ruled out by the final assumption defining a unitary network.

Given an SPN that satisfies our baseline stochastic assumptions (Assumption 2.1 in Section 2.2), plus assumptions (i) through (iii), it is meaningful to define P_{ij} as the probability that a class i item or job, upon completion of its service, next becomes a class j item or job, independent of all previous history. Then the *routing matrix* $P = (P_{ij})$ is an $I \times I$ substochastic matrix.

(iv) As the fourth assumption defining a unitary network, we require that P be transient (that is, $P^n \to 0$ as $n \to \infty$, or equivalently, the spectral radius of P is < 1), because otherwise stability is not achievable.

It is often assumed in the study of stochastic networks that $P_{ii} = 0$ for all $i \in \mathscr{I}$ (this is described as the case with *no self-feedback*), but that restriction is not generally required for our purposes. The matrix Γ, which was defined for a generic SPN model in Section 2.2, has the following special form for a unitary network (here prime denotes transpose as usual):

$$(2.35) \qquad\qquad \Gamma = P'.$$

One can study unitary networks in the context of either our basic SPN framework (Section 2.3) or the relaxed SPN framework (Section 2.4). The phrases "relaxed unitary network" and "unitary network with relaxed control" will both be used in reference to the latter setting. Unitary networks include both the queueing network models described later in this section and the bandwidth sharing models defined in Section 4.5.

The load vector ρ for a unitary network. Given a unitary network with routing matrix P, we define a corresponding K-vector ρ as follows. First, for each $i \in \mathscr{I}$, let α_i be the *total arrival rate* into buffer i, including both external arrivals and transitions of jobs into class i from other classes. Thus the vector $\alpha = (\alpha_1, \dots, \alpha_I)'$ satisfies the following system of linear equations, which are commonly referred to as *traffic equations*:

$$(2.36) \qquad\qquad \alpha = \lambda + P'\alpha.$$

Because P is transient by assumption, Theorem F.1 says that $(I - P')$ is nonsingular (here and in the equations that immediately follow, I denotes the identity matrix), and its inverse has the Neumann expansion

$$(2.37) \qquad\qquad (I - P')^{-1} = I + P' + (P')^2 + \cdots.$$

The unique solution of (2.36) is then

$$(2.38) \qquad\qquad \alpha = (I - P')^{-1}\lambda.$$

We now define $\rho = (\rho_1, \dots, \rho_K)'$ by setting

$$(2.39) \qquad\qquad \rho_k = \sum_{i \in \mathscr{I}} A_{ki}\alpha_i m_i,$$

which can be written more compactly as

$$(2.40) \qquad\qquad \rho = AM\alpha,$$

where

$$(2.41) \qquad\qquad M := \mathrm{diag}(m_1, \dots, m_I).$$

In our basic SPN model, m_i is unambiguously a mean service time, that is, the expected amount of time for which the required servers must be committed. Thus one interprets ρ_k as the total service effort required from pool k per time unit, in units like server-hours per hour, to complete the processing of all jobs passing through the network. In the relaxed SPN model, it is more natural to view m_i as a mean service *size*, in units like megabytes, and one interprets ρ_k as the average amount of server k capacity, in units like megabytes per second, required to process all jobs passing through the network.

In each case, ρ_k is expressed in units appropriate for comparison against the resource capacity b_k, and we call it the *load* on resource k. A key point is the following: in a unitary network, *this quantity does not depend on the system manager's control decisions*. In contrast, there may be discretion in a general SPN as to which processing activities are used, and the amount of work that must be done by a given resource will then depend on how that discretion is exercised.

Queueing networks. For purposes of this book, a *queueing network* is a unitary network whose capacity consumption matrix A has a single one in each column and the rest zeros. Thus, in a queueing network each job class is processed by servers from a single specified pool, and each such service is accomplished by a single server from that pool.

Note that our definition of a queueing network allows a many-to-one relationship between buffers and server pools (that is, we allow $I > K$), so it embraces what are usually called multiclass queueing networks (see Section 1.3); we delete the modifier simply to save words. In a queueing network context, when we say that a given buffer or job class is *associated* with a particular server pool, we mean that servers from that particular pool are able to process jobs from the given buffer. For a queueing network, the definition (2.39) of ρ_k takes the slightly simpler form

$$(2.42) \qquad \rho_k = \sum_{i \in \mathscr{I}(k)} \alpha_i m_i,$$

where $\mathscr{I}(k)$ is the set job classes (or buffers) associated with server pool k, that is,

$$(2.43) \qquad \mathscr{I}(k) := \{i \in \mathscr{I} : A_{ki} = 1\}.$$

A useful additional term for queueing networks is *station*, which is used in this book (and elsewhere) to mean a server pool and all of its associated buffers. A commonly cited control policy for queueing networks is *first-come-first-served* (FCFS), which directs servers at each station to process jobs in the order of their most recent arrival to that station. (In a multiserver context, it is not correct to describe this policy as first-in-first-out, because job A may start service later than job B but still leave earlier than B. Specifically, this can happen if the two jobs are processed by different servers and A happens to have a shorter service time than B.) One can call this control policy *station-level FCFS*, or *local FCFS*, to distinguish it from *network-level FCFS*, or *global FCFS*, in which jobs are processed at each station in the order of their arrival to the network.

It may be worth noting that Kelly (1979) used the term "queue" for what we have called a station. Not surprisingly, given the prominence of both that book and its author, one still encounters that usage occasionally, although it conflicts with the common understanding of "queue" as a group of waiting jobs. We generally try to avoid using "queue" in this book, either as a noun or as a verb, except in the references to "queueing theory" and "queueing networks."

2.7 More on the Concept of Class

Item class (or job class) is an informational concept, and modeling discretion is involved in the definition of classes. The class designation attached by a model builder to an item at a particular point in its processing history may communicate not only immutable characteristics of the item, but also whatever is relevant from the item's processing history for purposes of future decision making, and for assessing the likelihoods of future processing outcomes. In this regard, the following points deserve emphasis.

Interchangeability.   Items bearing the same class designation are interchangeable as inputs to processing activities. By assumption, the joint distribution of service time and outputs from a service of any given type does not depend on *which* particular item or job of a given class is used as input to that service.

Service types and job classes.   Modeling discretion is also involved in the definition of processing activities (that is, in the definition of service types). Activity definitions are intertwined with class definitions, and there may be more than one viable representation of a given physical system. Consider, for example, a single-server processing system in which red jobs and green jobs arrive according to independent Poisson processes, each with arrival rate $\lambda > 0$, and jobs queue up for service on an FCFS basis without regard to color. Suppose that service times for red jobs are i.i.d. and exponentially distributed with mean 1, while those for green jobs are i.i.d. and exponentially distributed with mean 2, and the two service time sequences are independent of one another and of the two Poisson arrival processes.

The total arrival process for jobs of both colors is Poisson at rate 2λ, and the service times for successive arrivals (ignoring their colors) are i.i.d. with a hyperexponential distribution (see Appendix D.8) having mean 1.5. Moreover, the Poisson arrival process and the service time sequence are independent of one another. Thus, if we model this system as one with a single undifferentiated job class and a single service type, the baseline stochastic assumptions imposed in Section 2.2 are all satisfied. On the other hand, we can model the system as one with two job classes (namely, red jobs and green jobs) and two activities or service types (namely, serving red jobs and serving green jobs), specifying that service is by order of arrival without regard to color. (See Section 3.3 for a more detailed discussion of essentially this same example, but with two servers rather than one.) Once again, our baseline stochastic assumptions are satisfied.

Defining job classes to achieve independence. Extending this example, now suppose that the single server just described is followed by another one in tandem fashion (see Figure 1.1), where jobs again queue for service on an FCFS basis without regard to color. Also, assume that the service times for each color at the second server are i.i.d. with the same exponential distribution as at the first server, but independent of the first-stage service times and of the arrival processes.

Suppose that we model this tandem system by defining just two job classes, namely, jobs seeking their first service and jobs seeking their second one, and define two activities in the obvious way. Then all of our baseline stochastic assumptions are satisfied *except* that first-stage service times and second-stage service times are *not* independent. On the other hand, if we define four job classes and four service types in the obvious way, specifying service by order of arrival at each server, without regard to color, then all of the baseline stochastic assumptions hold.

This example shows that the independence assumptions imposed in Section 2.2 are specific to a given set of class and activity definitions. By allowing the model builder to proliferate job classes and service types, we make it easier to satisfy the baseline stochastic assumptions, but the resulting class designations may not be observable for purposes of system control (see later in this section for elaboration).

Markov representations.   Proliferating job classes to achieve independence, as in the preceding example, is a variant of a common modeling practice in applied probability generally. The practice to which we refer is the inclusion of "supplementary variables" in the definition of system "state" so as to achieve a Markov representation. See Chapter 3 for examples, specifically Sections 3.4 and 3.8.

Class designations and control decisions.   A class designation may be a modeling artifice rather than an observable property of an item or job. Consider, for example, the tandem system described earlier in this section, in which some jobs are red and others are green. If job color is not observable by the system manager, or if we simply wish to forbid the use of color as a basis for prioritizing service, then the set of allowable control policies must be restricted accordingly, but we may still attach different class designations to jobs of different colors in order to satisfy our standard stochastic assumptions.

Class as a (route, stage) pair.   The classic book by Kelly (1979) is devoted primarily to queueing network models, and its use of the word "class" is consistent with, but narrower than, our use of that term in this book. Kelly treats models in which (i) each arriving job or customer follows one of finitely many deterministic routes, (ii) customers following the various routes arrive according to independent Poisson processes, and (iii) the "class" designation attached to a customer at any given time consists of the route it is following (an immutable characteristic) *and* the stage of service to which it has progressed (an indicator of current status).

On pages 57–58, Kelly observes that, thanks to the "random splitting" property of the Poisson process, this model is more general than it first appears. In particular, one can analyze a system with Poisson arrivals and complex stochastic routing *as if* the complete route of a customer were known at the moment of its arrival: each Poisson arrival stream in the original formulation is separated into independent Poisson substreams according to the customers' future routes, and the complete route of a customer is retained as part of its class designation thereafter.

Notational efficiency. Readers will find that the modeling style used in this book, whereby all relevant information about an item, job, or customer is summarized in a single class designation, leads to a compact and efficient notation. That is, our class-centered notational system allows a compact presentation of both SPN system equations and the corresponding fluid models.

2.8 Sources and Literature

The SPN formulation in this chapter is similar in spirit, and in many of its details, to the one proposed by Dai and Lin (2005), which is based in turn on work by Harrison (2000, 2003), although we diverge from the Dai–Lin treatment in several significant regards. In particular, we do not include the "input activities" of the Dai–Lin formulation, whose primary purpose was to allow the formulation of models with alternate routing and immediate commitment; Section 4.2 provides an alternative approach to formulation of such models.

The term "unitary network" (see Section 2.6) was used by Bramson and Williams (2003), but our definition differs substantially from theirs; in most regards, their definition is more general than ours, but they restrict attention to single-server pools, whereas we allow multiserver pools. Our conception of a queueing network (see Section 2.6) is essentially the one propounded by Harrison (1988), which is substantially more general than the classical formulations by Baskett et al. (1975) and by Kelly (1979); see pages 152–153 of Harrison (1988) for elaboration.

3

Markov Representations

Loosely speaking, a stochastic processing network (SPN) is said to be *stable* under a specified control policy if it approaches a statistical equilibrium as $t \to \infty$. (See Section 3.1 for a formal definition.) Of course, one would not expect such limiting behavior unless *both* the core stochastic elements of the SPN *and* the policy chosen by the system manager are in some sense time homogeneous, or time invariant. The first section of this chapter describes the approach taken in this book to ensure such time invariance, and to provide a tractable but still very general mathematical setting in which to address the substantive issue of stability.

Our key requirement is that the SPN be representable as a continuous-time Markov chain (CTMC), under the control policy that we wish to evaluate, by including enough "supplementary variables" in the definition of system state. Assumption 2.1 (baseline stochastic assumptions) plays an important role in meeting that requirement, but restrictions on the control policy are equally important. In the latter regard, our attention will be focused primarily on the "simply structured" control policies introduced in Chapter 2, but the analytical framework developed here can accommodate policies of a more general form.

Our general framework is described in Section 3.1, and an important result on SPN stability is proved in Section 3.2. Sections 3.3 and 3.4 present a series of simple, concrete examples for which CTMC representations are and are not achievable. Moving from the specific to the general, Section 3.5 develops a canonical CTMC representation for the basic and relaxed SPN models formulated earlier in Chapter 2, assuming that a simply structured policy is adopted. Sections 3.6 and 3.7 expand upon the CTMC framework assumed at the chapter's outset, and Section 3.8 discusses briefly the more general (nondiscrete)

Markov representations that may be needed if our baseline stochastic assumptions are relaxed.

3.1 General Framework and Definition of Stability

The following development applies equally well to the basic SPN model (Section 2.3) and to the relaxed model (Section 2.4). In it, we take as given a control policy, and consider the buffer contents process Z that is constructed from primitive stochastic elements and the given control policy. (See Appendix D for a review of definitions and basic theory related to continuous-time Markov chains.)

Assumption 3.1 (Markov representation). There exists an irreducible Markov chain $X = \{X(t), t \geq 0\}$, with countable state space $\mathscr{X}$, and a function $f : \mathscr{X} \to \mathbb{Z}_+^J \times \mathbb{Z}_+^I$ such that

$$(3.1) \qquad (N(t), Z(t)) = f(X(t)), \quad t \geq 0,$$

on every sample path. Also, the set $B(z) \subset \mathscr{X}$ is finite for each $z \in \mathbb{Z}_+^I$, where

$$(3.2) \qquad B(z) := \left\{ x \in \mathscr{X} : f(x) = (n, z) \text{ for some } n \in \mathbb{Z}_+^J \right\}.$$

Finally, there exists at least one state $x^* \in \mathscr{X}$ such that

$$(3.3) \qquad f(x^*) = (0, 0).$$

Remark 3.2. One may paraphrase (3.1) by saying that the service count process N and buffer contents process Z can be *embedded in an ambient Markov chain X*, or more simply, that the SPN can be embedded in an ambient Markov chain. Alternatively, one may say that the SPN is embedded in a *Markovian environment*.

Remark 3.3. The definitions and assumptions advanced in Chapter 2 imply the following: if $Z(t) = 0$ at any given time t, then it must also be that $N(t) = 0$. That is, there cannot exist any open services when all buffers are empty. Thus the first zero on the right side of (3.3) is actually redundant.

Hereafter, the state x^* in (3.3) will be called an *empty state*. Section 3.5 will consider an SPN with Poisson arrivals (as required by our baseline stochastic Assumption 2.1) and a simply structured control policy, and describe a "canonical" Markov representation in which the empty state is unique. In contrast, the more general case with a Markovian arrival process (MArP) will be considered in Section 4.1, and in that setting the empty state is typically *not* unique.

Of course, the assumed irreducibility implies that the empty state x^* in (3.3) is reachable from any other state. See Section 3.7 for further discussion of the irreducibility assumption.

Remark 3.4.   In Section 3.6, a mild additional assumption will be imposed (this is Assumption 3.8) regarding the Markov representation.

Initial state $X(0)$.   Assumption 3.1 remains in force throughout the remainder of this book, and henceforth the "initial state" of an SPN will be understood to mean an initial state $x = X(0) \in \mathcal{X}$ for the CTMC in which the SPN is embedded. From (3.1), we know that the vector $Z(0)$ is recoverable from $X(0)$. Typically, the joint distribution of processing variables for services that are open at time zero will be different for different initial states, but part of Assumption 2.1 is the stipulation that processing variables for services started *after* time zero be independent of $X(0)$.

Proposition 3.5 (equivalent definitions of stability).   *Under Assumption 3.1, the following statements are equivalent: (a) X is positive recurrent; (b) X has a unique stationary distribution π; (c) $Z(t)$ converges in distribution to a nondefective limit as $t \to \infty$. Moreover, when any one of the preceding conditions is satisfied, the following strong law of large numbers (SLLN) holds for any bounded function $h : \mathcal{X} \to \mathbb{R}$ and any initial distribution for $X(0)$:*

$$\mathbb{P}\left\{ \lim_{t \to \infty} \frac{1}{t} \int_0^t h(X(s))ds = \bar{h} \right\} = 1,$$

where $\bar{h} := \sum_{x \in \mathcal{X}} \pi(x)h(x)$.

Proof.   The equivalence of (a) and (b) follows from Theorem D.17 in Appendix D. Now we prove the equivalence of (a) and (c). If X is positive recurrent, then $\mathbb{P}\{X(t) = x\}$ converges to $\pi(x)$, a positive constant, as $t \to \infty$ for each $x \in \mathcal{X}$ (see part (i) of Theorem D.24 in Appendix D). For each $z \in \mathbb{Z}_+^I$, $B(z)$ is a finite set. It follows that

$$\mathbb{P}\{Z(t) = z\} = \mathbb{P}\{X(t) \in B(z)\} \to \sum_{x \in B(z)} \pi(x) > 0$$

as $t \to \infty$, which implies (c). Conversely, if X is either null recurrent or transient, it follows from Part (ii) of Theorem D.24 in Appendix D that

$$\mathbb{P}\{Z(t) = z\} = \mathbb{P}\{X(t) \in B(z)\} \to 0$$

as $t \to \infty$, and hence (c) cannot hold. The SLLN follows from Theorem D.25. □

Definition 3.6. An SPN satisfying Assumption 3.1 is said to be *stable* if the equivalent statements in Proposition 3.5 hold.

3.2 Sufficient Condition for SPN Stability

The lemma proved in this section will allow us to connect SPN stability with the fluid limit theory to be developed in Chapter 6. For an initial state $x \in \mathcal{X}$, following standard practice in Markov process theory, we use $\mathbb{E}_x$ to denote an expectation conditioned on $X(0) = x$. Also, for each $x \in \mathcal{X}$ we define

(3.4) $$|x| := |z| := \sum_{i \in \mathcal{I}} z_i,$$

where $z = f(x)$.

Lemma 3.7. *If there exists a $\delta > 0$ such that*

(3.5) $$\lim_{|x| \to \infty} \frac{1}{|x|} \mathbb{E}_x |Z(|x|\delta)| = 0,$$

then the ambient Markov chain X is positive recurrent.

Proof. By Theorem D.22, it suffices to prove that the skeleton DTMC $Y = \{Y_n : n \in \mathbb{Z}_+\}$ is positive recurrent, where for each $n \in \mathbb{Z}_+$

$$Y_n := X(n\delta).$$

By Assumption 3.1, X is irreducible. Therefore, by Lemma D.20, Y is irreducible. By Theorem C.27, to prove Y is positive recurrent, it suffices to prove that the state-dependent Foster–Lyapunov drift condition (C.24) is satisfied. Namely, there are functions $V(x) : \mathcal{X} \to \mathbb{R}_+$ and $n(x) : \mathcal{X} \to \mathbb{N}$, constants $a, b > 0$, and a finite set $C \subset \mathcal{X}$ such that

(3.6) $$\mathbb{E}_x V(Y_{n(x)}) \leq V(x) - \frac{1}{a} n(x) + b \mathbf{1}_C(x) \quad \text{for each } x \in \mathcal{X},$$

where $1_C(x)$ is the indicator function of the set C. In the rest of this proof, we check that (3.6) is satisfied. By assumption (3.5), there exists an integer $n_0 > 0$ such that

$$\frac{1}{|x|}\mathbb{E}_x\big|X(|x|\delta)\big| \leq 1/2$$

for $x \in \mathscr{X}$ with $|x| > n_0$. Equivalently,

$$\mathbb{E}_x\big|X(|x|\delta)\big| \leq |x| - \frac{1}{2}|x| \quad \text{for } |x| > n_0.$$

For each initial state $x \in \mathscr{X}$, the number of jobs in the system at time h is bounded by the initial number plus the number of external arrivals in $(0,h]$. Thus, for each $x \in \mathscr{X}$,

$$\mathbb{E}_x\big|X(\delta)\big| \leq |x| + c,$$

where

$$c := \sum_{i \in \mathscr{I}} \mathbb{E}\Big[E_i(\delta)\Big] < \infty.$$

Define

$$B_{n_0} := \{x \in \mathscr{X} : |x| \leq n_0\}.$$

By Assumption 3.1, B_{n_0} is a finite set. For $x \in \mathscr{X}$, define $V(x) := |x|$ and

$$n(x) := \begin{cases} |x| & \text{if } |x| > n_0 \\ 1 & \text{if } |x| \leq n_0 \end{cases}.$$

Then we have

$$\mathbb{E}_x\big[V(Y_{n(x)})\big] \leq V(x) - \frac{1}{2}n(x) + (c+1/2)1_{B_{n_0}},$$

proving (3.6) with $a = 2$, $b = (c+1/2)$, and $C = B_{n_0}$. $\qquad\square$

3.3 First Examples of Markov Representations

In the remainder of this chapter, we consider a variety of SPN models, progressing from simple examples to general model families; the treatment will then be further generalized in Chapter 4. Our purpose is to show how, for various control policies of interest, a Markov representation can be obtained under Assumption 2.1 (baseline stochastic assumptions).

To completely specify the transition structure of the relevant Markov chain X can be a daunting task, especially when dealing with a general model family, but such details are largely irrelevant for our purposes. To avoid an oppressive accumulation of mundane detail, our strategy is to introduce basic ideas via simple examples, demonstrate selectively how they apply in more general settings, and omit most of the detailed specification.

In preparation for the verbal descriptions of different control policies that follow, readers are reminded that, according to the linguistic convention established in Section 2.1, a "buffer" may contain both waiting jobs and in-service jobs at any given time; for us, "buffer i" is a place where all class i jobs reside, including both the ones that are currently in process and the ones that are waiting.

Temporary nonstandard use of semicolons. Throughout this section and the next one, semicolons will be used in expressions like $x = (n; z)$ and $x = (1, 0; 0, 1; 2, 3)$ to divide components of a vector into intuitively meaningful groups, or to emphasize that the items being separated in the specification of a vector are vectors themselves. Such semicolons are mathematically equivalent to commas. This practice will not be continued beyond Section 3.4.

Single-pool example with static buffer priorities. Figure 3.1 pictures a single-hop processing system with two job classes and a pool of two identical servers. For class $i = 1, 2$ the arrival process is assumed to be Poisson with rate λ_i, and class i service times form an i.i.d. sequence distributed $\exp(\mu_i)$; thus the mean service time for class i is $m_i = 1/\mu_i$. Also, the two arrival processes and the two service time sequences are assumed to be mutually independent.

If one or both of the servers are idle when a class i job arrives, that job is processed immediately by an idle server, and it leaves the system when its service is complete. (If both of the servers are idle at the time of an arrival, it is immaterial which one does the processing.) Otherwise, the job waits in buffer i. If a server completes the processing of a job and finds that only one buffer contains waiting jobs, the server takes whichever of the waiting jobs arrived first and immediately begins its service. If both buffers contain waiting jobs, we assume that the

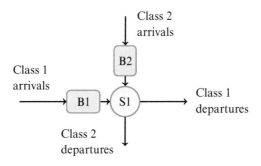

Figure 3.1 A multiclass single-pool system.

server chooses the first-arriving job that is waiting in buffer 2. This is an example of a (nonpreemptive) static buffer priority (SBP) policy. Specifically, it is the SBP policy that gives top priority to class 2.

Recall that N and Z denote the service count process and buffer contents process (or job count process), respectively, for a generic SPN. In our current context, there is a one-to-one correspondence between processing activities and job classes, so a "type i service" is interpreted simply to mean the service of a class i job, and hence $N_i(t)$ is interpreted as the number of class i services under way at time t. To implement the SBP policy that we have assumed, it is sufficient to know, at each time t, the current values $n = N(t)$ and $z = Z(t)$. Thus, given the memoryless character of the Poisson arrival processes and exponential service time distributions that we have assumed, $\{(N(t), Z(t)), t \geq 0\}$ is itself Markov under the assumed SBP policy, so a Markov state description is obtained by simply taking

$$(3.7) \qquad\qquad x := (n; z),$$

where $n \in \mathbb{Z}_+^2$ and $z \in \mathbb{Z}_+^2$. Each state $x = (n; z)$ must satisfy

$$(3.8) \qquad\qquad n \leq z, \quad n_1 + n_2 \leq 2,$$

and hence the continuous-time Markov chain $X = \{X(t), t \geq 0\}$ has state space

$$(3.9) \qquad \mathscr{X} = \left\{ (n; z) \in \mathbb{Z}_+^2 \times \mathbb{Z}_+^2 : (n; z) \text{ satisfies (3.8)} \right\}.$$

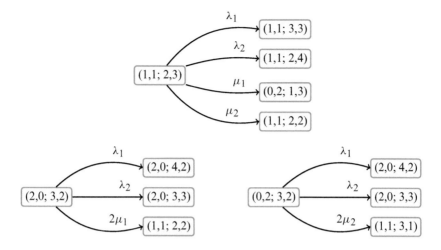

Figure 3.2 Transition rates under the SBP policy from three states.

To describe the transition rates of X in general, one must consider each state $x = (n;z) \in \mathcal{X}$, identify the set of "next" states $\tilde{x} = (\tilde{n};\tilde{z}) \in \mathcal{X}$ that are reachable from x in one jump, and specify the transition rate from x to $\tilde{x}$ for each such combination. This is done for three representative states in Figure 3.2. In each case, the reachable next states $\tilde{x}$ are shown on the right side of the figure and transition rates are indicated by symbols above the arrows. Trusting that these examples illustrate the general mechanisms at work in this example, we delete the comprehensive specification.

Single-pool example with FCFS control. Let us now consider the same system operating under the first-come-first-served (FCFS) control policy, that is, the policy that processes jobs in the order of their arrival. (As noted earlier in Section 2.6, it is not correct in a multiserver context to describe this policy as first-in-first-out.) Effectively, then, jobs in buffers 1 and 2 are merged into a single (logical) buffer according to the order in which they arrived.

The content of the merged buffer can be described by a finite string of class designations, such as $(1,2,2,1,2)$. This string indicates that there are five jobs either being processed or waiting to be processed by the two servers in the pool. The first-arriving job is from class 1, and it is being processed by one of the servers. The second-arriving job is from class 2, and it is being processed by the other server. The other three

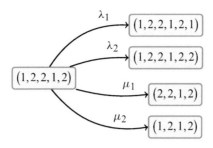

Figure 3.3 Transition rates under FCFS from an illustrative state.

jobs, in order of their arrival, are from classes 2, 1, and 2, respectively; they are all waiting in the buffer. In general, a Markov state description under FCFS control is obtained by taking

$$x := (c_1, c_2, \ldots, c_q),$$

where $q \in \mathbb{Z}_+$ and $c_\ell \in \{1, 2\}$ for $\ell = 1, \ldots, q$. Here q is the total number of jobs in the system, c_1 is the class of the one that arrived first, $\ldots$, and c_q is the class of the one that arrived last. The function $f : x \to z$ in (3.1) is given by

$$z_i = \sum_{\ell=1}^{q} 1_{\{c_\ell = i\}} \quad \text{for } i = 1, 2.$$

Figure 3.3 specifies transition rates from state $(1, 2, 2, 1, 2)$, which was considered at the beginning of this paragraph.

This FCFS control policy is *not* "simply structured," as we defined that term in Section 2.3, because the choice of a next service type to initiate at a decision time does not depend solely on the pair $(\hat{n}, \hat{z})$ for the decision time. In fact, this example and the one that follows it (which involves "global FIFO" control) are the only nonsimply structured policies that appear anywhere in this book, although it is easy to generate other examples of such policies.

Modified criss-cross network with global FIFO control. Figure 3.4 pictures a queueing network with two servers and three job classes. It is similar to the criss-cross network (Figure 1.2), but here it is the "downstream" server (server 2) that has two buffers to serve. We assume that classes 1 and 3 arrive according to independent Poisson processes

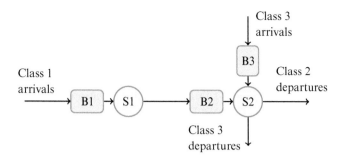

Figure 3.4 Modified criss-cross network.

at rates λ_1 and λ_3, respectively, and that class i service times are i.i.d. and distributed $\exp(\mu_i)$ for $i = 1, 2, 3$.

The only control decisions to be made for this system concern the order in which server 2 will take jobs from buffers 2 and 3. The policy we want to consider is global FCFS (which can also be called global FIFO, since this example has only single-server pools), which dictates that server 2 process jobs from buffers 2 and 3 in the order of their arrival to the *network*, *not* in the order of their arrival to that *server*. To implement this policy, it will suffice to track the *age* of each job currently in the system, defined as the elapsed time since the job arrived from the outside world. A Markov state description is obtained by defining

$$x := \big(a_1(1), \ldots, a_1(q_1);\ c(1), a_2(1), \ldots, c(q_2), a_2(q_2)\big),$$

where $q_1 \in \mathbb{Z}_+$ is the number of jobs in buffer 1, $a_1(1), \ldots, a_1(q_1)$ are the ages of those jobs, $q_2 \in \mathbb{Z}_+$ is the total number of jobs in buffers 2 and 3 together, and the pair $\big(c(\ell), a_2(\ell)\big)$ specifies the class designation and the age, respectively, of the ℓth job in the latter population. Unfortunately, the state space $\mathscr{X}$ is uncountable when states are defined in this way. In Section 3.8, we shall discuss Markov processes with general state space.

3.4 Examples with Phase-Type Distributions

In the preceding three examples, we have assumed exponential service time distributions and independent Poisson arrival processes. To achieve a Markovian state description under those assumptions, one starts with the basic processes $N(t)$ and $Z(t)$, then asks (only) what supplementary variables are needed to support the chosen control policy, such as the queue positions required for FCFS control in a multiclass environment. The following example shows how, with more general service time distributions, a Markov state description can still be achieved by appending more supplementary variables. Here and later, we assume the reader's familiarity with notation and terminology developed in Sections D.8 and D.9, such as "Erlang$(2,\alpha)$" and "service regulating Markov chain."

Single-pool example with phase-type distributions. Consider again the system pictured in Figure 3.1, with two job classes and a pool of two identical servers, but now suppose that service times for class 1 are distributed Erlang$(2,\alpha)$ and service times for class 2 have the two-phase hyperexponential distribution that is denoted $H_2(p,\beta)$ in Section D.8. Thus a class 1 service time v_1 has the representation

(3.10)
$$v_1 = \xi_1 + \xi_2 \quad \text{where} \quad \xi_1 \text{ and } \xi_2 \text{ are i.i.d. and distributed } \exp(\alpha),$$

and a class 2 service time v_2 has the representation

(3.11)
$$v_2 = \begin{cases} \zeta_1 & \text{with probability } p_1, \\ \zeta_2 & \text{with probability } p_2, \end{cases}$$

where ζ_1 and ζ_2 are independent random variables and ζ_s is distributed $\exp(\beta_s)$, $s = 1,2$. Therefore, the expected service time for class 1 jobs is $2/\alpha$, and the expected service time for class 2 jobs is

$$p_1/\beta_1 + p_2/\beta_2.$$

Both (3.10) and (3.11) are examples of phase-type service time distributions (see Section D.8). In the former case, the associated "service regulating Markov chain" Y begins in service phase 1 with probability 1, stays there for ξ_1 time units, then transitions to phase 2 and stays there for ξ_2 time units, at which point Y is absorbed into a terminal phase and the service is complete. In the latter case, Y begins in service phase s with probability p_s and stays there for ζ_s time units ($s = 1,2$), at which point Y is absorbed and the service is complete. (Here, as in Section D.8, we use the term "service phase," or just "phase," in preference to "state" when referring to service regulating Markov chains.)

Assume as before that the system is operating under the SBP policy that gives priority to class 2. Because service times of different jobs are independent by assumption, completion times for services in process at any given time are viewed as the absorbtion times of *independent* service regulating Markov chains. Then a Markov state description is the following:

(3.12) $$x := \big(\eta_1(1), \eta_1(2); \eta_2(1), \eta_2(2); z_1, z_2\big),$$

where $\eta_i(s)$ denotes the number of class i jobs whose service regulating Markov chains are in service phase s. The constraints (3.8) still apply with

$$n_1 = \eta_1(1) + \eta_1(2) \quad \text{and} \quad n_2 = \eta_2(1) + \eta_2(2).$$

Figure 3.5 specifies the transition rates from state $\big(1,0;0,1;2,3\big)$. Red coloration is used in this figure to indicate components of the state description that are changed by the transition.

A queueing network with dependence between service time and route. Figure 3.6 portrays a simple queueing network similar to the one pictured earlier in Figure 1.1, with two single-server pools arranged in tandem. Now, however, we suppose that routing is stochastic: some

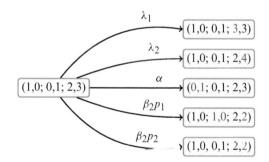

Figure 3.5 Transition rates with phase-type service time distributions.

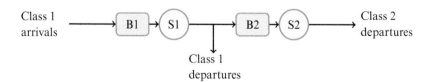

Figure 3.6 Tandem system with stochastic routing.

class 1 customers exit the system after their initial service, while others go on for a second service as class 2 customers. Modifying the notation that was used in Chapter 2, let us denote by v a service time for a generic class 1 customer, and denote by φ the output vector associated with that service. Thus the range of potential values for v is $[0, \infty)$ and the two possible values for φ are $(0,0)$ and $(0,1)$. To be more specific, suppose that each time server 1 begins processing a new customer, a fair coin is flipped to determine the service phase s of that customer: if a head is observed ($s = 1$), then the customer's service time is distributed $\exp(\beta_1)$ and it exits after completing service; if a tail is observed ($s = 2$), then the customer's service time is distributed $\exp(\beta_2)$ and it continues as a class 2 customer after completing its initial service. Thus the marginal distribution of v is hyperexponential, and the marginal distribution of φ gives equal probability to the two possible values identified earlier, but v and φ are *not* independent in general. Rather, v and φ have what is called a *joint phase-type distribution* in Section D.9. Completing the system specification, assume that the external arrival process into buffer 1 is Poisson with parameter λ_1 and service times for class 2 customers are distributed $\exp(\mu_2)$. A Markov state description is

$$x := (s; z_1, z_2),$$

where $z = (z_1, z_2)$ is the vector of buffer contents (or customer counts) as usual, and s is the service phase of the the customer currently being processed by server 1. (We can set $s = 0$ as a matter of convention when $z_1 = 0$.) Figure 3.7 shows transition rates from two illustrative states of this system.

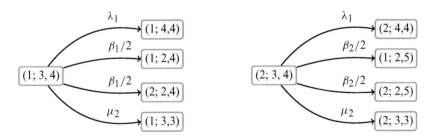

Figure 3.7 Transition rates from two illustrative states for the system in Figure 3.6.

3.5 Canonical Representation with a Simply Structured Policy

Throughout this section, we focus primarily on the basic SPN model formulated in Section 2.3, assuming a simply structured control policy. Under such a policy, one defines the updated service count vector $\hat{n}$ and updated buffer contents vector $\hat{z}$ at each decision time t as in Section 2.3, and the vector u of new service starts at t is $u = g(\hat{n}, \hat{z})$, where g is a given policy function in (2.20). That logic determines the number of open services of each type going forward from t, and all services proceed at full speed until they are completed.

We develop a Markov representation of the SPN model, hereafter called the *canonical* Markov representation, which has two parts: specification of the Markov state description, and specification of the corresponding generator matrix. In each case, we first treat the basic SPN model, as stated previously, and then indicate the minor changes that are required to treat the relaxed SPN model laid out in Section 2.4. Readers are reminded that a simply structured relaxed policy is defined by a function h. At each decision time t, one calculates a vector β of service rates or service intensities via $\beta = h(\hat{n}, \hat{z})$, which is (2.24), and the vector u of new service starts at t is determined from β via (2.25). Thus one can again write $u = g(\hat{n}, \hat{z})$, as in the basic SPN model, but in the relaxed SPN setting g is actually derived from h via (2.24) and (2.25).

Notation related to joint phase-type distributions. To begin, we introduce the following notation in conjunction with Assumption 2.1. Extending in an obvious way the notation of Section D.9, for each $j \in \mathcal{J}$ we denote by $(\mathcal{S}_j, \Delta_j, p_j, F_j, f_j)$ the phase space, output space, initial distribution, and transition rate functions of the joint phase-type distribution for type j services, and denote by s_j a generic element of $\mathcal{S}_j$. Thus $F_j(s_j, \tilde{s}_j)$ is the rate at which the service regulating Markov chain for a type j service makes silent transitions from s_j to $\tilde{s}_j$ $(s_j, \tilde{s}_j \in \mathcal{S}_j, \tilde{s}_j \neq s_j)$, and $f_j(s_j, \delta)$ is the rate at which it makes transitions from service phase s_j that terminate the service and trigger output vector δ $(s_j \in \mathcal{S}_j, \delta \in \Delta_j)$.

Markov state description. Under Assumption 2.1, the probability is zero that an external arrival and a service completion will ever occur simultaneously, or that two service completions will ever occur simultaneously. Given that service initiation decisions at each decision time are based solely on $\hat{n}$ and $\hat{z}$, a suitable Markov state description in either the basic or relaxed SPN setting is

(3.13) $x := (\eta, z),$

where elements of x are defined as follows. First,

(3.14) $\eta := \big(\eta_j(s_j), s_j \in \mathscr{S}_j, j \in \mathscr{J}\big),$

with $\eta_j(s_j)$ being the number of type j services currently open whose service regulating Markov chain is in service phase s_j ($j \in \mathscr{J}, s_j \in \mathscr{S}_j$); hereafter we shall refer to η as a *phase count vector*. Second, $z = (z_1, \ldots, z_I)$ is the buffer contents vector as usual. With the possible exception of irreducibility (see Section 3.7 for elaboration), all aspects of Assumption 3.1 are satisfied with this definition of system state. Also, denoting by $n = (n_1, \ldots, n_J)$ the service count vector as usual, one sees that n can be recovered from η as follows:

(3.15) $n_j = \sum_{s_j \in \mathscr{S}_j} \eta_j(s_j), j \in \mathscr{J}.$

Unique empty state. Recall from (3.3) and Remark 3.3 that an *empty state* is one for which $z = 0$. In our current setting, that further implies $\eta = 0$, so we conclude that there is exactly one empty state in the canonical Markov representation, namely, $x^* = (0, 0)$.

We turn now to specifying the generator matrix Λ (that is, specifying nonzero transition rates) for the ambient Markov chain X. As in Section 3.1, the state space of X is denoted $\mathscr{X}$. Here a generic element of Λ will be denoted $\Lambda(x, \tilde{x})$, as opposed to the subscripts used to denote elements of Λ in Appendix D, because the states $x, \tilde{x}$ that we wish to consider are themselves vectors with subscripted components.

Transition categories and basic accounting relationships. Transitions fall into the three categories listed in the tabular display that immediately follows. For each category, we denote by (n, z) the service count vector and buffer contents vector immediately before the transition, and indicate in the tabular display the updated values $(\hat{n}, \hat{z})$ after new arrivals and service completions are accounted for, but before accounting for new service starts; this extends in an obvious way notation that was established earlier in Section 2.3. Of course, the pair (n, z) does not constitute a complete state description immediately before the decision time; in this display and the one that follows it, we are merely noting some "accounting" relationships that are helpful in describing state transitions.

Silent transition of a service regulating
    Markov chain: $\qquad\qquad\qquad\qquad (\hat{n}, \hat{z}) = (n, z)$

External arrival into buffer i ($i \in \mathscr{I}$): $\qquad (\hat{n}, \hat{z}) = (n, z + e^i)$

Type j service completion triggering
    output vector $\delta \in \Delta_j$ ($j \in \mathscr{J}$): $\qquad (\hat{n}, \hat{z}) = (n - e^j, z - B^j + \delta)$

Finally, denoting by $(\tilde{n}, \tilde{z})$ the service count vector and buffer contents vector *after* we account for new services that may be initiated at the transition time, one has the following accounting relationships for the three transition categories we have identified.

Silent transition of a service regulating
    Markov chain: $\qquad\qquad\qquad\qquad (\tilde{n}, \tilde{z}) = (\hat{n}, \hat{z})$

External arrival: $\qquad\qquad\qquad\qquad (\tilde{n}, \tilde{z}) = (\hat{n} + g(\hat{n}, \hat{z}), \hat{z})$

Service completion: $\qquad\qquad\qquad\qquad (\tilde{n}, \tilde{z}) = (\hat{n} + g(\hat{n}, \hat{z}), \hat{z})$

Everything said in this paragraph applies equally well to the basic and relaxed SPN model formulations, with the understanding that g is derived from h in the latter setting via (2.24) and (2.25).

Transition rates in the basic SPN setting. In the basic SPN setting, transition rates $\Lambda(x, \tilde{x})$ for the ambient Markov chain X are determined by two types of model data: the arrival rates λ_i for the various classes $i \in \mathscr{I}$; and the transition rate functions for the model's service regulating Markov chains. Consider, for example, the first transition category in the preceding tables. More specifically, consider a silent transition from s_j to $\tilde{s}_j$ for a type j service regulating Markov chain. This takes the ambient chain from state $x = (\eta, z)$ to state $\tilde{x} = (\tilde{n}, z)$, where $\tilde{\eta}_j(s_j) = \eta_j(s_j) - 1$, $\tilde{\eta}_j(\tilde{s}_j) = \eta_j(\tilde{s}_j) + 1$, and all other components of $\tilde{\eta}$ equal those of η. The transition rate from s_j to $\tilde{s}_j$ is $F_j(s_j, \tilde{s}_j)$ for each of the $\eta_j(s_j)$ type j service regulating chains now in state s_j, so the overall transition rate is $\Lambda(x, \tilde{x}) = \eta_j(s_j) F_j(s_j, \tilde{s}_j)$.

    Transition rates are more complicated for the second transition category, but still straightforward in principle. Let us consider, for example, state $x = (\eta, z)$ for the ambient Markov chain, modified by an external arrival into buffer i; such arrivals occur at rate λ_i. As noted earlier, the vector u of new service starts is $u = g(\hat{n}, \hat{z})$, and we shall consider the specific scenario where $g(\hat{n}, \hat{z}) = e^\ell$; this means that a single new service of type ℓ is dictated by the policy under consideration.

Completing specification of that transition scenario, let us suppose that the service regulating Markov chain for a type ℓ service has a random initial state (that is, its initial distribution p_ℓ puts positive mass on two or more states), and let us consider a particular state $s_\ell \in \mathscr{S}_\ell$ of the service regulating chain such that $p_\ell(s_\ell) > 0$. Then the new state of the ambient Markov chain will be $\tilde{x} = (\tilde{\eta}, \hat{z})$, where $\tilde{\eta}_\ell(s_\ell) = \eta_\ell(s_\ell) + 1$ and all other elements of $\tilde{\eta}$ equal those of η; the associated transition rate is simply $\Lambda(x, \tilde{x}) = \lambda_i p_\ell(s_\ell)$. If we modify the scenario by supposing that the decision function g calls for initiation of two or more new services, all of whose service regulating Markov chains might have random initial states, then the specification of $\Lambda(x, \tilde{x})$ will be more involved, but it is not conceptually subtle or difficult.

Turning finally to the third transition category, let us consider a state $x = (\eta, z)$ for the ambient Markov chain, modified by a service-completing transition of a type j service regulating Markov chain. More specifically, suppose that the transition is from service phase $s_j \in \mathscr{S}_j$, and that the service completion triggers output vector $\delta \in \Delta_j$. The overall rate at which such transitions occur from the hypothesized state $x = (\eta, z)$ is $\eta_j(s_j) f_j(s_j, \delta)$, the updated system status vector $(\hat{n}, \hat{z})$ is as stated in the "accounting" relationships described previously, and the vector u of new service starts is $u = g(\hat{n}, \hat{z})$. We shall consider the specific scenario where $g(\hat{n}, \hat{z}) = e^\ell$ (that is, the service completion triggers a single service start that is of type ℓ) and $\ell \neq j$, and consider a particular service phase $s_\ell \in \mathscr{S}_\ell$ such that $p_\ell(s_\ell) > 0$. Then the new state of the ambient Markov chain will be $\tilde{x} = (\tilde{\eta}, \hat{z})$, where $\tilde{\eta}_j(s_j) = \eta_j(s_j) - 1$, $\tilde{\eta}_\ell(s_\ell) = \eta_\ell(s_\ell) + 1$, and all other elements of $\tilde{\eta}$ equal those of η; the associated transition rate is simply $\Lambda(x, \tilde{x}) = \eta_j(s_j) f_j(s_j, \delta) p_\ell(s_\ell)$.

Transition rates in the relaxed SPN setting. In a relaxed SPN model, transition intensities $\Lambda(x, \tilde{x})$ involve not only the model data indicated earlier, but also the service rates $\beta = (\beta_j, j \in \mathscr{J})$ that are calculated at each decision time via the formula $\beta = h(\hat{n}, \hat{z})$. Specifically, for the first and third transition categories identified earlier, one must multiply both $F_j(s_j, \tilde{s}_j)$ and $f_j(s_j, \delta)$ by β_j wherever they appear in a transition intensity formula for the basic model; no other changes are needed.

The finite space H of phase count vectors. In either the basic or relaxed SPN formulation, we denote by H the set of all phase count vectors η that can occur in the canonical Markov state description (3.13) under a simply structured control policy. It follows from (3.15) and the service count bound (2.34) that this is a finite set.

3.6 A Mild Added Restriction on the CTMC Representation

Recall from (2.3) that we use the capital Greek letter Ψ to denote the complete set of initial processing variables for an SPN model. Ψ contains a number of pairs (v, φ) that have the following role in our model. First, v is a positive random variable representing the remaining service time, or residual service time, for a service of some type $j \in \mathscr{J}$ that is open at $t = 0$. Second, φ is the output vector for that same service, so it takes values in the output space Δ^j identified in Section 2.1. We call Ψ the model's *IP set*, and its elements (v, φ) are called *IP pairs*.

As a supplement to Assumption 3.1, we add the following very mild restriction, with justification to follow. This assumption will be used in Section 6.3 to establish the finiteness of a certain set Π^0, leading to the stochastic bound (6.36), which is needed in our analysis of fluid limits.

Assumption 3.8. As x ranges over the countable state space $\mathscr{X}$, there are just finitely many possible conditional distributions of IP pairs (v, φ) in Ψ, given that $X(0) = x$.

To make clear the content of this assumption, it may be helpful to consider again the first example discussed in Section 3.4. The network structure for that example is pictured in Figure 3.1, where the buffers occupied by the two job classes are labeled B1 and B2, and the single pool of two identical servers is labeled S1. To begin, we consider the same static buffer priority policy (hereafter, *SBP policy*) that was assumed earlier, but first-come-first-served processing will be considered shortly. There is a one-to-one correspondence between job classes and service types in this example, so we shall speak solely in terms of job classes in the discussion that immediately follows.

Service times for class 1 are distributed Erlang$(2, \alpha)$, and those for class 2 have the two-phase hyperexponential distribution that is denoted $\Pi_2(p, \beta)$ in Section D.8. Thus, for each class $i = 1, 2$ the current service phase of a class i job is either $s = 1$ or $s = 2$, and the conditional distribution of a job's residual service time is determined by its current service phase. Specifically, denoting by $\mathscr{D}(i, s)$ the residual service time distribution for a class i job in service phase s, we have the following: $\mathscr{D}(1, 1) = \text{Erlang}(2, \alpha)$, $\mathscr{D}(1, 2) = \exp(\alpha)$, $\mathscr{D}(2, 1) = \exp(\beta_1)$, and $\mathscr{D}(2, 2) = \exp(\beta_2)$.

The SBP policy that we have assumed is an example of a simply structured control policy, so we can adopt the canonical Markov state description $x = (\eta, z)$ that was proposed in Section 3.5. No more than two services can be open at $t = 0$, and Ψ consists of the residual service times for those open services, plus their associated output vectors, which in this example are simply zero vectors, because all jobs exit after completing the single service they require. Given any initial state $X(0) = x = (\eta, z)$, each of the residual service time random variables included in Ψ has one of the four conditional distributions listed in the previous paragraph, so Assumption 3.8 is satisfied.

Let us consider now the same example, but assuming that jobs are processed on a FCFS basis by the two servers, irrespective of class, giving us a control policy that is *not* simply structured. Combining the discussion of FCFS representations in Section 3.3 with that of phase-type representations in Section 3.4, one sees that a suitable Markov state description for the current example is

$$(3.16) \qquad x := (c_1, \ldots, c_q, s_1, s_2),$$

with the following interpretations: q is the total number of jobs currently in the system (that is, residing in either B1 or B2); c_1 is the class of the one that arrived first, ..., and c_q is the class of the one that arrived last; s_1 is the service phase of the older job currently in service (that is, the one that arrived first from the outside world), and s_2 is the service phase of the younger job currently in service; if $q = 1$ (just one job in the system, hence just one job in service, that being of class c_1 and in service phase s_1), then we set $s_2 = 0$, and the conditional distribution of the residual service time for the one job in service is $\mathcal{D}(c_1, s_1)$; and if $q = 0$ (empty system, so no jobs in service), then we set $s_1 = s_2 = 0$. If $q \geq 2$ (two services open at $t = 0$), then the conditional distributions of the residual service times for the older and younger of the jobs currently in service, given that $X(0) = x$, are $\mathcal{D}(c_1, s_1)$ and $\mathcal{D}(c_2, s_2)$, respectively. Thus Assumption 3.8 is again satisfied for the FCFS case.

All of the Markov representations encountered in this book have a structure similar to (3.16), as follows: the system state x contains phase indicators for the services that are open at $t = 0$, and the distribution of initial processing variables depends on x only through those phase indicators. More specifically, the joint distribution of residual service time v and output vector φ for a service open at $t = 0$ is determined by that service's phase indicator, and each service time distribution is of phase type with just finitely many phases. In this way, one can verify that Assumption 3.8 holds for each of the control policy families identified in this book, using the Markov representation that we specify for that family.

Is Assumption 3.8 actually needed? Based on the preceding paragraph, a reader might plausibly conclude that Assumption 3.8 is not just "very mild" but actually unnecessary, being implied by our other assumptions. Strictly speaking, that is not true, because we have never committed to any specific Markov representation in our general development, simply assuming that one meeting certain minimal requirements exists and is adopted (Assumption 3.1). We assume Poisson arrivals, phase-type service time distributions, etc., to ensure existence of a phase-oriented Markov representation for each of the specially structured models we analyze, but a user of our general theory might want to adopt a completely different representation. Alternatively, a user might want to weaken one or more of our baseline stochastic assumptions (see Section 2.2), and in either of those cases there is another restriction to be met, namely the one expressed here as Assumption 3.8. As an example, when a model includes service time distributions with countably many phases, the assumption is not satisfied.

3.7 Sufficient Condition for Irreducibility

Part of Assumption 3.1 requires that the ambient Markov chain X be irreducible. This assumption restricts somewhat the class of control policies to which our theory applies, but the policies that it excludes are seldom if ever of practical interest. In this section, we provide a simple sufficient condition for irreducibility, which is satisfied by most models one encounters in the literature, followed by an example showing how that condition might fail to hold.

Consider an SPN operating under a given control policy, denoting by N and Z its service count process and buffer contents process, respectively. Let $X = \{X(t), t > 0\}$ be a CTMC with countable state space $\mathcal{X}$ such that (3.1) and (3.3) hold for some function f and an empty state $x^* \in \mathcal{X}$. (Uniqueness of the empty state is *not* assumed initially.) Let us further suppose that, for any starting state $x \in \mathcal{X}$, there is a finite time $t > 0$ such that

$$(3.17) \qquad \mathbb{P}_x\{X(t) = x^*\} > 0.$$

In words, this means that the empty state x^* is reachable from any other state. Now we can redefine the state space of X to be the subset $\mathcal{X}^* \subset \mathcal{X}$ consisting only of states that are reachable from x^*. Then X

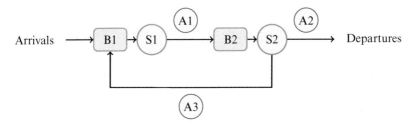

Figure 3.8 Tandem queueing network with three activities.

confined to the state space $\mathscr{X}^*$ is an irreducible CTMC for which (3.1) and (3.3) again hold.

Having seen that (3.17) ensures irreducibility, we now assume uniqueness of the empty state, as in the setting of Section 3.5, and show that the following condition is sufficient for (3.17) to hold: for each state $x \in \mathscr{X}$, there exists a time t such that

$$(3.18) \quad \mathbb{P}_x\{Z(t) = 0 | \tau > t\} > 0, \quad \text{where} \quad \tau = \inf\{t > 0 : E(t) \neq E(0)\}.$$

In words, τ is the time of the first external arrival, and (3.18) says there exists a time $t > 0$ such that, if external arrivals are turned off during $[0, t]$, the network will be empty at time t with positive probability.

Proposition 3.9. *Assume that the empty state x^* is unqiue. Under assumption (3.18), and given our baseline stochastic assumptions, the empty state x^* is reachable from any other state $x \in \mathscr{X}$.*

Proof. Fix a state $x \in \mathscr{X}$ and a time $t > 0$. Then

$$(3.19) \quad \begin{aligned} \mathbb{P}_x\{X(t) = x^*\} &= \mathbb{P}_x\{Z(t) = 0\} \\ &= \mathbb{P}_x\{\tau > t\}\mathbb{P}_x\{Z(t) = 0 | \tau > t\}. \end{aligned}$$

Under our baseline stochastic assumptions, τ has an exponential distribution not depending on the initial state x, so $\mathbb{P}_x\{\tau > t\} > 0$ for each $t \geq 0$. Combining that with (3.19) proves that (3.17) holds for any t satisfying (3.18). $\qquad\qquad\square$

A negative example. Consider the network pictured in Figure 3.8, which has two job classes, two servers, and three activities or service types: activity 1 (denoted A1 in the figure) is performed by server 1 (S1), and it transfers class 1 jobs to class 2; activity 2 (A2) is performed by server 2 (S2), and it removes class 2 jobs from the network; and activity 3 (A3) is also performed by S2, but it transfers class 2 jobs back to class 1. Service times for activity j are i.i.d. random variables distributed $\exp(\mu_j)$, where $\mu_j > 0$ $(j = 1, 2, 3)$, and all services must be carried through to completion without interruption once they have begun. External arrivals into class 1 follow a Poisson process with rate $\lambda > 0$, and there are no external arrivals into class 2.

As usual, "decisions times" are times at which either an external arrival occurs or a service is completed. We assume that S1 processes class 1 jobs in the order of their arrival without insertion of idleness. To specify the control policy used by S2, consider a decision time t, and as in Section 2.3, let $\hat{z}$ and $\hat{n}$ be the updated job count vector and updated service count vector, respectively, at t. If S2 has a decision to make at t, meaning that $\hat{n}_2 = \hat{n}_3 = 0$, then S2 uses the following rule: begin a type 3 service (which will send a class 2 job back to class 1) if $\hat{z}_2 = 1$, begin a type 2 service (which will remove a class 2 job from the network) if $\hat{z}_2 \geq 2$, and remain idle if $\hat{z}_2 = 0$.

For this network, an adequate Markov state description is $X(t) = (N(t), Z(t))$, and the unique empty state is $x^* = (0, 0)$. The rule specified for S2 ensures that $Z_1(t) + Z_2(t) \geq 1$ for all times t after the first external arrival occurs. That is, the empty state can never again be reached once there are any jobs in the system. One may argue that this example is artificial, featuring a control policy rather obviously contrived to prevent emptiness, but still the specified policy is nonidling and simply structured. Moreover, it can be shown that the network is stable (meaning that the CTMC X with a suitably restricted state space is positive recurrent) under the "obvious" load condition $\lambda < \mu_1 \wedge \mu_2$. In Chapter 12, where the modeling framework includes a discrete time parameter, a more subtle example will be presented in which emptiness is unachievable after a brief initial period; see specifically Remark 12.19.

3.8 Markov Representations with General State Space

Perhaps the simplest nontrivial example of a stochastic processing network is the $G/G/1$ queueing system that was introduced in Section 1.5. If both the interarrival time distribution and the service time distribution are exponential, then the buffer contents process Z is a continuous time Markov chain, but that is no longer true if those distributions are general. A Markov state description with general distributions is

$$x := (z, u),$$

where $u = (u_0, u_1)$ is a pair of "supplementary variables" that can be defined as follows (this is not the only choice, as we explain later): u_0 is the time that will elapse before the next arrival occurs, referred to hereafter as the *remaining interarrival time*; and u_1 is the time that will elapse before the service currently under way is completed (with $u_1 = 0$ by convention when $z = 0$, say), referred to hereafter as the *remaining service time*.

A similar augmentation can be used to obtain Markov state descriptions of more general systems. Consider, for example, the single-hop system pictured in Figure 3.1, which has two job classes and two servers, assuming for concreteness that jobs are scheduled via the SBP policy that gives priority to class 2. With general interarrival and service time distributions (as opposed to the exponential distributions that we originally assumed), a Markov state description for that system is

$$(3.20) \qquad\qquad x := (n, z, u),$$

where $n = (n_1, n_2)$ is the service count vector as usual, $z = (z_1, z_2)$ is the buffer contents vector (or job count vector) as usual, and u is a four-vector that specifies the remaining interarrival times for the two arrival processes and the remaining service times for each of the services currently under way. (Various conventions are possible for matching class designations to the remaining service time components of u, using information carried by the vector n, but that detail is irrelevant for current purposes.) That is, the corresponding state process

$$X := \left\{ \big(N(t), Z(t), U(t) \big), t \geq 0 \right\}$$

is Markov, but because the components of $U(t)$ take values in a continuum, X is an example of a Markov process with a *general* (nondiscrete) *state space*. Indeed, X falls in a category that Davis (1984) called *piecewise deterministic Markov processes* (PDMPs).

The foundational theory for Markov processes with general state space is considerably more complex than that for Markov chains. For example, instead of positive recurrence, which has a straightforward and elementary definition in the discrete state setting, one must use the more recondite notion of *positive Harris recurrence* in the general state setting. For that reason, and given the approximation theorems that are cited in Appendices D and E (specifically, the phase-type approximation theorem cited in Section D.8 and the MArP approximation theorem cited in the preamble to Appendix E), we have chosen to focus in this book on settings where Markov chain models can be employed.

Several additional comments are in order. (i) The fluid model approach to stability that will be developed in later chapters applies with equal ease to the general state setting, if one is willing to accept certain well-known sufficient conditions for positive Harris recurrence. Dai (1995a) and Bramson (2008) undertake just such an analysis for multiclass queueing networks with general distributions. (ii) For some service policies, such as global FIFO (see the last example in Section 3.3), a general state space is still needed to properly describe the system's dynamics *even when all underlying distributions are exponential.* (iii) We have described the components of u in (3.20) as remaining interarrival times and remaining service times. Such a state, while well defined mathematically, may not correspond to what is observable by a system manager. Alternatively, one can define the components of u as the "ages" of the arrival processes and of the ongoing services, that is, the time that has elapsed since the last arrival into each buffer and since the start of each service that is currently under way. The resulting theory is similar.

3.9 Sources and Literature

An important special class of queueing networks (see Section 2.6 for a general definition of that term) are open Jackson networks, the theory of which was developed in the pioneering paper by Jackson (1957). Open Jackson networks are defined by the following two restrictions, in addition to the assumptions that define a queueing network. First, there is a one-to-one correspondence between job classes and server pools; that is, $I = K$ and each server pool processes jobs from just one buffer. Second, all external arrival processes are Poisson, and all service time distributions are exponential. Thus, for a Jackson network the buffer contents process $Z = \{Z(t), t \geq 0\}$ is a CTMC. Under the standard load condition that will appear later as (5.1), Jackson proved the positive recurrence of Z by explicitly constructing a stationary distribution. (Theorem D.17 shows that existence of a stationary distribution implies positive recurrence.) Motivated by applications to computer systems and communication networks, there was an explosion of research in the 1970s extending this approach to more general queueing networks. Two important papers in that stream are Baskett et al. (1975) and Kelly

(1975), the content of which is summarized in the influential book by Kelly (1979). In additon to covering the classical theory, Serfozo (1999) covers work that appeared in the following 20 years, including some for models falling outside the queueing network framework.

The definition of a generalized Jackson network is the same as that of an open Jackson network except that service time distributions are allowed to be general (that is, not necessarily exponential) and interarrival time distributions for external arrival processes are also general. Such models can be studied in the setting of Section 3.8, where stability is defined as positive Harris recurrence of the ambient Markov process, the state space of which may be nondiscrete. Borovkov (1986) proved stability of a generalized Jackson network in that sense, given the standard load condition referred to previously, when all service time distributions and interarrival time distributions satisfy a strong Cramér condition. That condition essentially means that the tails of the interarrival time and service time distributions are asymptotically exponential. In particular, it implies that all such distributions have finite moments of all orders. Borovkov's strong condition was subsequently relaxed in independent work by Sigman (1990) and by Foss (1991). Meyn and Down (1994), retaining Borovkov's strong assumption on service times, were able to achieve stronger stability results, and their proofs used Lyapunov functions and newly developed stability criteria of Foster–Lyapunov type for Markov processes with general state space; see Meyn and Tweedie (2009) for an account of the latter theory. Compared with Jackson's short, elegant proof, the proofs in these papers are highly complex. The fluid methodology developed in Dai (1995a) and recapitulated in this book covers the generalized Jackson network as a special case; see Corollary 8.19.

4

Extensions and Complements

The first two sections of this chapter describe alternative model assumptions that generalize or extend the SPN formulations developed earlier in Chapter 2. Sections 4.3 through 4.5 concern network models with *processor sharing*, which can only be treated within our SPN framework after a standard model transformation. Section 4.6 describes two attractive control policies for queueing networks, and Section 4.7 identifies a family of specially structured models called "parallel-server systems," which have been the subject of numerous studies in recent years.

Section 4.8 discusses several alternative models of a small-scale example, thereby introducing the family of *fork-and-join networks*. That discussion also serves to illuminate some modeling issues that frequently arise in SPN applications. Finally, Section 4.9 provides brief comments on sources and literature.

4.1 Markovian Arrival Processes

Consider an SPN that satisfies Assumption 2.1 (baseline stochastic assumptions), operating under a control policy such that Assumption 3.1 (Markov representation) also holds. Part of the former assumption says that components of the external arrival process E are independent Poisson processes, and part of the latter assumption says that the network processes N and Z can be embedded in an irreducible CTMC called the ambient Markov chain, here denoted $X_0 = \{X_0(t), t \geq 0\}$, with generic state $x_0 \in \mathcal{X}_0$. (The reason for using a subscript zero in this notation will become apparent shortly.) When we say that N and Z can be "embedded" in X_0, this means there is a projection function f_0 such that

$$(4.1) \qquad (N(t), Z(t)) = f_0(X_0(t)), \quad t \geq 0.$$

Assumption 3.1 also ensures the existence of an "empty state" $x_0^* \in \mathcal{X}_0$ such that

(4.2) $$f_0(x_0^*) = (0,0).$$

Now consider that same model modified in just one regard, as follows: E is a Markovian arrival process (MArP), as defined in Appendix E, whose arrivals-regulating Markov chain (ARMC) $Y = \{Y(t), t \geq 0\}$ has generic state $s \in \mathcal{S}$. In this setting, X_0 may no longer have the Markov property when considered in isolation, and hence (4.1) may not provide a Markovian embedding. Rather, it is now natural to consider the *augmented ambient Markov chain* $X := (X_0, Y)$ with generic state $x = (x_0, s) \in \mathcal{X} := \mathcal{X}_0 \times \mathcal{S}$. (Note that the dynamic behavior of X_0 depends on that of Y, but not vice versa.) To embed N and Z in X, we can define a projection function $f : \mathcal{X} \to \mathbb{Z}_+^J \times \mathbb{Z}_+^I$ by setting

(4.3) $$f(x_0, s) := f_0(x_0),$$

so that

(4.4) $$f(X(t)) = f_0(X_0(t)) = (N(t), Z(t)), \quad t \geq 0.$$

Having successfully developed the embedding (4.4), let $s \in \mathcal{S}$ be any state of the ARMC and define $x^* := (x_0^*, s)$. Then $x^* \in \mathcal{X}$, and moreover $f(x^*) = (0,0)$ by (4.2) and (4.3). Thus x^* is an empty state for the augmented ambient Markov chain X. We conclude that X may have multiple empty states in the MArP setting, one for each state s of the ARMC Y.

To verify that the augmented ambient Markov chain X satisfies Assumption 3.1, it must be shown that X is irreducible. Following the construction in Section 3.7, one sufficient condition for irreducibility is existence of at least one empty state x^* that is reachable from any other state; under mild restrictions on the system manager's control policy, it can be shown that this is indeed the case.

Finally, in terms of generalizing the fluid model methodology to be developed in Chapter 6, the only characteristic of the Markovian arrival process that will be used in stability analysis is the vector λ of long-run arrival rates that appears in the associated SLLN, which is part (a) of Proposition E.7.

4.2 Alternate Routing with Immediate Commitment

In Section 1.4, we introduced "alternate routing" decisions by means of an example, pictured in Figure 1.3, in which one of the two arrival streams feeds into buffer 3, and those class 3 items can be processed by either server 1 or server 2; the routing decision (that is, which server will process the item) can be postponed until processing is actually to begin. That model fits neatly within the basic SPN framework laid out in Section 2.3.

Consider now the alternative model pictured in Figure 4.1, which is identical in all regards except that the routing decision must be made at the moment of arrival. In many applications, including the task allocation problem described later in Chapter 11, this requirement of "immediate commitment" is realistic, but it cannot be captured in the framework of Section 2.3. We therefore develop in this section an "augmented" SPN formulation whose only added feature is its allowance of alternate routing with immediate commitment. The development applies equally well to either the basic SPN model formulated in Section 2.3, or the relaxed model in Section 2.4; in that sense, this section formulates augmented versions of both the basic and relaxed SPN models.

In our augmented formulation, there are external arrivals from several different *sources*, and arrivals from any given source must be immediately directed into one of several acceptable buffers. For the example pictured in Figure 4.1, "source 1" arrivals are simply the external arrivals into class 1 that featured in our original example. "Source 2" arrivals are the ones that can be directed to either of the

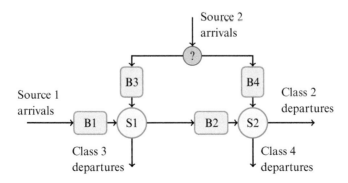

Figure 4.1 Alternate routing with immediate commitment.

two servers. Formal representation of this system requires that we create two different buffers in which source 2 arrivals may be stored, depending on how they are routed. That is, in the formal representation of the example we create two distinct item classes, identified as class 3 and class 4 in Figure 4.1, and each source 2 arrival is immediately converted to one of those two classes.

Formal development of the augmented SPN model proceeds as follows. First, recall that our buffers (or item classes) are indexed by $i \in \mathscr{I}$. Instead of viewing the I-dimensional arrival process E as primitive, we now take as primitive an L-dimensional process $U = \{U(t), t \geq 0\}$ of *uncommitted arrivals*. (Components of U correspond to the different "sources" referred to earlier.) Hereafter, let $\mathscr{L} := \{1, \ldots, L\}$. In the obvious way, part (a) of Assumption 2.1 (baseline stochastic assumptions) is replaced by the following: the uncommitted arrival processes $U_1, \ldots, U_L$ are independent Poisson processes with strictly positive arrival rates $v_1, \ldots, v_L$. From that, we have the following strong law of large numbers for uncommitted arrivals: with probability one, for each $\ell \in \mathscr{L}$,

$$(4.5) \qquad \frac{1}{t} U_\ell(t) \to v_\ell \quad \text{as } t \to \infty.$$

Less restrictively, one can allow U to be an arbitrary MArP; see Section 4.1 and Appendix E. In that case, by part (a) of Proposition E.7 (the SLLN for a MArP), there exists a vector $v = (v_\ell)$ such that (4.5) continues to hold; (E.5) and (E.6) express the vector v of long-run arrival rates in terms of MArP data, but using notation that conflicts badly with the notation of this section, and (4.18) provides an alternative, less explicit expression for v. Stability analysis is largely the same as in the case where uncommitted arrivals follow independent Poisson processes, except that one must use a more comprehensive Markov representation of the system state (this is elaborated later in this chapter).

In addition to the data and assumptions related to uncommitted arrivals, there is given an $L \times I$ *source-buffer matrix* $G = (G_{\ell i})$ whose entries are zeros and ones that specify the buffers to which uncommitted arrivals can be directed at the moment of their arrival. That is, the system manager can direct an arrival from source ℓ to any buffer i such that $G_{\ell i} = 1$, and he or she must direct that arrival to one of those buffers without delay. Mathematically, this can be formalized by requiring that the system manager specify, as part of his or her chosen control, an $L \times I$ matrix-valued process $V = \{V(t), t \geq 0\}$ with components $V_{li}(t)$ that are nondecreasing, right-continuous, and integer valued and further satisfy the following:

(4.6)
$$V_{\ell i}(\cdot) \equiv 0 \text{ if } G_{\ell i} = 0 \quad \text{and} \quad \sum_{i \in \mathscr{I}} V_{\ell i}(t) = U_{\ell}(t) \text{ for all } \ell \in \mathscr{L} \text{ and } t \geq 0.$$

One interprets $V_{\ell i}(t)$ as the cumulative number of arrivals from source ℓ that are routed to buffer i over the time interval $[0, t]$, and we define the I-dimensional process E of class-level external arrivals in the obvious way:

(4.7)
$$E_i(t) = \sum_{\ell \in \mathscr{L}} V_{\ell i}(t) \quad \text{for all } i \in \mathscr{I} \text{ and } t \geq 0.$$

An admissible control is now a pair (V, τ), where V has the properties specified immediately above and τ satisfies the restrictions laid out in Section 2.3. (Note that the restrictions on τ are expressed in terms of E, which is expressed in terms of V.) Our previous formulation corresponds to the special case where $L = I$ and G is the identity matrix.

MArP arrivals. In all future discussion of alternate routing with immediate commitment, we take the uncommitted arrival process U to be a MArP, as in Section 4.1, because that added generality is important for the model's application to task allocation in data centers, on which we focus in Chapter 11. Accordingly, we impose the modified version of Assumption 3.1 where (3.1) is replaced by (4.4), plus Assumption 3.8 as before; those assumptions guarantee the existence of a Markov representation with certain properties. The arrivals-regulating Markov chain underlying U is denoted $Y = \{Y(t), t \geq 0\}$, and the (finite) batch space for U is denoted Δ, as in Appendix E.

Simply structured routing policies. Also, attention is restricted hereafter to *simply structured routing policies*, which means the following: if t is an arrival time for the uncommitted arrival process U, and the jump of U at that time is $\Delta U(t) = \delta$, then the corresponding jump of the routing process V is

(4.8)
$$\Delta V(t) = g(Z(t-), \delta),$$

where $g : \mathbb{Z}_+^I \times \Delta \to \mathbb{Z}_+^{L \times I}$ is a policy function satisfying

(4.9)　　　$g_{\ell i}(\cdot, \cdot) \equiv 0$ if $G_{\ell i} = 0$　and　$\sum_{i \in \mathscr{I}} g_{\ell i}(z, \delta) = \delta_\ell$

for all $z \in \mathbb{Z}_+^I$ and $\delta \in \Delta$.

Of course, (4.9) is necessary for the simply structured policy to satisfy (4.6). In words, (4.8) means that the allocation of new arrivals to job classes depends only on the buffer contents vector immediately before the arrival event and the composition of the arrival batch.

For the proof of the following proposition, readers are reminded that (4.4) is assumed to hold for some function f, where $X = \{X(t), t \geq 0\}$ is an ambient CTMC. This proposition will be used later in the proof of Corollary 5.5, which sets the stage for our analysis of task allocation models in Chapter 11. (Specifically, Corollary 5.5 is cited in the proof of Lemma 11.2).

Proposition 4.1 (SLLN for class-level arrivals). *Consider an SPN with alternate routing and immediate commitment. Under the previously stated assumptions, and further assuming the SPN is stable (that is, the ambient Markov chain X is positive recurrent), there exists a matrix $\phi \in \mathbb{R}_+^{L \times I}$ and a vector $\lambda \in \mathbb{R}_+^I$ such that*

(4.10)　　$\mathbb{P}\left\{ \lim_{t \to \infty} \frac{1}{t} V(t) = \phi \right\} = 1$　*and*　$\mathbb{P}\left\{ \lim_{t \to \infty} \frac{1}{t} E(t) = \lambda \right\} = 1.$

Moreover,

(4.11)　　$\phi_{\ell i} = 0$ *if* $G_{\ell i} = 0, \quad \sum_{i \in \mathscr{I}} \phi_{\ell i} = v_\ell$ *for all* $\ell \in \mathscr{L},$

and

(4.12)　　　　$\lambda_i = \sum_{\ell \in \mathscr{L}} \phi_{\ell i}$ *for all* $i \in \mathscr{I}..$

Proof.　Fix a state $x \in \mathscr{X}$. Define $T_x^{(0)} := 0$, $\tau_x^{(0)} := \inf\{t > 0 : X(t) \neq x\}$, and for $n = 1, 2, \ldots,$

$$T_x^{(n)} := \inf\left\{ t > \tau_x^{(n-1)} : X(t) = x \right\},$$

$$\tau_x^{(n)} := \inf\left\{ t > T_x^{(n)} : X(t) \neq x \right\}.$$

Define $\sigma_n := T_x^{(n)}$ for $n = 0, 1, \ldots$. Because X is positive recurrent by hypothesis, $\sigma_n < \infty$ almost surely for each n, and moreover, $0 \leq \mathbb{E}(\sigma_{n+1} - \sigma_n) < \infty$ for all $n = 1, 2, \ldots$. By the strong Markov property of X (Theorem D.13) and Corollary D.14, the sequence

$$\left\{ \sigma_n - \sigma_{n-1}, \left(X(s), \sigma_{n-1} \leq s < \sigma_n \right) : n = 1, 2, \ldots \right\}$$

is i.i.d. Recall from Section 4.1 that the arrival process $\{U(t), t \geq 0\}$ is generated using an arrivals-regulating Markov chain Y, which is a component of the CTMC X, plus some Poission processes that are independent of X. Therefore, the sequence

(4.13)
$$\left\{ \sigma_n - \sigma_{n-1}, \left((Z(s), U(s) - U(\sigma_{n-1})), \sigma_{n-1} \leq s < \sigma_n \right) : n = 1, 2, \ldots \right\}$$

is i.i.d. For $n = 1, 2, \ldots$, define

$$\eta(n) := \sigma_n - \sigma_{n-1},$$
$$\xi(n) := V(\sigma_n) - V(\sigma_{n-1}),$$
$$\zeta(n) := U(\sigma_n) - U(\sigma_{n-1}).$$

It follows from (4.13) and (4.8) that the sequence

$$\left\{ (\eta(n), \xi(n), \zeta(n)) : n = 1, 2, \ldots \right\} \text{ is i.i.d.}$$

Thus, by the strong law of large numbers, with probability one,

(4.14)
$$\lim_{n \to \infty} \frac{\sigma_n}{n} = \lim_{n \to \infty} \frac{\sum_{\ell=1}^n \eta(\ell)}{n} = \mathbb{E}(\eta(1)) = \mathbb{F}_{ix}(\sigma_1) < \infty,$$

(4.15)
$$\lim_{n \to \infty} \frac{V(\sigma_n)}{n} = \lim_{n \to \infty} \frac{\sum_{\ell=1}^n \xi(\ell)}{n} = \mathbb{E}(\xi(1)) = \mathbb{E}_x(V(\sigma_1)),$$

(4.16)
$$\lim_{n \to \infty} \frac{U(\sigma_n)}{n} = \lim_{n \to \infty} \frac{\sum_{\ell=1}^n \zeta(\ell)}{n} = \mathbb{E}(\zeta(1)) = \mathbb{E}_x(U(\sigma_1)).$$

For each $t \geq 0$, let $n(t) := \sup\{n \geq 0 : \sigma_n \leq t\}$. In words, $n(t)$ is the number of reentries into state x that occur over the interval $[0, t]$. Thus $\{n(t), t \geq 0\}$ is a piecewise-constant, right-continuous process, and

$$\sigma_{n(t)} \leq t < \sigma_{n(t)+1}.$$

Fixing t for the moment and setting $n := n(t)$ to compactify notation, it follows that

(4.17)
$$\frac{n}{\sigma_{n+1}} \frac{V(\sigma_n)}{n} = \frac{1}{\sigma_{n+1}} V(\sigma_n) \le \frac{1}{t} V(t)$$
$$\le \frac{1}{\sigma_n} V(\sigma_{n+1}) = \frac{n+1}{\sigma_n} \frac{V(\sigma_{n+1})}{n+1}.$$

Because (4.14) holds, $n = n(t) \to \infty$ with probability one as $t \to \infty$. Therefore, it follows from (4.14), (4.15), and (4.17) that, with probability one,

$$\lim_{t \to \infty} \frac{1}{t} V(t) = \frac{\mathbb{E}_x(V(\sigma_1))}{\mathbb{E}_x(\sigma_1)} := \phi.$$

This proves the first part of (4.10). Then defining λ in terms of ϕ via (4.12), the second part of (4.10) follows from the definition (4.7) of E. Using (4.14), (4.16), and a pair of "bracketing" inequalities similar to those in (4.17), one obtains, with probability one,

(4.18)
$$\lim_{t \to \infty} \frac{1}{t} U(t) = \frac{\mathbb{E}_x(U(\sigma_1))}{\mathbb{E}_x(\sigma_1)} := v,$$

which is the strong law of large numbers (4.5) for the Markovian arrival process U. Finally, (4.11) follows from (4.6), (4.10), and (4.5). □

A more elaborate example.   Before concluding this section, it may be helpful to provide a second, slightly more elaborate example of alternate routing with immediate commitment. Figure 4.2 portrays a four-server network in which source 1 arrivals can be directed to either buffer 1 or buffer 3. In the former case, they will be processed first by server 1 and then by server 2, perhaps with intermediate storage, and in the latter case they will be processed first by server 3 and then by server 4. In similar fashion, source 2 arrivals must be committed to one of two routes upon arrival: either server 1 followed by server 3, or else server 2 followed by server 4. The point illustrated by this example is that an initial routing decision may not only commit the new arrival to immediate processing by a particular means, but also constrain or dictate subsequent processing options for that arrival.

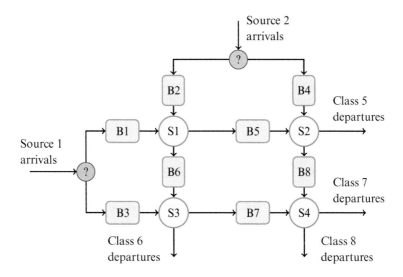

Figure 4.2 Four-server network with alternate routing and immediate commitment.

4.3 PS Networks

We consider in this section a family of models defined by three features: *relaxed control* (see Section 2.4), which includes the assumption of single-server pools; a *unitary* network structure (see Section 2.6); and a *processor sharing* service discipline (see the following subsection). These will be called *processor sharing networks*, or *PS networks* for brevity. A simple example of a PS network is the classical processor sharing queueing model introduced by Kleinrock (1967), which was mentioned briefly in Chapter 2 (see Section 2.5) and will be discussed again at the end of Section 4.4.

Recall that in a unitary network we have $J = I$ (just one processing activity per job class), and B is the $I \times I$ identity matrix. This means that activity i processes (only) jobs of class i, and there is no other way to process those jobs. Accordingly, both service types and job classes will be indexed in this section by $i \in \mathcal{I}$. Columns of the $K \times I$ capacity consumption matrix A may have more than one positive entry, meaning that more than one server may be involved in processing any given job class.

The unitary network assumption further stipulates that jobs change class in Markovian fashion after completing service. As in Section 2.6, we denote by $P = (P_{ij})$ the $I \times I$ *routing matrix*, assuming P to be substochastic and transient (that is, $P^n \to 0$ as $n \to \infty$); one interprets P_{ij} as the probability that a class i item or job, upon completion of its service, next becomes a class j item or job, independent of all previous history.

Processor sharing service discipline. In queueing theory, the term "service discipline" refers to the order in which servers process jobs or customers. Two familiar examples are first-in-first-out service and last-in-first-out service. PS is yet another service discipline, and it actually contradicts certain assumptions of our relaxed SPN model formulation in Section 2.4. For that reason, the following paragraphs provide a complete and self-contained description of relaxed control with processor sharing.

First, each time a service is completed and a job changes class, its next service is opened immediately, and in similar fashion, the service of each external arrival is opened immediately after its arrival. Thus, at each time t, rather than having at most one open service of each type, the number of open services of any type i equals the number of jobs occupying buffer i.

Decision times are as described in Section 2.4, and in the current model the updated service count vector $\hat{n}$ and updated job count vector $\hat{z}$ (these quantities were originally defined in Section 2.3) are identical at every decision time. At each decision time t, we have

$$(4.19) \qquad \beta = h(\hat{z}),$$

where β is the I-vector of service rates for different job classes going forward from t, and $h : \mathbb{Z}_+^I \to \mathbb{R}_+^I$ is a policy function that satisfies the capacity constraint

$$(4.20) \qquad Ah(\cdot) \leq b.$$

Finally, for each $i \in \mathscr{I}$, the service rate or service effort β_i at any given time is *divided equally* among the type i services that are then open; expressing that in slightly different words, one may say that the total service effort devoted to class i jobs at any given time is divided equally among the class i jobs then in the system.

To repeat, the family of network models described earlier (that is, unitary networks with relaxed control and processor sharing) will be called *PS networks* for brevity. PS networks do *not* fit within the relaxed SPN framework developed in Section 2.4, because in that formulation the system manager cannot open a new service of any given type until previously opened services of that type have all been completed. However, given our assumption that service time distributions are of phase type (see Assumption 2.1), the following will be explained in Section 4.4: a PS network is equivalent in a distributional sense to another SPN model that meets all of our standard assumptions (that is, the assumptions imposed in Sections 2.1, 2.2, and 2.4), and hence can be analyzed using the methods developed in this book.

Markov representation. To specify a PS network in mathematical terms, one can adopt the relatively simple Markov state description

$$(4.21) \qquad \eta := \big(\eta_i(s_i), s_i \in \mathscr{S}_i, i \in \mathscr{I}\big),$$

where $\eta_i(s_i)$ is the number of type i services currently open whose service regulating Markov chain is in service phase s_i $(i \in \mathscr{I}, s_i \in \mathscr{S}_i)$; this is a simplified version of the Markov state description (3.13) and (3.14) that was adopted in Section 3.5. In the current context, one can describe η as a vector of *refined job counts*, taking the view that jobs are sorted into a large number of *refined job classes*, each refined class corresponding to a different (i, s_i) pair. As noted earlier in (3.15), one can extract from η the vector n of service counts, which is equivalent in the current model to the vector z of buffer contents. That is,

$$(4.22) \qquad n_i = z_i = \sum_{s_i \in \mathscr{S}_i} \eta_i(s_i), \quad i \in \mathscr{I}.$$

In the obvious way, we denote by $\eta = \{\eta(t), t \geq 0\}$ the vector process of refined job counts in our PS model, and denote by $\eta_i(s_i, t)$ its (i, s_i)th component process. Thus (4.22) can be rewritten in process terms as follows:

$$(4.23) \qquad N_i(t) = Z_i(t) = \sum_{s_i \in \mathscr{S}_i} \eta_i(s_i, t), \quad i \in \mathscr{I}, t \geq 0.$$

We assume that the buffer contents process Z is observable, but not that η is, noting that the service rate vector β in (4.19) depends on the current state η only through z.

Now consider an arbitrary time t, denoting by η the corresponding system state, and by β the vector of service rates determined at the decision time immediately preceding t. Also, let $\beta_i(s_i)$ be the portion of β_i that is allocated to class i jobs undergoing service phase s_i, or equivalently stated, the portion of β_i devoted to *refined* job class (i, s_i). The PS service discipline dictates that

$$(4.24) \qquad \beta_i(s_i) = \begin{cases} \beta_i \frac{\eta_i(s_i)}{z_i} & \text{if} \quad z_i > 0, \\ 0 & \text{if} \quad z_i = 0. \end{cases}$$

That is, in our PS model the total service rate β_i is divided among the refined job classes $(i, s_i), s_i \in \mathscr{S}_i$, in proportion to their current population sizes $\eta_i(s_i), s_i \in \mathscr{S}_i$. But beyond that, the PS service discipline further dictates that the allocation $\beta_i(s_i)$ be divided equally among the $\eta_i(s_i)$ individual jobs that currently have the refined class designation (i, s_i).

Defining stability for a PS network. As noted earlier, a PS network is not an SPN as we have defined that term in Chapter 2, because in a PS network two or more jobs of a given class can receive service simultaneously, which is not allowed in our relaxed SPN formulation. Strictly speaking, then, the definition of SPN stability in Section 3.1 does not apply to a PS model, but we can and do adopt the following essentially identical definition.

Definition 4.2. A PS network is said to be *stable* if the CTMC η defined by (4.21) is positive recurrent.

4.4 Equivalent Head-of-Line Model for a PS Network

Continuing discussion of a PS network, let us consider now the stochastic evolution of its ambient Markov chain η. The crucial observation is the following: the transition intensities that comprise the generator matrix for η remain the same if we assume that all of the service effort allocated to a particular refined job class is devoted to just one of the jobs having that refined class designation. The justification for this claim is twofold: first, the duration of any particular service phase s_i for any particular job class i is by definition exponentially distributed; and second, the minimum of ℓ independent random variables, each exponentially distributed with mean ℓm, is exponentially distributed with mean m.

Hereafter we focus on an alternative model that is identical to the PS model except for the following: the service effort allocated to each refined job class is *devoted entirely to the job of that refined class that first made the transition to its current service phase.* Imagining that jobs of each refined class form a line in the order to their arrival, the alternative model can be described as one in which service effort is allocated within each refined class on a *head-of-line* (HL) basis, or oldest-service-first basis. As noted in the previous paragraph, this alternative model gives rise to an ambient Markov chain $\tilde{\eta}$, where $\tilde{\eta}_i(s_i, t)$ is the number of jobs in refined class (i, s_i) at time t that has the same generator matrix as the Markov state process η for our original PS model.

Hence, $\tilde{\eta}$ and η have the same distribution when their initial conditions (that is, the distributions of their initial processing variables) are identical. Hereafter, we use the term *equivalent HL model*, or *EHL model* for brevity, when referring to the alternative head-of-line model introduced in the preceding paragraph, and also when referring to the associated Markov chain $\tilde{\eta}$. Both processes and parameters associated with the EHL model will be denoted with tildes.

Conformance with the relaxed SPN formulation. To establish that the EHL model satisfies all the assumptions of our relaxed SPN formulation (Section 2.4), it will be helpful to describe it in slightly different terms, introducing a modest amount of additional notation for that purpose. Recall from Appendix D.8 that the phase-type service time distribution for any job class in our original PS model is characterized by a triple (p, γ, P), where p is an initial distribution over transient service phases, γ is a vector of exit rates from transient service phases, and P is a Markov transition matrix among transient service phases. In this section, the service time parameters for a particular class $i \in \mathcal{I}$ in the original PS model will be denoted (p^i, γ^i, P^i). In contrast, the letter P with no superscript will denote hereafter the routing matrix for our original PS model; that is, P_{ij} is the probability that a class i job transitions to class j after completing all phases of its class i service.

In the EHL model, we have a finite collection $\mathcal{C}$ of refined job classes. That is, each $c \in \mathcal{C}$ corresponds to a pair (i, s_i), where $i \in \mathcal{I}$ and $s_i \in \mathcal{S}_i$; we shall write $c = (i, s_i)$ when it is necessary to speak explicitly about that correspondence. External arrivals into refined classes $c \in \mathcal{C}$ occur according to independent Poisson processes, and we denote by $\tilde{\lambda}(c)$ the external arrival rate into refined class c; of course, one may have $\tilde{\lambda}(c) = 0$ for some $c \in \mathcal{C}$. Service times for jobs of refined class c are exponentially distributed with mean $\tilde{m}(c) > 0$, and jobs switch from one refined class to another in Markovian fashion after completing service. We denote by $\tilde{P} = \big(\tilde{P}(c,d)\big)$ the associated routing matrix.

The Poisson arrival rates $\tilde{\lambda}(c)$ for our EHL model are derived from the external arrival rates λ_i in the original PS model, plus the initial distributions p^i for that model's phase-type service time distributions, and the mean service time $\tilde{m}(c)$ are derived from the exit rates γ^i. Specifically, if $c = (i, s_i)$, then $\tilde{\lambda}(c) = \lambda_i p^i(s_i)$ and $\tilde{m}(c) = 1/\gamma^i(s_i)$. The routing probabilities $\tilde{P}(c,d)$ for the EHL model are similarly derived from the initial distributions p^i, the phase-transition matrices P^i, and the routing matrix P of the original PS model as follows: for $c = (i, s_i)$ and $d = (j, s_j)$,

$$(4.25) \qquad \tilde{P}(c,d) = \begin{cases} P^i(s_i, s_j) & \text{if } i = j, \\ \big(1 - \sum_{s_i' \in \mathscr{S}_i} P^i(s_i, s_i')\big) P(i,j) p^j(s_j) & \text{if } i \neq j. \end{cases}$$

The factor $1 - \sum_{s_i' \in \mathscr{S}_i} P^i(s_i, s_i')$ is the probability that completion of service phase s_i for a class i job actually completes the entire service of that job. The term $P(i,j) p^j(s_j)$ is the probability that this completed class i job becomes a class j job next, and that its service begins in phase s_j. The substochastic matrices $P^1, \ldots, P^I$ and P are all transient, and it follows from this that $\tilde{P}$ is transient as well. The resource consumption matrix $\tilde{A}$ for the EHL model is given by $\tilde{A}_{kc} = A_{ki}$ for each server $k \in \mathscr{K}$ and each refined class $c = (i, s_i)$ with $i \in \mathscr{I}$ and $s_i \in \mathscr{S}(i)$.

To establish conformance with assumptions of our relaxed SPN formulation, one must identify as "decision times" in the EHL model all times t at which either an external arrival occurs or a service *phase* is completed, because the service rate allocations to different refined job classes generally change at those times. Let us agree to call these *refined decision times* for the EHL model.

We denote by $\hat{\eta}(c)$ the number of jobs occupying refined job class c in the EHL model immediately after a refined decision time t, that is, after the refined job counts have been updated to reflect external arrivals, silent transitions among service phases, and service phase completions resulting in either departures from the network or transitions from one job class to another, but before they have been updated to reflect new service starts. This use of a hat to denote an updated quantity immediately after a refined decision time follows a practice that originated in Section 2.3, but it should be noted that here $\hat{\eta}(\cdot)$ denotes an updated version of the EHL phase count vector $\tilde{\eta}(\cdot)$, rather than an updated version of the original PS phase count vector $\eta(\cdot)$.

A control policy for the EHL model is specified by a *refined decision function* $\tilde{h}$ that determines service rate allocations going forward from t, and it will be defined in terms of the decision function h that is inherited from the original PS model. In preparation, we set

$$(4.26) \qquad \hat{z}_i := \sum_{c \in \mathscr{C}(i)} \hat{\eta}(c) \quad \text{for } i \in \mathscr{I},$$

where $\mathscr{C}(i)$ is the set of all $c \in \mathscr{C}$ such that $c = (i, s_i)$ for some $s_i \in \mathscr{S}_i$. Thus $\hat{z}_i$ is the number jobs in our original "unrefined" class i immediately after the refined decision time referred to previously.

To write out the EHL policy function $\tilde{h}$ in terms of h, we shall take as given an updated phase count vector $\hat{\eta} = (\hat{\eta}(c), c \in \mathscr{C})$ and define a corresponding vector $\tilde{\beta} = (\tilde{\beta}(c), c \in \mathscr{C})$, finally setting $\tilde{h}(\hat{\eta}) = \tilde{\beta}$. Specifically, we define $\hat{z}$ in terms of $\hat{\eta}$ via (4.26), set $\beta = h(\hat{z})$ as in (4.19), and then define $\tilde{\beta}(c)$ via the following obvious modification of (4.24) for $c = (i, s_i)$:

$$(4.27) \qquad \tilde{\beta}(c) := \begin{cases} \beta_i \dfrac{\hat{\eta}(c)}{\hat{z}_i} & \text{if} \quad \hat{z}_i > 0, \\ 0 & \text{if} \quad \hat{z}_i = 0. \end{cases}$$

Equation (4.27) expresses in mathematical form the EHL service mechanism that was described verbally earlier. With its policy function $\tilde{h}$ defined in this way, the EHL model conforms to all structural requirements of the relaxed SPN formulation in Section 2.4, and its stochastic elements (Poisson arrivals, exponential service time distributions, and Markovian switching between refined job classes) satisfy our baseline stochastic assumptions (Section 2.2) as well.

Equivalent load vector computations. To repeat, job classes in the EHL model have the form $c = (i, s_i)$, where $i \in \mathscr{I}$ and $s_i \in \mathscr{S}_i$. The number of such classes is

$$\tilde{I} := \sum_{i \in \mathscr{I}} \#\{\mathscr{S}_i\},$$

where $\#\{\cdot\}$ denotes the number of elements in a finite set. The EHL model is a unitary network, as defined in Section 2.6, having $\tilde{I}$ classes and an equal number of processing activities. In the preceding paragraphs, we have specified the following data for this unitary network, expressing each in terms of data for the PS network from which it was derived: an $\tilde{I}$-vector $\tilde{\lambda}$ of external arrival rates, a $K \times \tilde{I}$ capacity consumption matrix $\tilde{A}$, an $\tilde{I} \times \tilde{I}$ routing matrix $\tilde{P}$, and a mean service time $\tilde{m}(c)$ for each class c. Let us now define an $\tilde{I}$-dimensional diagonal matrix $\tilde{M}$ by analogy with (2.41), and let $\tilde{\rho}$ be the K-dimensional load vector for the EHL model, as defined in Section 2.6. Then formulas (2.40) and (2.38) together give

(4.28) $$\tilde{\rho} = \tilde{A}\tilde{M}\tilde{\alpha}, \quad \text{where} \quad \tilde{\alpha} = (I - \tilde{P}')^{-1}\tilde{\lambda}.$$

In similar fashion, one can define a load vector $\rho = (\rho_1, \ldots, \rho_K)'$ for the original PS network by setting

(4.29) $$\rho = AM\alpha, \quad \text{where} \quad \alpha = (I - P')^{-1}\lambda.$$

Components of the vector ρ in (4.29) are interpreted as in Section 2.6, namely, ρ_k is the total amount of service effort required from server k per time unit, expressed in units like server-hours per hour. Of course, $\tilde{\rho}_k$ has that same interpretation in the EHL model, so the following proposition is "obvious" in a sense; its formal proof simply confirms that no mistake has been made in the development to this point.

Proposition 4.3. $\tilde{\rho} = \rho$.

Proof. From the definition (4.28) of the total arrival rate vector $\tilde{\alpha}$, one can verify that the column vector $\big(\tilde{\alpha}(i, s_i), s_i \in \mathscr{S}(i)\big)$ is given by

$$\alpha_i \big(I - (P^i)'\big)^{-1} p^i.$$

Thus

$$\begin{aligned}
\tilde{\rho}_k &= \sum_{i \in \mathscr{I}} A_{ki} \sum_{s_i \in \mathscr{S}(i)} \tilde{\alpha}(i, s_i)/\gamma^i(s_i) \\
&= \sum_{i \in \mathscr{I}} A_{ki}\alpha_i e' \big[\operatorname{diag}(\gamma^i)\big]^{-1} \big(I - (P^i)'\big)^{-1} p^i \\
&= \sum_{i \in \mathscr{I}} A_{ki}\alpha_i m_i = \rho_k,
\end{aligned}$$

where e is a column vector of ones and the second equality follows from (D.41). $\qquad\square$

Two equivalent notions of stability. Having established that the EHL model is a relaxed SPN, as that term was defined in Section 2.4, we can and do define stability of the EHL model as in Section 3.1. However, the following equivalence is immediate from the fact that η (the ambient Markov chain for our original PS model) and $\tilde{\eta}$ (the ambient Markov chain for the corresponding EHL model) have the same generator, and hence have the same recurrence classification.

Proposition 4.4. *A processor sharing network is stable in the sense of Definition 4.2 if and only if its equivalent head-of-line model is stable in the sense of Definition 3.6.*

Classical processor sharing queue. In the final paragraphs of Section 2.5, mention was made of the PS queueing model introduced by Kleinrock (1967). In it, one has a single server and independent Poisson arrivals into each of I different customer classes, each of which has its own phase-type distribution of service times. At each time $t \geq 0$, the server divides its capacity equally among all jobs present in the system, so Kleinrock's model is a special case of the general PS network formulation propounded in Section 4.3.

Its corresponding EHL model has routing among a number of refined customer classes (typically the number of them is larger than I), each of those refined classes has an exponential service time distribution, the server divides its capacity among the refined classes in proportion to their population sizes at any given time, and the service rate allocated to any one refined class is all directed to the oldest member of that class present. This service discipline is called head-of-line proportional processor sharing (HLPPS); see Section 4.6 for further discussion.

4.5 Bandwidth Sharing Networks

We consider in this section a class of stochastic processing networks that generalize the bandwidth sharing (BWS) example presented in Section 1.4. These *bandwidth sharing networks* are primarily of interest for modeling the dynamic behavior of Internet flows, in which context they are also called *connection-level models* or *flow-level models*. The standard BWS model is introduced in the following subsection. It is a single-hop special case of a PS network, as described in Section 4.3, having I different job classes, $J = I$ processing activities or service types (that is, a one-to-one correspondence between job classes and service types), and K distinct servers or processing resources.

In the canonical application to Internet modeling, the "servers" represent links of a communication network, "jobs" represent files requiring transfer over one or more of those links, and the "route" of a job is by definition the set of links used to transfer it from its point of origin to its ultimate destination. (Thus the word "route" is used to mean a collection of "servers" throughout this section, which is different from the meaning ascribed to the word in Section 2.6 and elsewhere in this book.) By assumption, a job (or the file it represents) is transferred over all links on its route simultaneously, without intermediate buffer storage, so the execution of a "service" (that is, a file transfer) requires simultaneous capacity allocations on each of the links that constitute the route for that job. This simultaneous resource possession is the model's salient feature, and the assumption of simultaneous transfer over all links on the route, without intermediate buffering, is what gives the model its single-hop character.

In the standard BWS model described in the following subsection, each job class has a unique associated route. One may also consider more complex models with *multipath routing*, in which there may be more than one route available for serving jobs of any given class, and hence a many-to-one relationship between service types and job classes; see Section 4.9.

The standard model. The standard BWS model is a special case of the PS network formulated in Section 4.3. As stated previously, it is a single-hop model, meaning that jobs simply leave the system when their one service is complete. It should be emphasized, however, that the "equivalent head-of-line model" introduced in the next subsection is *not* single-hop.

There are jobs of classes $1,\ldots,I$ that arrive according to independent Poisson processes at rates $\lambda_1,\ldots,\lambda_I$, and each class i arrival has a *size* that is drawn from a class-specific distribution with mean $m_i > 0$ ($i = 1,\ldots,I$). The job sizes for the different classes form I mutually independent sequences of independent and identically distributed (i.i.d.) random variables. In the usual way, let $\mu_i := m_i^{-1}$ ($i = 1,\ldots,I$). We assume the job size distribution for each class to be of phase type.

The system's processing resources are *links* numbered $1,\ldots,K$, which have capacities $b_1,\ldots,b_K$, respectively; the significance of link capacities will be explained shortly. The processing of a job is accomplished by allocating a *flow rate* to it over time: a job departs from the system when the integral of its allocated flow rate equals its size. In general, the processing of a job consumes the capacity of several different resources simultaneously, as follows. There is given a nonnegative $K \times I$ matrix

$A = (A_{ki})$, and each unit of flow allocated to type i jobs consumes A_{ki} units of link k capacity ($k \in \mathcal{K}$ and $i \in \mathcal{I}$). Thus, denoting by $\beta = (\beta_1, \ldots, \beta_I)'$ the vector of total flow rates allocated to the various job classes at a given time, $A\beta$ is the corresponding capacity consumption vector. It follows that β must satisfy the capacity constraint $A\beta \leq b = (b_1, \ldots, b_K)'$. Hereafter we use the terms "capacity allocation" and "flow rate allocation" interchangeably.

Job sizes are measured in units like megabytes, so flow rates and link capacities are expressed in units like megabytes per second, and the capacity consumption rate A_{ik} is positive or zero depending on whether link k is or is not part of the route used by jobs of class i. One naturally thinks in terms of the case where elements of A are binary (that is, $A_{ki} = 1$ if link k is part of the route used by class i jobs, and $A_{ki} = 0$ otherwise), but imposing that restriction does not actually simplify analysis of the model.

In the preceding model description, we have denoted by β_i the total flow rate allocated to class i jobs at a particular time, without regard to how that allocation is divided among individual jobs. To resolve that ambiguity, an *equal sharing* rule is assumed hereafter, which means that the class i flow rate allocation is divided equally among all class i jobs present. This assumption is standard in the literature, and it corresponds at least roughly with actual practice in Internet flow management. As stated earlier, our standard BWS model is therefore a special case of the PS network formulation in Section 4.3. Because the BWS model is single-hop, its $I \times I$ routing matrix contains only zeros.

Equivalent head-of-line model. From the development in Section 4.4, we know that the standard BWS model described in the preceding subsection is equivalent to a head-of-line model that satisfies all the assumptions of our relaxed SPN formulation (Section 2.4). There are two steps involved in the transition from our original model to the EHL model: first, we describe system state in terms of *refined job classes*, and then we convert from equal sharing to head-of-line capacity allocation *within* each refined job class. The job count process Z of the original BWS network has the same distribution as an *aggregated* job count process $\hat{Z}$ for the EHL model; each component of $\hat{Z}$ is a sum of job counts in refined classes (i, s_i) that have a common first component, as in (4.26). As noted in Section 4.4, the EHL model is a unitary network.

4.6 Queueing Networks with HLSPS and HLPPS Control

Hereafter we use the term "relaxed queueing network" to mean a queueing network model, as defined in Section 2.6, subject to relaxed control, as defined in Section 2.4. It is a requirement of our relaxed SPN formulation that such a network have single-server stations, and as before we denote by $\mathscr{I}(k)$ the set of buffers (or job classes) that are processed by server $k \in \mathscr{K}$. We introduce in this section two examples of relaxed queueing networks, the stability of which will be studied later in Sections 8.4 and 8.2, respectively.

HLSPS control.   One type of relaxed control policy is identified by the phrase *head-of-line static processor sharing* (HLSPS). Here the word "static" contrasts with the "dynamic" sharing of server capacity that occurs under the HLPPS policy described in the following subsection. Associated with each HLSPS policy is a *proportion vector* $\gamma = (\gamma_1, \ldots, \gamma_I) > 0$ that satisfies

$$(4.30) \qquad \sum_{i \in \mathscr{I}(k)} \gamma_i = 1$$

for each server k. When the network operates under such a policy, all nonempty buffers simultaneously receive service. Specifically, job class $i \in \mathscr{I}(k)$ receives a fraction γ_i of pool k capacity. Thus the service rate received by class i is $b_k \gamma_i$.

In accordance with the definition of a simply structured relaxed control policy (Section 2.4), all of this service goes to the oldest job in buffer i when the buffer is nonempty. (The term "head-of-line" derives from imagining jobs in each buffer as lined up in the order of their most recent arrival to that buffer, so the oldest job in a buffer is the one at the head of the line.) When buffer i is empty, the capacity allocated to that class is lost (wasted) by assumption. A special case of interest is the *egalitarian* HLSPS policy, which has $\gamma_i = \gamma_j$ whenever classes i and j are processed by the same server.

To describe this relaxed control policy in mathematical terms, one can say that the vector β of class-level service rates going forward from a decision time t is $\beta = h(\hat{z})$, where $\hat{z}$ is the updated buffer contents vector as usual, and the policy function h is defined as follows: for each $k \in \mathscr{K}$ and $i \in \mathscr{I}(k)$,

$$(4.31) \qquad h_i(\hat{z}) = \begin{cases} b_k \gamma_i & \text{if} \quad \hat{z}_i > 0, \\ 0 & \text{if} \quad \hat{z}_i = 0. \end{cases}$$

HLPPS control. HLPPS is another simply structured relaxed control policy under which all nonempty buffers receive service simultaneously. Specifically, for a job class $i \in \mathscr{I}(k)$, the fraction of server k capacity that is allocated to class i at time t is $\gamma_i(Z(t))$, where

(4.32) $$\gamma_i(z) = \frac{z_i}{\sum_{j \in \mathscr{I}(k)} z_j},$$

with the convention that the fraction equals zero if the denominator is zero. Thus the total service rate allocated to class i is $b_k \gamma_i(t)$, all of which goes to the oldest job in buffer i. Note that the aggregate service rate granted to classes in the set $\mathscr{I}(k)$ is b_k (the full capacity of server k), regardless of system status.

Two appealing aspects of the HLPPS policy are the following: (a) its implementation does not require knowledge of "unobservable" system parameters, such as service rates, external arrival rates, and routing probabilities; and (b) the allocation of server k capacity at any given time is based solely on "local" information, by which we mean the contents of buffers that are processed by that server.

Queueing networks with PS control. The final paragraphs of Section 4.4 described a single-server queueing system with the classical processor sharing service discipline, assuming that all service time distributions are of phase type. There we explained how, by exploiting the phase structure of services, one can construct an EHL model whose single server uses an HLPPS control policy. Exactly the same idea applies to a queueing network in which each server uses a PS control policy, sharing its capacity equally at each point in time among all the customers then present for which it is responsible. Thus *a queueing network with a PS control policy is equivalent, in a distributional sense, to another queueing network using the HLPPS control policy described earlier*. Data of the latter network are derived from those of the former network as described in Section 4.4.

4.7 Parallel-Server Systems

A *parallel-server system* is an SPN, either basic (Section 2.3) or relaxed (Section 2.4), that satisfies the following additional assumptions. (i) Each column of the capacity consumption matrix A contains a single one and the rest zeros, and the same is true of the material requirements matrix B. Thus each processing activity consists of a single server from some specified pool serving or processing a single item or job from

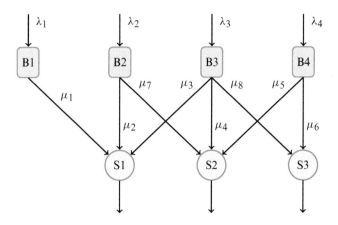

Figure 4.3 Example of a parallel-server system.

some specified class. (ii) Each of the output vectors $\varphi^j(\ell)$ is identically equal to zero, which means that each arriving job receives just one service before it exits. Such networks are often described as *single-hop* in communication theory, as opposed to *multihop* networks where some jobs visit two or more buffers before exiting.

In a parallel-server system, it may be possible to process jobs in a particular buffer (that is, to process jobs of a particular class) using a server from any one of several different pools. In that case, the service time distribution may depend on the server pool chosen. Figure 4.3 pictures a parallel-server system that has four job classes, three server pools, and a total of eight activities. The first-order data portrayed in the figure are an arrival rate λ_i for each job class i, and a mean service rate (reciprocal of mean service time) μ_j for each activity j.

4.8 Example Involving Fork-and-Join Jobs

This section is devoted to a rather elaborate example, with three objectives in mind. First, it introduces a family of SPN models that have many potential applications, including high-fidelity representation of distributed computing paradigms such as MapReduce (see Section 11.1 and the commentary that follows). Second, it further illustrates subtleties in the notion of "job class" or "item class." Finally, it reinforces the following obvious but important point: what constitutes an acceptable definition of "class" depends on the control policy that the model builder wishes to analyze.

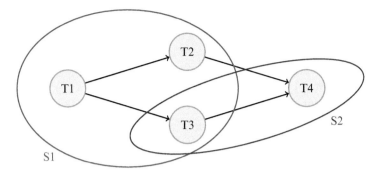

Figure 4.4 Precedence constraints among tasks.

Consider a processing network in which new arrivals are "jobs" that each comprise four constituent "tasks." Those tasks are performed subject to *precedence constraints* portrayed graphically in Figure 4.4, where tasks are represented by circles (T1 denotes task 1, and so forth) and an arrow from one task to another means that the former task must be completed before the latter one can be started. (The significance of the elliptical regions labeled S1 and S2 will be explained later.) That is, task 1 must be completed before either task 2 or task 3 can be started, and both task 2 and task 3 must be completed before task 4 can be started. Our discussion of this example will consider two distinct scenarios that differ with regard to the number of servers available and the breadth of those servers' capabilities. An "activity" or "service type" corresponds to a particular server executing or performing one of the task types indexed by $\ell = 1, 2, 3, 4$.

Fork-and-join networks. This example is representative of a model family where each arriving job consists of some fixed number of tasks, and precedence constraints among the tasks are expressed by means of a directed acyclic graph. These are often called *fork-and-join* networks: the term "fork" refers to a task or operation whose completion allows two or more other tasks to proceed in parallel, while a "join" refers to a task or operation that cannot proceed until two or more antecedent tasks have been completed.

For example, if one seeks a high-fidelity representation of distributed computing and related system management decisions, the natural result is some variant of a fork-and-join network: fork operations are used to represent the decomposition of a job into constituent tasks that are parceled out to different servers and may call upon different data chunks, and join operations represent the stepwise assembly of partial results into a final resolution. (This application domain will be explored later in Chapter 11, but attention is focused there on a highly simplified model having a parallel-server structure, that is, without any join operations.)

Similar mixtures of parallel and sequential tasks occur in the health care arena, specifically in managing the flow of patients, tissue samples, medical records, etc., in hospitals. (It might be, for example, that a patient cannot see a doctor until her test results and her medical records are also ready.) As the following discussion suggests, high-fidelity models of such systems quickly become intractable; there is a premium to formulating simplified approximate models that capture the essence of system behavior without getting bogged down in details.

Dedicated server scenario.   Initially, let us suppose that there are four servers in our example, each of which is dedicated to one of the four task types, and that each server processes arrivals on a FIFO basis. This gives rise to the SPN model pictured in Figure 4.5, where servers are denoted S1, S2, etc., and buffers are denoted B1, B2, etc. The items stored in buffer 1 are new arrivals awaiting execution of task 1 (by server 1), and completion of task 1 for any given job creates a pair of "tokens" that authorize server 2 to proceed with task 2 for the job in question, and server 3 to proceed with task 3. Figuratively speaking, such tokens are the items that queue up on a FIFO basis in buffers 2 and 3, respectively. Completion of task 2 for a given job creates a token (one might call this a partial authorization) that we envision as being stored in buffer 4, and completion of task 3 creates a similar token that is stored in buffer 5; server 4 is authorized to proceed with task 4 for a given job only when

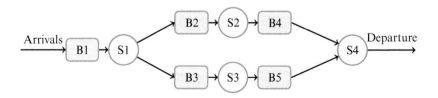

Figure 4.5 Fork-and-join network with a dedicated server performing each task type.

tokens verifying completion of both task 2 and task 3 are present in buffers 4 and 5, respectively.

The SPN pictured in Figure 4.5 has four associated processing activities, which we number 1, 2, 3, 4 in the obvious way: activity ℓ consists of server ℓ performing type ℓ tasks ($\ell = 1, 2, 3, 4$). Thus activity 1 requires a single input item and produces two output items, while activities 2 and 3 each require a single input item and each produce a single output item. Finally, activity 4 requires two input items and results in a job's departure from the system under study.

Flexible server scenario. Consider now an alternative scenario with two *flexible* servers, by which we mean that each server can execute several task types. Specifically, let us suppose that the server capabilities are as pictured in Figure 4.4, that is, S1 can execute tasks of type 1, 2, and 3, while S2 can execute tasks of type 3 and 4. Let us further suppose that, when a type 4 task is undertaken in this alternative scenario, it is essential that the two "tokens" being "joined" actually correspond to the same original job. In our original scenario, with one dedicated server for each task type, that "synchronization problem" was solved by simply specifying that items in each buffer be processed on a FIFO basis. Now, however, with flexible servers that may perform tasks of different types, something more is needed to ensure synchronization.

To illustrate that point, suppose that job α arrives to the system shortly before job β. If server 1 (S1) tends to be faster than server 2 (S2), the following sequence of events can occur. First, S1 executes task 1 for job α, causing tokens to be deposited in both buffer 2 (B2) and buffer 3 (B3), and then it begins executing task 1 for job β while S2 begins execution of task 3 for job α. Suppose S1 finishes execution of task 1 for job β, task 2 for job α, and task 3 for job β, all before S2 finishes task 3 for job α. In that case, the earliest arriving token in B4 corresponds to job α, while the earliest arriving token in B5 corresponds to job β. Thus, if S2 later withdraws the earliest arriving token in each of those two buffers in order to execute a type 4 task, it will be wrongly joining results from two different jobs.

How can synchronization be assured in the alternative scenario with flexible servers? One obvious approach is to label each of the tokens referred to in this subsection with the number of the job to which it corresponds (by order of arrival, say), and require that when S2 undertakes a type 4 task it uses as inputs two tokens (one generated after completion of a type 2 task and the other generated after completion

Figure 4.6 Job classes and transition structure with rigid sequential processing.

of a type 3 task) that are labeled with the same job number. This approach essentially creates countably many item classes (or buffers) and countably many activities, with each item class and each activity having a different job number specified in its definition. Such a detailed model structure has very limited value, because it does not naturally suggest parsimonious measures of system status on which to base dynamic decision making.

Rigid sequential processing. At the other extreme, one could eliminate the synchronization problem altogether by imposing an artificial constraint that task 2 be executed before task 3. This leads to the *rigid sequential* flow pattern pictured in Figure 4.6. (Neither servers nor activities are portrayed in this figure.) Here the "items" occupying each buffer are new or partially completed jobs, and the associated class number is simply the next task to be executed. Because both servers are able to execute type 3 tasks, we identify the following five distinct processing activities:

Activity number	Input class	Server used	Task performed	Output class
1	1	S1	T1	2
2	2	S1	T2	3
3	3	S1	T4	4
4	3	S2	T3	4
5	4	S2	T4	Exit

Flexible sequential processing. A less extreme solution for the synchronization problem is the *flexible sequential* flow pattern portrayed in Figure 4.7, where sequential execution of tasks 2 and 3 is again required, but now either one of them may be executed first. The "items" occupying each buffer are again new or partially completed jobs, but now labeled with the list of constituent tasks that have yet to be

Figure 4.7 Job classes and transition structure with flexible sequential processing.

performed. (Here again, neither servers nor activities are portrayed in the figure.) Those labels constitute five distinct "class" designations for jobs. Newly arrived jobs are of class 1234, and after task 1 is completed for such a job it transitions to class 234. A job of class 234 may next transition to either class 24 or class 34, depending on whether it is selected first for execution of task 3 or for execution of task 2. A job of class 34 transitions next to class 4, because only task 3 can be chosen next for execution, and similarly for class 24. With this configuration, we identify a different processing activity for each feasible combination of job class, next task to be executed, and server to do the execution. That gives a total of eight activities, as follows:

Activity number	Input class	Server used	Task performed	Output class
1	1234	S1	T1	234
2	234	S1	T2	34
3	234	S1	T3	24
4	234	S2	T3	24
5	24	S1	T2	4
6	34	S1	T3	4
7	34	S2	T3	4
8	4	S2	T4	Exit

Each activity involves a single server, consumes a single "input item," and generates either one "output item" or a departure from the system.

In summary, the system configurations pictured in Figures 4.6 and 4.7 both result from the imposition of artificial constraints on processing activities, the purpose of those constraints being to simplify the "bookkeeping" required for synchronization. The only difference between the flexible sequential arrangement in Figure 4.7 and our original system specification is that the former forbids simultaneous execution of tasks that could, in principle, be done in parallel. The rigid sequential arrangement in Figure 4.6 limits scheduling freedom much more severely. Consider, for example, a server that wishes to proceed with the execution of task 3 for a job on which only task 1 has been completed. That server must wait not only through the time required for execution of task 2, but also through any queueing delay that may precede the execution of task 2.

4.9 Sources and Literature

As noted in Appendix E, our standard reference on Markovian arrival processes (Section 4.1) is (Asmussen, 2003, chapter XI). The alternate routing example pictured in Figure 4.2 is due to Laws and Louth (1990). The processor sharing service discipline discussed in Section 4.3 plays a prominent role in the now-classical theory of product-form queueing networks, which was pioneered by Baskett et al. (1975) and Kelly (1979); Serfozo (1999) provides a summary of that work and its later extensions.

The bandwidth sharing model described in Section 4.5 was proposed by Massoulié and Roberts (2000), and the literature treating such networks has developed rapidly, due to both their elegance and their role as Internet models. Massoulié (2007) explored an equivalent head-of-line model for the case with phase-type file size distributions, as in Section 4.5.

Perhaps surprisingly, a BWS network with multipath routing (see Section 4.5) can be reduced to an equivalent standard model by redefining "resources" in an appropriate manner; see (Kang et al., 2009, section 5.5). This is an important theoretical insight. However, there is another, more direct approach that one can take when analyzing a BWS network with multipath routing: represent the network as an SPN having a many-to-one relationship between service types and job classes (that is, identify one activity or service type for each combination of job class and route), and then proceed directly with the analysis of that model.

A round-robin version of the HLSPS control policy (Section 4.6) was studied by Bertsekas and Gallager (1992), and the HLPPS policy was introduced by Bramson (1996b). The term "parallel-server system" (Section 4.8) originated in the work of Harrison and López (1999), who considered only models with single-server pools; other authors use the term "flexible server system" to describe essentially the same model family. The example with fork-and-join jobs discussed in Section 4.8 is adapted from the work of Pedarsani et al. (2014).

5

Is Stability Achievable?

Up to this point, the term "stochastic processing network (SPN)" has been used in a way that includes specification of a control policy. Throughout most of this chapter, however, we consider an SPN as consisting only of data, including both structural data like the resource consumption matrix A and material requirements matrix B, and parameter values like external arrival rates and expected outputs from different service types, *without* fixing a control policy. Taking that view, a natural question to ask is the following: does there exist *any* control policy under which the SPN is stable?

To a large extent, that is equivalent to asking whether the server pools have enough capacity to handle the load imposed on them by external arrivals, as expressed through first-order system data (average arrival rates, mean service times, and expected outputs from different service types). We begin in Section 5.1 by defining the "standard load condition" for a unitary network, as defined in Section 2.6. The standard load condition will later be shown necessary for existence of a stable policy, and to be sufficient under certain additional assumptions.

The manager of a unitary network has no decisions to make about which services to undertake, only about their order, and that lack of discretion makes capacity analysis particularly simple. The corresponding first-order analysis for a general SPN involves consideration of alternative ways in which arriving jobs might be processed. The analysis takes the form of a *static planning problem* (SPP) whose decision variables are long-run average activity rates. That is the subject of Sections 5.2 and 5.3.

In our formulation of the static planning problem, the objective is to minimize the maximum utilization rate for any of the network's K server pools. That minimum value is denoted γ^*. The system is said to be *subcritical* if $\gamma^* < 1$, *critical* if $\gamma^* = 1$, and *supercritical* if $\gamma^* > 1$. For a unitary network, subcriticality is equivalent to the standard load condition referred to previously, and one might plausibly conjecture that a stable policy exists for a general SPN if and only if it is subcritical. The "only if" part of that conjecture is correct: it will be shown in Section 5.4 that no stable policy can exist in the critical or supercritical cases. The "if" part of the conjecture is *not* generally true, as we show by example in Sections 5.5 and 5.6. That is, the examples presented in those sections show that certain network structures may prevent stability even in the subcritical case.

In Section 5.7, we introduce the notion of a *maximally stable* control policy for a given SPN, which means that the specified policy has two properties: first, its implementation does not require knowledge of the arrival rate vector λ, and second, for any λ such that a stable control policy exists, the specified policy is stable. Later chapters will identify particular policies that are maximally stable for particular model families, such as queueing networks or bandwidth sharing networks. As usual, the chapter concludes with comments about sources and literature.

5.1 Standard Load Condition for a Unitary Network

Recall that the load vector ρ for a unitary network was defined in Section 2.6 by (2.40); in our basic SPN model, ρ_k is interpreted as the total service effort required from pool k per time unit, in units like server-hours per hour, and in the relaxed model it is interpreted as the average amount of server k capacity, in units like megabytes per second, required to process all jobs passing through the network. In either case, a natural conjecture is that the following *standard load condition* is necessary and sufficient for existence of a stable control policy:

$$(5.1) \qquad\qquad\qquad \rho < b.$$

It will be seen shortly that (5.1) is indeed necessary for existence of a stable policy, but is not sufficient in general.

5.2 Defining Criticality via the Static Planning Problem

Considering first the relaxed SPN formulation developed in Section 2.4, suppose we denote by $x \in \mathbb{R}_+^J$ the long-run average value of the process $\beta = \{\beta(t), t \geq 0\}$ defined by (2.24). Thus x_j represents the long-run average level of effort, or long-run average total service rate, devoted to type j services per time unit. From (2.27), we see that x must satisfy the capacity constraints

$$(5.2) \qquad\qquad Ax \leq b,$$

where b is the K-vector of server capacities. For the basic SPN formulation developed in Section 2.3, let x denote instead the long-run average value of the service count process $N = \{N(t), t \geq 0\}$. Then x_j can again be described as the long-run average level of effort devoted to type j services per time unit. We have from (2.11) that x must again satisfy (5.2), where b is now the vector of server pool sizes. Furthermore, in either the basic or relaxed model context, x must satisfy certain material balance constraints.

To express those constraints compactly, first recall from Section 2.3 that we denote by m_j the mean service time for activity $j \in \mathcal{J}$, by B the $I \times J$ matrix whose (i,j)th element is the number of class i items consumed by a type j service, and by $\Gamma = (\Gamma_{ij})$ the $I \times J$ matrix whose jth column is the expected output vector from a class j service. Defining $M = \mathrm{diag}(m_1, \ldots, m_J)$ as in Section 2.6, we then set

$$(5.3) \qquad\qquad R := (B - \Gamma)M^{-1}.$$

Thus R_{ij} represents the long-run average rate at which activity j depletes the content of buffer i, expressed in items removed per unit of effort devoted to that activity. (A negative value is interpreted to mean that activity j *increases* the content of buffer i on average.) With this notation, the aforementioned material balance constraints are expressed as follows:

$$(5.4) \qquad\qquad Rx = \lambda.$$

These equalities demand that, for each buffer i, the long-run average net removal rate $(Rx)_i$ be just adequate to counterbalance the external arrival rate λ_i.

To recapitulate, a plausible necessary condition for existence of a policy achieving long-run stability is that there exist $x \in \mathbb{R}_+^J$ satisfying (5.2) and (5.4). Given the SPN model data (R, A, λ, b), an efficient way to check for the existence of such a vector x is by solving the following linear program (LP):

(5.5)   Min γ

(5.6)   Subject to $Rx = \lambda,$

(5.7) $Ax \leq \gamma b,$

(5.8) $x \geq 0.$

We call this a *static planning problem* (SPP), because it uses only first-order system data, suppressing the stochastic variability that motivates dynamically varying service rates. The decision variables of the SPP are a J-vector x of average service rates, as explained previously, and a scalar γ that upper bounds the utilization rate for servers in each of the system's K server pools. The SPP objective is to minimize γ. That is, we seek a static processing plan x that minimizes the maximum utilization of any server pool, consistent with material balance requirements.

Let us denote by γ^* the optimal objective value for the SPP. Obviously, $\gamma^* \leq 1$ if and only if there exists an $x \in \mathbb{R}_+^J$ satisfying (5.2) and (5.4), and similarly, $\gamma^* < 1$ if and only if there exists an x that satisfies $\lambda = Rx$, $Ax < b$, and $x \geq 0$. We say that the SPN is *critically loaded*, or just *critical*, if $\gamma^* = 1$, and the terms *subcritical* and *supercritical* refer to the cases $\gamma^* < 1$ and $\gamma^* > 1$, respectively. Also, a supercritical SPN will sometimes be described as *overloaded*, for obvious reasons.

For a unitary network, formula (5.3) specializes to $R = (I - P')M^{-1}$, where P is the routing matrix, so it follows from (2.38) that the material balance equation (5.6) has a unique solution, namely, $x = M\alpha$. Thus $Ax = \rho$ by (2.40), and hence the optimal solution of the static planning problem (5.5) through (5.8) has

(5.9) $\gamma^* = \min\{\rho_1/b_1, \ldots, \rho_K/b_K\}.$

This brings us to the following conclusion.

Proposition 5.1. *A unitary network is subcritical if and only if it satisfies the standard load condition* (5.1).

Henceforth the ratios appearing on the right side of (5.9) will be called *load factors* for the unitary network's server pools.

Alternate routing with immediate commitment.   Consider now the augmented SPN formulation described in Section 4.2, where uncommitted arrivals from L different sources are assumed to follow a Markovian arrival process (MArP), and the associated L-vector of long-run average arrival rates is denoted by $\nu = (\nu_\ell)$. In that formulation, arrivals from each source ℓ must be allocated or routed without delay to an eligible job class, and the system manager's routing policy manifests itself in a nonnegative $L \times I$ matrix ϕ, where $\phi_{\ell i}$ is the long-run average rate at which arrivals from source ℓ are routed into class i. Recall from Section 4.2 that ϕ must satisfy

$$(5.10) \qquad \phi_{\ell i} = 0 \text{ if } G_{\ell i} = 0, \quad \sum_{i \in \mathscr{I}} \phi_{\ell i} = \nu_\ell \text{ for all } \ell \in \mathscr{L},$$

where G is an $L \times I$ matrix of zeros and ones that defines available routing options (this is problem data). Denoting by $\lambda = (\lambda_i)$ the I-vector of long-run average arrival rates into the various classes under the chosen routing policy, we then have

$$(5.11) \qquad \lambda_i = \sum_{\ell \in \mathscr{L}} \phi_{\ell i} \quad \text{for all } i \in \mathscr{I}.$$

In this setting, we expand the decision variables of our static planning problem to include not only the J-vector x of average activity rates but also the $L \times I$ matrix ϕ, arriving at the following variant of the original problem (5.5) through (5.8), which is also a linear program:

$$(5.12) \qquad \text{Min} \quad \gamma$$
$$(5.13) \qquad \text{Subject to (5.10), (5.11)} \quad \text{and} \quad Rx = \lambda,$$
$$(5.14) \qquad Ax \le \gamma b,$$
$$(5.15) \qquad \phi, x \ge 0.$$

As before, we denote by γ^* the optimal objective value, and the SPN is said to be subcritical, critical, or supercritical depending on whether $\gamma^* < 1$, $\gamma^* = 1$, or $\gamma^* > 1$.

5.3 The Subcritical Region

Let us return now to the basic and relaxed SPN models formulated in Chapter 2. For such a model, the arrival rate vector $\lambda \in \mathbb{R}_+^I$ has thus far been treated simply as data, but it will now be viewed as a vector of variable system parameters, and we write $\gamma^*(\lambda)$ to indicate the optimal value of the static planning problem (5.5) through (5.8) as a function of λ. Given the other system data (R, A, b) involved in the SPP, we then define the *subcritical region*

$$(5.16) \qquad \Lambda := \{\lambda \in \mathbb{R}_+^I : \gamma^*(\lambda) < 1\}.$$

It is left as an exercise to verify the following equivalent characterization:

(5.17) $\quad \Lambda = \{\lambda \in \mathbb{R}_+^I : Rx = \lambda \text{ for some } x \in \mathbb{R}_+^J \text{ satisfying } Ax < b\}.$

It will be shown in the next section that subcriticality is a *necessary* condition for stability (this is Theorem 5.2). That is, there can only exist a stable control policy for a given SPN if $\lambda \in \Lambda$.

To give a concrete example of a subcritical region, let us consider the SPN pictured in Figure 5.1, which is closely related to the alternate routing model depicted earlier in Figure 1.3. Here we have $I = 4$ buffers or job classes, $J = 5$ activities or service types, and $K = 3$ single-server processing stations. The five activities are as follows: S1 processing jobs from B1 (this is activity 1), S2 from B1 (activity 2), S2 from B2 (activity 3), S3 from B3 (activity 4), and S3 from B4 (activity 5). Outputs from these activities follow the routes shown on the diagram deterministically. As indicated on the diagram, S1 service times have mean duration 3, and service times for S2 and S3 have mean duration 1, regardless of the job class being served. It is left as an exercise for the reader to determine the 4×5 input–output matrix R and the 3×5 capacity consumption matrix A for this SPN model.

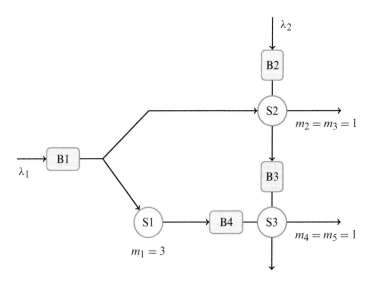

Figure 5.1 Three-station network with alternate routing.

As mentioned in the previous section, it is true in general that $\gamma^*(\lambda) < 1$ if and only if there exists an $x \in \mathbb{R}^I_+$ such that $Rx = \lambda$ and $Ax < b$. For the example pictured in Figure 5.1, where $b = e$ (the three-vector of ones), it can be verified that such an x exists if and only if the following both hold:

(5.18) $$\lambda_1 + \lambda_2 < 4/3 \quad \text{and} \quad \lambda_1 + 2\lambda_2 < 2.$$

(The first inequality says that the total number of services required per time unit from S1 and S2 together is less than their maximum combined output rate. The second inequality similarly compares the total number of services required per time unit from S2 and S3 together with their maximum combined output rate; the term $2\lambda_2$ reflects the fact that each class 2 arrival must be served by both S2 and S3.) The lines $\lambda_1 + \lambda_2 = 4/3$ and $\lambda_1 + 2\lambda_2 = 2$ intersect at the point $(\lambda_1, \lambda_2) = (2/3, 2/3)$, so the subcritical region Λ, which consists of all (λ_1, λ_2) satisfying (5.18), is the polytope pictured in Figure 5.2, excluding the sloped boundary lines that are colored red in the diagram.

Assuming that λ lies within that subcritical region, what control policy can one use to stabilize this system? Presumably there are many feasible options, but one concrete candidate is the so-called back-pressure policy described in Chapter 9: with attention restricted to the basic SPN formulation where server splitting and service interruptions

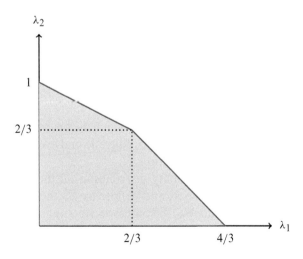

Figure 5.2 Subcritial region for the three-station network.

are disallowed, we show in Section 9.7 that the back-pressure policy is stable for a certain class of networks, assuming $\lambda \in \Lambda$, and that class includes the system pictured in Figure 5.1.

5.4 Only Subcritical Networks Can Be Stable

Throughout the remainder of this book, attention is essentially restricted to subcritical networks, for which the question of stability will be studied via fluid model analysis. In this section, we show that subcriticality is a *necessary* condition for stability. In Theorem 5.2, that result is established for both the basic and relaxed SPN models formulated in Chapter 2, under our baseline stochastic assumptions. Corollary 5.4 extends the theorem to allow a general Markovian arrival process.

Theorem 5.2. *Consider either the basic SPN model formulated in Section 2.3, or the relaxed model formulated in Section 2.4. If Assumptions 2.1 (baseline stochastic assumptions) and 3.1 (Markov representation) are satisfied, and if the SPN is stable, then the arrival rate vector λ must lie in the subcritical region Λ defined via (5.16).*

Remark 5.3. For the basic SPN model, we do *not* restrict attention to simply structured policies. For the relaxed model, however, that restriction is implicit, because only simply structured policies are discussed in Section 2.4.

Proof. Let us first consider the basic SPN model. Assumption 3.1 assures the existence of an ambient Markov chain X, with irreducible state space $\mathscr{X}$, and a function $f : \mathscr{X} \to \mathbb{Z}_+^J \times \mathbb{Z}_+^I$ such that

$$(5.19) \qquad\qquad (N(t), Z(t)) = f(X(t)), \quad t \geq 0.$$

We have assumed that the SPN is stable, which means by definition that X is positive recurrent. Because X is irreducible and positive recurrent, it has a unique stationary distribution π.

We denote by $\mathbb{P}_\pi$ the probability distribution on the path space of X when $X(0)$ has distribution π, and denote by $\mathbb{E}_\pi$ the corresponding expectation operator. Under $\mathbb{P}_\pi$, the processes $\{X(t), t \geq 0\}$, $\{N(t), t \geq 0\}$ and $\{Z(t), t \geq 0\}$ are all stationary. Also, $N(t)$ is bounded by (2.34), so $\mathbb{E}_\pi[N(t)]$ exists and is independent of t. Let

$$(5.20) \qquad\qquad x := \mathbb{E}_\pi[N(0)].$$

The following relationships will be proved, establishing the theorem via (5.17):

(5.21) $$\lambda = Rx \quad \text{and}$$

(5.22) $$Ax < b.$$

To prove (5.22), recall first that $AN(0) \leq b$ on each sample path by (2.11), and that $Z(0) = 0$ implies $N(0) = 0$ by (2.12). Also, the last sentence of Assumption 3.1 ensures that $\mathbb{P}_\pi\{Z(0) = 0\} > 0$. Thus $\mathbb{P}_\pi\{N(0) \neq 0\} < 1$, implying that

$$Ax = A\mathbb{E}_\pi[N(0)] = \mathbb{E}_\pi[AN(0)]$$
$$= \mathbb{E}_\pi[AN(0); N(0) \neq 0] \leq b\mathbb{P}_\pi\{N(0) \neq 0\} < b.$$

Turning now to the proof of (5.21), we shall rely on the following sample path representation of the buffer contents process Z, which is a composite of (2.9) and (2.10):

(5.23) $$Z(t) = Z(0) + E(t) + \sum_{j \in \mathscr{J}} \Phi^j\left(F_j(t)\right) - BF(t), \quad t \geq 0.$$

Let us assume for convenience that the initial state has $Z(0) = N(0) = 0$ (Assumption 3.1 guarantees the existence of such a state), denoting by $\mathbb{P}(\cdot)$ the corresponding probability distribution on the path space of X. The last step in our proof will be to establish the following SLLN for the cumulative service completion process F:

(5.24) $$\mathbb{P}\left\{\lim_{t \to \infty} \frac{1}{t}F(t) = M^{-1}x\right\} = 1,$$

where $M = \mathrm{diag}(m_1, \ldots, m_J)$ as usual. This will be combined with the SLLN (2.14) for arrivals and the SLLN in (2.15) for output vectors. Specifically, we divide both sides of (5.23) by t, let $t \to \infty$, and invoke (2.14), (2.15), and (5.24) to obtain the following:

(5.25) $$\mathbb{P}\left\{\lim_{t \to \infty} \frac{1}{t}Z(t) = \lambda + (\Gamma - B)M^{-1}x\right\} = 1.$$

Defining

(5.26) $$a := \lambda + (\Gamma - B)M^{-1}x,$$

it is clear from (5.25) that $a \geq 0$. If it were true that $a_i > 0$ for some $i \in \mathscr{I}$, that would imply $|Z(t)| \to \infty$ with probability one, contradicting part (c) of Proposition 3.5. Therefore, we have proved that $a = 0$. Combining that with (5.26) and the definition (5.3) of R gives (5.21), completing the proof of the theorem.

It remains only to prove (5.24). The service effort process T is defined via (2.22), so a direct application of the SLLN for Markov chains (Theorem D.25) gives the following:

$$(5.27) \qquad \mathbb{P}\left\{ \lim_{t \to \infty} \frac{1}{t} T(t) = x = \mathbb{E}_\pi[N(0)] \right\} = 1.$$

To make connection between this SLLN for T and the desired SLLN for F, first suppose the system is such that $N_j(t) \leq 1$ for all $t \geq 0$ and $j \in \mathscr{J}$, that is, no more than one service of any given type can be open at any given time. This is the case, for example, if each pool contains a single server ($b_k = 1$ for all k) and each type of service is conducted by a single specified server (each column of A contains a single 1). In that case, we have the simple relationship

$$(5.28) \qquad F_j(t) = \max\{\ell \geq 1 : v_j(1) + \cdots + v_j(\ell) \leq T_j(t)\}$$

for all $t \geq 0$ and $j \in \mathscr{J}$. Combining this with (5.27) and the SLLN for type j service times in (2.15) gives

$$(5.29) \qquad \mathbb{P}\left\{ \lim_{t \to \infty} \frac{1}{t} F_j(t) = m_j^{-1} x_j \right\} = 1 \quad \text{for all } j \in \mathscr{J},$$

which is equivalent to (5.24). Equation (5.28) is not generally valid, but as we show in the next chapter (Lemma 6.8), it provides an adequate asymptotic approximation, and (5.29) continues to hold in the general case. This completes our treatment of the basic SPN model.

For the relaxed SPN model, the proof of the theorem is very similar, but now attention centers on the J-dimensional service rate process $\beta = \{\beta(t), t \geq 0\}$ defined by a policy function h via (2.24). In place of (5.19), we can write

$$(\beta(t), Z(t)) = f(X(t)), \quad t \geq 0,$$

for some function $f : \mathscr{X} \to \mathbb{R}_+^J \times \mathbb{Z}_+^I$, where X is the canonical ambient Markov chain identified in Section 3.5. Our earlier definition (5.20) is replaced by

$$(5.30) \qquad \qquad x = \mathbb{E}_\pi[\beta(0)],$$

where π again denotes the unique stationary distribution of the ambient Markov chain X. The cumulative service effort process T is now defined in terms of β via (2.28), and so the the SLLN for Markov chains (Theorem D.25) gives

$$(5.31) \qquad \mathbb{P}\left\{\lim_{t\to\infty} \frac{1}{t}T(t) = x = \mathbb{E}_\pi[\beta(0)]\right\} = 1.$$

We need to prove that (5.21) and (5.22) hold with x redefined via (5.30). The latter is proved as before, using the capacity constraint $A\beta(0) \le b$ and the fact that $\mathbb{P}_\pi\{Z(0) = 0\} > 0$. The proof of (5.21), using (5.31) in place of (5.27), is actually simpler than before, because our relaxed model formulation includes the constraint $N_j(t) \le 1$ for all $t \ge 0$ and $j \in \mathscr{J}$, and hence (5.28) holds exactly. All other steps of the proof are identical. □

Corollary 5.4 (Markovian arrival process). *Theorem 5.2 remains valid in the setting of Section 4.1, where Assumption 2.1 is weakened to allow a Markovian arrival process. In that case, λ is the I-vector of long-run arrival rates defined in part (a) of Proposition E.7 (the SLLN for Markovian arrival processes).*

Proof. Assuming a MArP, rather than the independent Poisson arrival streams that are part of Assumption 2.1, we adopt an ambient Markov chain X such that (4.4) holds. The proof then proceeds exactly as before, except that Proposition E.7 (the SLLN for MArPs) is used in place of (2.14). □

Corollary 5.5 (Alternate routing with immediate commitment). *Consider an SPN with alternate routing and immediate commitment, maintaining the assumptions stated in Section 4.2. If the network is stable, then it must be subcritical, as that term was defined in the last paragraph of Section 5.2. That is, the optimal objective value for the modified static planning problem (5.12) through (5.15) must satisfy $\gamma^* < 1$.*

Proof. First, as an analog of (5.17), readers can easily verify the following: $\gamma^* < 1$ if and only if there exist $\phi \in \mathbb{R}_+^{L\times I}$, $\lambda \in \mathbb{R}_+^I$, and $x \in \mathbb{R}_+^J$ such that (5.10), (5.11), and the following all hold:

$$(5.32) \qquad Rx = \lambda \quad \text{and} \quad Ax < b.$$

Assuming that the SPN is stable, one can define $\phi \in \mathbb{R}_+^{L \times I}$ and $\lambda \in \mathbb{R}_+^I$ as in the proof of Proposition 4.1 (the SLLN for class-level arrivals), which ensures that (5.10) and (5.11) both hold. We then define x exactly as in the proof of Theorem 5.2, observing that $Rx = \lambda$ and $Ax < b$ as before. $\qquad\square$

5.5 Instability with Multiresource Activities

In Section 1.4, we introduced a first example of a so-called bandwidth sharing network (see Figure 1.4), and Figure 5.3 pictures a related model having three servers (or links) instead of two. In the latter example, there are three input processes (or traffic sources), and jobs of each class require simultaneous possession of two servers for their processing, as indicated in the diagram. For concreteness, let us assume that service times for each of the three job types are exponentially distributed with mean 1, and that $\lambda_1 = \lambda_2 = \lambda_3 = r > 0$. Because each server is involved in processing two of the three job types, there are $2r$ hours of work for each server arriving per hour on average.

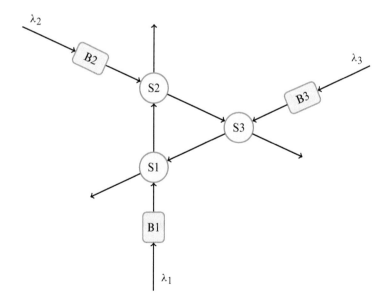

Figure 5.3 Three-link bandwidth sharing network.

It follows that the static planning problem of Section 5.2 has optimal value $\gamma^*(r) = 2r$, and hence the system is subcritical if and only if $r < 1/2$.

A general formulation of a bandwidth sharing network was introduced in Section 4.5, assuming our relaxed formulation is appropriate (that is, services can be conducted at fractional rates and server capacities can be divided among multiple jobs), and it will be shown later (see Section 8.2) that there does indeed exist a stable policy whenever the system is subcritical. Specifically, we show that a control policy called *proportionally fair bandwidth allocation* is stable for such a model if $\lambda \in \Lambda$.

Now suppose that the physical configuration of the system pictured in Figure 5.3 prevents dividing the capacity of any server between two or more activities. (In a telecommunications context, this would mean that each network link can participate in the transmission of just one file or message at any given time, although it may be able to transmit that file or message at a high rate.) Thus we adopt the basic SPN formulation laid out in Section 2.3. Because each type of service involves two of the system's three servers, and no server can split its capacity between multiple activities, no more than one service can be ongoing at any given time, so the long-run maximum number of services that can be completed per time unit is 1. Thus there cannot exist any stabilizing control policy unless $r \leq 1/3$.

Two important conclusions emerge from consideration of this example. First, there may be substantial differences between our basic and relaxed SPN formulations with regard to stability. Second, subcriticality does not guarantee existence of a stable control policy, at least for the basic SPN model where capacity splitting is forbidden.

5.6 Instability with Multiinput Activities

Our next example, pictured in Figure 5.4, shows that the second conclusion cited in the previous paragraph is not unique to the basic SPN model. Here we have two input processes that deposit items called *components* into separate buffers, and a single activity that requires one component from each buffer as input. For obvious reasons, the activity is called *assembly*. The salient feature of this example is that no service can begin until one component is available in each buffer. Thus, recalling that E_1 and E_2 denote the external arrival processes into the two buffers, while Z_1 and Z_2 are the corresponding buffer

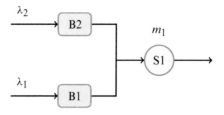

Figure 5.4 Assembly operation.

content processes, one sees after a bit of thought that $Z_1(t) + Z_2(t) \geq |E_1(t) - E_2(t)|$ for all $t \geq 0$. That is, the total number of components in the system at time t, either waiting in inventory or in the process of assembly, is at least as large as the input imbalance $|E_1(t) - E_2(t)|$.

Under our baseline stochastic assumptions (Section 2.2), E_1 and E_2 are independent Poisson processes, so assuming that λ_1 and λ_2 are not both equal to zero, it follows from the central limit theorem that

$$\mathbb{P}\{|E_1(t) - E_2(t)| > a\} \to 1 \text{ as } t \to \infty$$

for each $a > 0$. Thus we conclude that this assembly system is intrinsically unstable: its ambient Markov chain cannot be positive recurrent, no matter how small the mean service time m_1 might be, and regardless of whether "relaxed control" is allowed. That is, the assembly model is only stable in the trivial case $\lambda = (0,0)$, but as readers can easily verify, the subcritical region for this SPN is the line segment (here $\mu_1 = 1/m_1$ as usual)

$$\Lambda = \{(\lambda_1, \lambda_2) : \lambda_1 = \lambda_2 < \mu_1\}.$$

The intrinsic instability of the preceding example derives from the multiinput nature of its single processing activity: because an assembly operation requires one component of each type, the two input processes effectively "serve one another," and the logic of our static planning problem (Section 5.2) does not detect the difficulty arising from that feature. However, the presence of a multiinput activity does not automatically cause SPN instability.

Consider, for example, the system pictured in Figure 5.5. Here server 1 (S1) conducts assembly operations as in the previous example, but there is a second server (S2) that is able to process units of component 2 for direct sale in a complementary "side business." If

(5.33) $\lambda_2 > \lambda_1, \quad \lambda_1 m_1 < 1 \quad$ and $\quad (\lambda_2 - \lambda_1)m_2 < 1,$

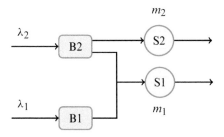

Figure 5.5 Assembly operation with complementary side business.

then the following policy is stable: S1 initiates new services (that is, initiates new assembly operations) whenever there are components available in both buffers, and S2 initiates new "side business" services whenever $Z_1 < Z_2$ (that is, whenever there are "unmatched" components in B2). Given that the three inequalities in (5.33) hold, the stability of that policy will be proved later (see Theorem 8.21 in Section 8.5) using the fluid model approach developed in the next three chapters.

It can be shown, moreover, that all three of the strict inequalities in (5.33) are *necessary* for existence of a stable policy. Thus stability is only achievable for the example pictured in Figure 5.5 if λ satisfies those three inequalities. That is, stability is only achievable for this example if λ lies in the shaded parallelogram in Figure 5.6, excluding both the boundary line segments colored blue in the diagram and the one colored red. As readers can easily verify, the subcritical region Λ is the same, except that it includes the line segment colored red, on which $\lambda_1 = \lambda_2$. This provides another example where stability is *not* achievable for all λ in the subcritical region Λ.

5.7 Stability Region and Maximally Stable Policies

The important notion of "maximal stability" arises in the following context. First, we take the view that an SPN, which may also be called a "network model," consists only of its data, as described in the first paragraph of this chapter. Second, we view the SPN not as a single model, but rather as a family of models parameterized by the vector λ of external arrival rates. For any one model in that family, that is, for any given $\lambda \in \mathbb{R}_+^I$, we use the term "stable control policy," or just "stable policy" for brevity, to mean a control policy such that (a) Assumption 3.1 (Markov representation) is satisfied, and (b) the

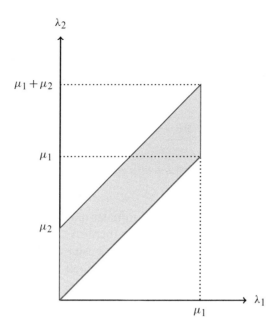

Figure 5.6 Subcritical and stability regions for the system in Figure 5.5.

network with arrival rate vector λ is stable under the specified policy, in the sense of Definition 3.6.

For a given SPN, we denote by Λ^* the set of all $\lambda \in \mathbb{R}_+^I$ such that a stable control policy exists, calling this the network's *stability region*. In the previous section, for example, we considered the system pictured in Figure 5.5, and it was asserted (but not actually proved) that Λ^* consists of all two-vectors λ satisfying the three strict inequalities in (5.33). A control policy is said to be *maximally stable* for the given SPN if (a) its implementation does not depend on the underlying arrival rate vector λ, and (b) it is a stable policy for each $\lambda \in \Lambda^*$.

In general, determining the stability region for a given SPN is a deep mathematical problem, but we know from Theorem 5.2 that $\Lambda^* \subset \Lambda$; that is, no stable policy can exist unless λ lies in the subcritical region. Thus the following corollary, which provides the basis for all our future proofs of maximal stability, is immediate from Theorem 5.2.

Corollary 5.6. *For a given SPN, consider a control policy whose implementation does not depend on λ. If the policy is stable for all $\lambda \in \Lambda$, then it is maximally stable.*

5.8 Sources and Literature

The static planning problem formulated in Section 5.2 first appeared in the paper by Harrison (2000). Theorem 5.2, in its current generality, has not appeared previously in the literature. The negative example in Section 5.5 is a modified version of one that appeared in the paper by Dai and Lin (2005). The intrinsic instability of the assembly operation pictured in Figure 5.4 was first demonstrated by Harrison (1973).

6

Fluid Limits, Fluid Equations, and Positive Recurrence

This chapter develops a fluid model methodology for stability analysis. The results apply both to our basic SPN model (see Section 2.3) and to our relaxed model (Section 2.4) operating under a simply structured control policy, given the baseline stochastic assumptions enunciated in Section 2.2. We denote by $X = \{X(t), t \geq 0\}$ the ambient Markov chain for the SPN under study, which has the explicit form specified in Section 3.5, and other notation established in Chapters 2 and 3 is similarly carried forward. The main results of this chapter are summarized in the next section.

The results developed here can be generalized on any or all of three dimensions: model structure, control policy, and stochastic assumptions. One such generalization will be developed in Section 7.3, where we analyze a queueing network with first-come-first-served scheduling at each station, which is not a simply structured control policy, as noted earlier in Section 3.3.

Another generalization will be developed in Chapter 11, where we analyze a task allocation model that violates in several respects the assumptions laid out in Chapter 2. In all cases, however, our approach to fluid-based stability analysis requires (a) strong laws of large numbers for both external arrivals and processing variables, which follow either from the baseline stochastic assumptions of Section 2.2 or from some suitable substitute, and (b) the Markov representation assured by Assumption 3.1.

6.1 Overview

In Section 6.4, we shall define in precise mathematical terms what is meant by the fluid limit of an SPN. In general, it is a set of four-tuples $(\hat{D}, \hat{F}, \hat{T}, \hat{Z})$ whose components are absolutely continuous functions on $[0, \infty)$. The functions that make up such a four-tuple correspond to the stochastic processes (D, F, T, Z) in our original SPN formulation. A single four-tuple $(\hat{D}, \hat{F}, \hat{T}, \hat{Z})$ belonging to the set is called a *fluid limit path*, and each component of a fluid limit path has the same dimension as its analog in the original stochastic model. Fluid limit paths are obtained as limits of the corresponding SPN processes under SLLN scaling (see Section 6.4), and they will be shown to satisfy (see Theorem 6.5 in Section 6.4) the following basic relationships, each of which corresponds to a stochastic model relationship that was displayed earlier in Chapter 2:

(6.1) $$\hat{Z}(t) = \hat{Z}(0) + \lambda t + \Gamma \hat{F}(t) - \hat{D}(t), \quad t \geq 0,$$

(6.2) $$\hat{Z}(t) \geq 0, \quad t \geq 0,$$

(6.3) $$\hat{D}(t) = B\hat{F}(t), \quad t \geq 0,$$

(6.4) $$\hat{F}(t) = M^{-1}\hat{T}(t), \quad t \geq 0,$$

(6.5) $$\hat{T} \text{ is nondecreasing with } \hat{T}(0) = 0,$$

(6.6) $$A\big(\hat{T}(t) - \hat{T}(s)\big) \leq b(t - s) \quad \text{for } 0 \leq s < t.$$

The relationships (6.1) through (6.6) are customarily called *fluid equations*. We shall continue that terminology in this book, even though two of the six relationships are not actually expressed as equalities. Because (6.4) defines F in terms of T, (6.3) defines D in terms of F, and (6.1) defines Z in terms of F and D, we see that D and F could actually be eliminated from (6.1) through (6.6). One reason for maintaining the fluid equations in their current spread-out form is that references to the processes D and F are often necessary, or at least convenient, in the mathematical development to come. Another reason is that the structure of (6.1) thorough (6.6) provides a reminder of the SPN relationships from which they are derived, which is an aid to intuition.

Definition 6.1. The fluid limit of an SPN is said to be *stable* if there exists a constant $\gamma > 0$ such that, for every fluid limit path $(\hat{D}, \hat{F}, \hat{T}, \hat{Z})$, one has $\hat{Z}(t) = 0$ for all $t \geq \gamma |\hat{Z}(0)|$.

To paraphrase this definition, one may say that the fluid limit of an SPN is stable if fluid limit paths are *uniformly attracted to the origin*. The following theorem, to be proved in Section 6.4, is the fulcrum that supports all other results developed in this book. When combined with ancillary results of the kind discussed after Theorem 6.2, it can be used to prove the stability of SPN models.

Theorem 6.2. *If the fluid limit of an SPN is stable, then the SPN is also stable (that is, its ambient Markov chain X is positive recurrent).*

To prove stability of the fluid limit for a given SPN, thus establishing stability of the original stochastic model via Theorem 6.2, one naturally relies on the fluid equations (6.1) through (6.6). However, each of those relationships holds regardless of the control policy chosen, so to establish the desired result, one invariably needs to derive one or more additional fluid equations (that is, relationships satisfied by all fluid limit paths) that capture the key characteristics of the chosen policy; those are the "ancillary results" referred to in the previous paragraph.

For an example, let us consider a queueing network operating under a control policy that is nonidling. (Again readers are reminded that each statement made in this chapter applies equally to our basic and relaxed formulations.) For each $k \in \mathscr{K}$, let

$$\mathscr{I}(k) := \{i \in \mathscr{I} : A_{ki} = 1\};$$

that is, $\mathscr{I}(k)$ denotes the set of buffers whose contents are processed by servers from pool k. In Section 7.1, it will be shown that for such a network model, every fluid limit path $(\hat{D}, \hat{F}, \hat{T}, \hat{Z})$ satisfies the following relationship for each $k \in \mathscr{K}$:

$$(6.7) \qquad \sum_{i \in \mathscr{I}(k)} \hat{Z}_i(t) > 0 \quad \text{implies} \quad \frac{d}{dt}\left(\sum_{i \in \mathscr{I}(k)} \hat{T}_i(t)\right) = b_k.$$

Combining this added "fluid equation" with (6.1) through (6.6), the following will then be shown in Chapter 8: *if the queueing network is feedforward* (see Section 1.3 for the meaning of that term), and if the load factor for server pool k (see Section 5.1) is strictly less than b_k, then the fluid limit is stable, which implies stability of the queueing network by Theorem 6.2.

The previous paragraph describes an analytical process with two distinct steps. First, the relationship (6.7) is established through analysis of the queueing network's fluid limit. Second, real analysis is used to show that any four-tuple of functions $(\hat{D}, \hat{F}, \hat{T}, \hat{Z})$ that satisfies the fluid equations (6.1) through (6.7) must be stable in the sense of Definition 6.1. The fluid limit is a continuous analog of the original stochastic model, and (6.1) through (6.7) describe it precisely enough to establish stability under the conditions stated earlier.

It is more or less obvious in this example that the displayed fluid equations do not *uniquely* determine the fluid limit, because there are many nonidling policies, and all of them have fluid limits satisfying those equations. This pattern will be repeated often in the following chapters: fluid equations will be derived for a policy or family of policies, and those equations will be used to prove stability of the stochastic model's fluid limit; there is no guarantee that the fluid equations uniquely characterize the fluid limit, but fortunately, that is not a matter of concern for us.

6.2 Fluid Models

Throughout the remainder of this book, the term *fluid model* will be used to mean a set of fluid equations that have been shown to hold for all fluid limit paths of some SPN or family of SPNs, and a four-tuple $(\hat{D}, \hat{F}, \hat{T}, \hat{Z})$ that satisfies the stated fluid equations will be called a *fluid model solution*. Thus, in our usage, equations (6.1) through (6.7) constitute a fluid model of a queueing network operating under a nonidling policy, and the results cited earlier guarantee that every fluid limit path for such a queueing network is also a fluid model solution. The following corresponds in a natural way to our earlier definition of fluid limit stability.

Definition 6.3. A fluid model is said to be *stable* if there exists a constant $\gamma > 0$ such that, for every fluid model solution $(\hat{D}, \hat{F}, \hat{T}, \hat{Z})$, one has $\hat{Z}(t) = 0$ for all $t \geq \gamma |\hat{Z}(0)|$.

For many researchers in applied probability, the term "fluid model" has a more restrictive connotation, referring to a deterministic, continuous analog of a given stochastic model. In that description, the word "deterministic" means that the fluid model is uniquely defined, or to put that another way, the fluid model has just one sample path for a given initial condition. In the remainder of this section, we discuss two examples, elaborating on the matter of uniqueness and foreshadowing important issues to come.

Regular points and dot notation for derivatives. Consider a function $f : \mathbb{R}_+ \to \mathbb{R}^d$ for some integer $d > 0$. A point $t > 0$ is said to be a *regular point* for f if f is differentiable at t. When the function f is clear from the context, we sometimes say simply that t is a regular point, without explicitly mentioning f. We use $\dot{f}(t)$ to denote the derivative of f at a regular point t. By convention, whenever the symbol $\dot{f}(t)$ is used, t is assumed to be a regular point. Sometimes, as in (6.7), we also use $\frac{d}{dt}f(t)$ to denote the derivative of f at t.

Unique fluid model solution for the criss-cross network. Consider the criss-cross network pictured in Figure 1.2, which is a feedforward queueing network with two single-server stations. We consider the nonidling control policy that gives priority to class 3 jobs at station 1. The fluid model of this network is governed by the following equations (here and in the rest of this subsection, we drop the hat notation from each fluid model solution $(D, F, T, Z))$: for $t \geq 0$,

(6.8) $Z_1(t) = Z_1(0) + \lambda_1 t - \mu_1 T_1(t),$

(6.9) $Z_2(t) = Z_2(0) + \mu_1 T_1(t) - \mu_2 T_2(t),$

(6.10) $Z_3(t) = Z_3(0) + \lambda_3 t - \mu_3 T_3(t),$

(6.11) $Z_i(t) \geq 0, \quad i = 1, 2, 3,$

(6.12) $T_i(0) = 0, \quad T_i(\cdot)$ is nondecreasing, $\quad i = 1, 2, 3,$

(6.13) $(T_1(t) + T_3(t)) - (T_1(s) + T_3(s)) \leq t - s$ and
$$T_2(t) - T_2(s) \leq t - s \quad \text{for } 0 \leq s \leq t,$$

(6.14) $Z_1(t) + Z_3(t) > 0$ implies $\dot{T}_1(t) + \dot{T}_3(t) = 1,$ and
$$Z_2(t) > 0 \text{ implies } \dot{T}_2(t) = 1,$$

(6.15) $Z_3(t) > 0$ implies $\dot{T}_3(t) = 1.$

Equations (6.8) through (6.13) simply specialize the general fluid equations (6.1), (6.2), and (6.5) to the example at hand, and similarly, (6.14) specializes the general relationship (6.7) that characterizes nonidling policies. Finally, (6.15) characterizes the specific nonidling policy that gives priority to class 3 at station 1. As mentioned earlier, all fluid model equations must be justified through a fluid limit procedure. Equation (6.14) will be justified in Section 7.1, and (6.15) will be justified in Section 7.2.

We shall argue shortly that for any initial state $Z(0) \in \mathbb{R}_+^3$, the fluid model solution (T, Z) to (6.8) through (6.15) is unique. Assuming that the load condition (1.2) is satisfied, this fluid model is also stable. Indeed, starting from any initial buffer contents vector $Z(0)$, the fluid model solution is piecewise linear and is easily constructed by an iterative process. To illustrate that process, let us consider $Z(0) = (0, 0, 1)$. There is an initial time interval $[0, t_1]$ during which server 1 spends all of its capacity serving buffer 3 (that is, processing fluid from buffer 3), as dictated by the priority policy. During this interval, buffer 3 decreases linearly (that is, the fluid level in buffer 3 decreases linearly), buffer 1 increases linearly, and buffer 2 remains empty. To be specific, the duration of the initial time interval is

$$t_1 := \frac{1}{\mu_3 - \lambda_3}.$$

At t_1, the buffer contents vector is

$$Z(t_1) = (\lambda_1 t_1, 0, 0).$$

In the period immediately after t_1, server 1 is 100% busy, spending a fraction $\lambda_3 m_3$ of its capacity to keep buffer 3 empty, and spending the remaining fraction $1 - \lambda_3 m_3$ of its capacity to linearly decrease buffer 1. The content of buffer 1 reaches zero at time t_2, where

$$t_2 - t_1 = \frac{\lambda_1 t_1}{\mu_1(1 - \lambda_3 m_3) - \lambda_1}.$$

After t_2, both buffer 1 and buffer 3 remain empty, as server 1 spends a fraction $\lambda_1 m_1$ of its capacity serving buffer 1, and spends a fraction $\lambda_3 m_3$ of its capacity serving buffer 3; a fraction $1 - \lambda_1 m_1 - \lambda_3 m_3$ of server 1 capacity remains unused, or idle. The dynamics of server 2 and buffer 2 can be described similarly.

We now argue that for any initial state $Z(0) \in \mathbb{R}_+^3$, the fluid model solution (T, Z) to (6.8) through (6.15) is unique. For $t \geq 0$, define

$$Y_1(t) := t - T_1(t) - T_3(t),$$
$$Y_2(t) := t - T_2(t),$$
$$Y_3(t) := t - T_3(t).$$

By (6.12) and (6.13), each of these functions $Y_i(\cdot)$ is nondecreasing. Also, one can check that (Y_3, Z_3) satisfies

(6.16) $Z_3(t) = Z_3(0) + \lambda_3 t - \mu_3 t + \mu_3 Y_3(t),$

(6.17) $Z_3(t) \geq 0,$

(6.18) $Y_3(0) = 0$ and $Y_3(\cdot)$ is nondecreasing,

(6.19) $\displaystyle\int_0^\infty Z_3(t) dY_3(t) = 0,$

where (6.19) follows from (6.15) and the definition of Y_3. Together, equations (6.16) through (6.19) identify $(\mu_3 Y_3, Z_3)$ as a solution to what is called a *one-dimensional Skorohod problem*. That term is used, for example, in section IX.2 of Asmussen (2003). In section 2.2 of Harrison (2013), it is shown that the unique solution $(\mu_3 Y_3, Z_3)$ of the problem (6.16) through (6.19) is obtained by applying the *one-sided reflection mapping* to the function $X_3 = \{X_3(t), t \geq 0\}$ that is defined as follows:

$$X_3(t) := Z_3(0) + (\lambda_3 - \mu_3)t \quad \text{for } t \geq 0.$$

Therefore, we have established that (T_3, Z_3) is uniquely determined from X_3.

We next argue that (T_1, Z_1) is uniquely determined. To begin, observe that

(6.20) $m_1 Z_1(t) + m_3 Z_3(t)$

$$= m_1 Z_1(0) + m_3 Z_3(0) + (\lambda_1 m_1 + \lambda_3 m_3)t - t + Y_1(t),$$

(6.21) $m_1 Z_1(t) + m_3 Z_3(t) \geq 0,$

(6.22) $Y_1(0) = 0$ and $Y_1(\cdot)$ is nondecreasing,

(6.23) $\displaystyle\int_0^\infty \big(m_1 Z_1(t) + m_3 Z_3(t)\big) dY_1(t) = 0,$

where (6.20) follows from (6.8) and (6.10), and (6.23) follows from the first part of (6.14). Applying the results in section 2.2 of Harrison (2013) a second time, we have that $(Y_1, m_1 Z_1 + m_3 Z_3)$ are uniquely determined from $X_1 = \{X_1(t), t \geq 0\}$, where

$$X_1(t) = m_1 Z_1(0) + m_3 Z_3(0) + (\lambda_1 m_1 + \lambda_3 m_3)t - t, \quad t \geq 0.$$

To be specific, the unique solution $(Y_1, m_1 Z_1 + m_3 Z_3)$ is obtained by applying the one-sided reflection mapping to X_1. As a consequence, (T_1, Z_1) is uniquely determined from X_1 and (T_3, Z_3).

Finally, one can argue similarly that $(Z_2, \mu_2 Y_2)$ is uniquely determined from $X_2 = \{X_2(t), t \geq 0\}$ via the one-sided reflection mapping, where

$$X_2(t) := Z_2(0) + \mu_1 T_1(t) - \mu_2 t \quad t \geq 0.$$

Nonunique fluid model solution for the Rybko–Stolyar network. Consider again the Rybko–Stolyar network (Figure 1.9), where classes 2 and 4 have priority at their respective stations. Assume that $\lambda_1 = \lambda_3 = 1$

and the standard load condition (1.8) is satisfied, but condition (1.9) is violated, namely,

$$(6.24) \qquad m_2 + m_4 \geq 1.$$

Given that $\lambda_1 = \lambda_3 = 1$ by assumption, the fluid model for this system is governed by the following equations: for each $t \geq 0$,

$$(6.25) \quad Z_1(t) = Z_1(0) + t - \mu_1 T_1(t),$$
$$(6.26) \quad Z_2(t) = Z_2(0) + \mu_1 T_1(t) - \mu_2 T_2(t),$$
$$(6.27) \quad Z_3(t) = Z_3(0) + t - \mu_3 T_3(t),$$
$$(6.28) \quad Z_4(t) = Z_4(0) + \mu_3 T_3(t) - \mu_4 T_4(t),$$
$$(6.29) \quad Z_i(t) \geq 0 \text{ for } i = 1,2,3,4,$$
$$(6.30) \quad T_i(0) = 0 \text{ and } T_i(\cdot) \text{ nondecreasing for } i = 1,2,3,4,$$
$$(6.31) \quad (T_1(t) + T_4(t)) - (T_1(s) + T_4(s)) \leq t - s \text{ and}$$
$$(T_2(t) + T_3(t)) - (T_2(s) + T_3(s)) \leq t - s \text{ for } 0 \leq s < t,$$
$$(6.32) \quad Z_1(t) + Z_4(t) > 0 \text{ implies } \dot{T}_1(t) + \dot{T}_4(t) = 1, \text{ and}$$
$$Z_2(t) + Z_3(t) > 0 \text{ implies } \dot{T}_2(t) + \dot{T}_3(t) = 1,$$
$$(6.33) \quad Z_2(t) > 0 \text{ implies } \dot{T}_2(t) = 1, \text{ and } Z_4(t) > 0 \text{ implies } \dot{T}_4(t) = 1.$$

Equations (6.25) through (6.31) are the usual SPN fluid equations (6.1) through (6.6), specialized to the Rybko–Stolyar network, while (6.32) is a specialization of the nonidling equation (6.7), and (6.33) holds under the policy that gives priority to classes 2 and 4. Equation (6.32) will be justified in Section 7.1 by a fluid limit procedure, and (6.33) will be justified in Section 7.2.

Assuming the initial buffer contents to be $Z(0) = (1,0,0,0)$, we now construct multiple solutions (T,Z) for the fluid model equations (6.25) through (6.33).

Divergent fluid model solution. Figure 6.1 portrays graphically the dynamics of one such solution, showing the divergent cycles described in the passage that follows. In the figure, the four different shadings in a cycle represent fluid levels in the system's four buffers. The fluid level in a buffer is the vertical distance between the upper boundary and the lower boundary of the shaded region. Here we list the changes of state at certain times. Setting $t_1 := 1/(\mu_1 - 1)$, one has

$$Z(t_1) = (0, (\mu_1 - \mu_2)t_1, t_1, 0).$$

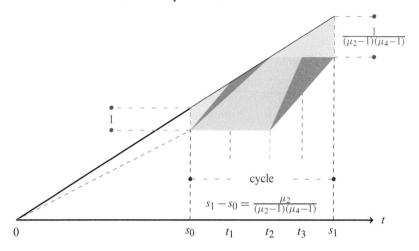

Figure 6.1 The dynamics of a divergent Rybko–Stolyar fluid network.

At $t_2 := 1/(\mu_2 - 1)$, one has

$$Z(t_2) = (0, 0, t_2, 0).$$

At t_3, where $t_3 - t_2 = t_2/(\mu_3 - 1)$, one has

$$Z(t_3) = \big(t_3 - t_2, 0, 0, (\mu_3 - \mu_4)(t_3 - t_2)\big).$$

At t_4, where

$$t_4 - t_2 = t_2/(\mu_4 - 1) = \frac{1}{(\mu_2 - 1)(\mu_4 - 1)},$$

one has

$$Z(t_4) = (t_4 - t_2, 0, 0, 0).$$

It follows that

$$t_4 = \frac{\mu_2}{(\mu_2 - 1)(\mu_4 - 1)}.$$

and

$$t_4 - t_2 \geq 1 \quad \text{if and only if (6.24) holds.}$$

By a scaling argument, we have

$$Z(s_n) = \left(\left(\frac{1}{(\mu_2 - 1)(\mu_4 - 1)}\right)^n, 0, 0, 0\right),$$

where $s_0 = 0$ and

$$s_n = s_{n-1} + \mu_2 \left(\frac{1}{(\mu_2 - 1)(\mu_4 - 1)} \right)^n.$$

It is obvious, then, that the following holds when $m_2 + m_4 > 1$:

$$s_n \to \infty \quad \text{and} \quad |Z(s_n)| \to \infty \quad \text{as} \quad n \to \infty,$$

demonstrating the divergence of this fluid model solution. If $m_2 + m_4 = 1$, one obtains a periodic behavior, namely,

$$Z(n\mu_2) = (1, 0, 0, 0) \quad \text{for all } n = 1, 2, \ldots.$$

In either case, the fluid model for this priority policy is unstable.

Convergent fluid model solution. We shall now construct another fluid model solution (T, Z) that is convergent, starting from the same initial state $Z(0) = (1, 0, 0, 0)$. This construction assumes the following mean service times, which satisfy (6.24):

$$m = (0.1, 0.4, 0.1, 0.8).$$

Let $d_i(t) = \mu_i \dot{T}_i(t)$, interpreting this as the departure rate from buffer i at time t ($i = 1, 2, 3, 4$ and $t \geq 0$). Noting that buffers 2, 3, and 4 are initially empty, we would like to construct a solution that keeps those buffers permanently empty, while buffer 1 drains to emptiness in finite time and remains empty thereafter. Thus, we would like our fluid solution to satisfy

(6.34) $\quad d_2(t) = d_1(t) = \mu_1 \dot{T}_1(t), \quad d_3(t) = 1, \quad d_4(t) = \mu_4 \dot{T}_4(t) = d_3(t)$

for $t \in [0, t_1]$, where t_1 is the first time that buffer 1 empties. Combining (6.34) and the nonidling property (6.32), one has

$$d_1(t) = \mu_1(1 - m_4) = 2,$$
$$\dot{T}_1(t) + \dot{T}_4(t) = 1,$$
$$\dot{T}_2(t) + \dot{T}_3(t) = m_2 d_1(t) + m_3 = 0.9$$

for $t \in [0, t_1]$. That is, server 2 is 90% busy in the time interval $[0, t_1]$, using 10% of its capacity on class 2 fluid and 80% on class 3, while server 1 is 100% busy, using 80% of its capacity on class 4 fluid and

20% on class 1. One can then verify that (T, Z) are as described in the following table:

$t \in [0, t_1]$	$t \in [t_1, \infty)$
$T_1(t) = .2t,$	$T_1(t) = T_1(t_1) + .1(t - t_1),$
$T_2(t) = .1t,$	$T_2(t) = T_2(t_1) + .4(t - t_1),$
$T_3(t) = .8t,$	$T_3(t) = T_3(t_1) + .1(t - t_1),$
$T_4(t) = .8t,$	$T_4(t) = T_4(t_1) + .8(t - t_1),$
$Z_1(t) = 1 - t,$	$Z_1(t) = 0,$
$Z_i(t) = 0, \quad i = 2, 3, 4,$	$Z_i(t) = 0, \quad i = 2, 3, 4$

This is a fluid model solution, because equations (6.25) through (6.33) are satisfied. It has $t_1 = 1$ and $Z(t) = 0$ for $t \geq t_1$.

Infinitely many fluid model solutions.   We have now constructed two fluid model solutions for the Rybko–Stolyar network, one of them divergent and the other convergent. However, one can actually construct infinitely many fluid model solutions by suitably mixing of the two preceding constructions. For example, for any $x \in (0, 1)$, follow the second construction until the time $s_1 \in (0, t_1)$ when the fluid level in buffer 1 falls to x. From that time onward, use the first construction, generating a sequence of divergent cycles that end at times s_n whose corresponding buffer content vectors are

$$Z(s_n) = \left(\left(\frac{x}{(\mu_2 - 1)(\mu_4 - 1)} \right)^n, 0, 0, 0 \right).$$

In this way, uncountably many divergent fluid model solutions can be generated.

6.3 Standard Setup for Study of Fluid Limits

To develop the notion of fluid limits with a minimum of technical difficulty, it will be convenient to have many versions of a single SPN defined simultaneously on a common probability space $(\Omega, \mathscr{F}, \mathbb{P})$, with different versions corresponding to different initial states x of the ambient Markov chain X. To be precise about that standard setup, let us first recall from Section 2.1 that the primitive stochastic elements of an SPN are divided into its *core stochastic elements* on the one hand, and its *initial processing variables* on the other. The core stochastic elements consist of an I-dimensional external arrival process $E = \{E(t), t \geq 0\}$ and a collection of processing variable pairs

$$\Pi := \{(v_j(\ell), \varphi^j(\ell)), j \in \mathscr{J}, \ell = 1, 2, \ldots\}.$$

(This use of the letter Π to denote the complete set of processing variables is new.) In the standard setup, we define a triple (E, Π, Π^0) on a single probability space $(\Omega, \mathscr{F}, \mathbb{P})$, where (i) E and Π are as described earlier; (ii) Π^0 is a finite collection of *potential initial processing variables*, to be explained shortly; and (iii) E, Π, and Π^0 are mutually independent.

Constructing sample paths. Recall from (2.3) that we use the capital Greek letter Ψ to denote the complete set of initial processing variables for an SPN model, calling this the model's *initial processing (IP) set*, and calling its constituent pairs (v, ϕ) *IP pairs*. Extending that previous notation, let us denote by Ψ^x the IP set for (or corresponding to) initial state $x \in \mathscr{X}$. That is, Ψ^x consists of initial processing variables (residual service times and their associated output vectors) that are used to construct network processes for initial state x.

It follows from Assumption 3.8, from the assumed independence of different IP pairs (see Section 2.1), and from the service count bound (2.34) that all of the IP sets $\{\Psi^x, x \in \mathscr{X}\}$ can be drawn from (that is, can be defined as subsets of) a single finite set Π^0 that is independent of both E and Π. Consider, for example, the basic SPN model operating under a simply structured policy, using the Markov state description $x = (\eta, z)$ as in Section 3.5. One can then proceed as follows.

First, let κ be the smallest integer satisfying (2.34), and for each $j \in \mathscr{J}$ and $s_j \in \mathscr{S}_j$ let $\mathscr{D}(j, s_j)$ be the joint phase-type distribution of residual service time v and associated output vector ϕ that corresponds to a type j service beginning in service phase s_j. One can take Π^0 to be a collection of mutually independent IP pairs such that κ of them have distribution $\mathscr{D}(j, s_j)$ for each combination of $j \in \mathscr{J}$ and $s_j \in \mathscr{S}_j$. With that setup, to construct Ψ^x for an initial state $x = (\eta, z)$, one can select from Π^0, for each combination of $j \in \mathscr{J}$ and $s_j \in \mathscr{S}_j$, just $\eta_j(s_j)$ mutually independent IP pairs with distribution $\mathscr{D}(j, s_j)$.

Using the policy mechanics specified in Sections 2.3 and 2.4 (for the basic model and relaxed model, respectively), we can and do construct a complete array of network processes (D, F, T, Z) for each initial state $x \in \mathscr{X}$, using as inputs to that construction the primitive stochastic elements (E, Π, Ψ^x). To indicate the dependence of a constructed variable or process on the initial state, we attach a superscript x to the notation developed in Chapter 2.

Thus, for example, $D^x = \{D^x(t), t \geq 0\}$ is the I-dimensional cumulative departure process for the SPN with initial state x. Similarly, the notation $v_j^x(\ell)$ will be used for negative values of ℓ in (6.36) and (6.47), emphasizing by means of the superscript that initial processing variables depend on the system's initial state. Before proceeding, it will be useful to restate the bound (2.34) using our superscript notation, as follows:

$$(6.35) \qquad N_j^x(t, \omega) \leq \kappa \quad \text{for all } j \in \mathscr{J}, x \in \mathscr{X}, \omega \in \Omega, \text{ and } t \geq 0.$$

Readers will see that (6.35) plays a critical role in justifying the fluid equation (6.3).

Remark 6.4. An important point is that the network processes for different initial states are all constructed from a common set of core stochastic elements. That is, the external arrival process E and processing variables $(v_j(\ell), \varphi^j(\ell))$ for positive values of ℓ, which correspond to services initiated after time zero, do *not* depend on the initial state x, and thus they are denoted as before, without a superscript x.

A stochastic bound. Let us denote by $\zeta(\omega)$ the largest of the service times v that occur in the set Π^0. Thus $\zeta(\omega)$ is the largest residual service time that can be associated with any of the services that are open at time zero in any of our constructions. Because Π^0 is finite, $\zeta(\omega) < \infty$ for almost all ω, and we have the following bound, expressed using the superscript notation introduced previously:

$$(6.36) \quad v_j^x(\ell, \omega) \leq \zeta(\omega) \quad \text{for all } j \in \mathscr{J}, \ell \in \mathscr{L}_j^0, \omega \in \Omega, \text{ and } x \in \mathscr{X},$$

where $\mathscr{L}_j^0$ is defined via (2.1).

6.4 Definition and Properties of Fluid Limits

This section develops the fluid limit theory that is needed for the statement and proof of Theorem 6.2, whose significance was discussed earlier in Section 6.1. Fluid limits are obtained after network processes are subjected to law of large numbers (LLN) scaling, in which time and space are scaled by a common large parameter.

For a state $x = (\eta, z) \in \mathscr{X}$, let

$$|x| := |z| := \sum_{i \in \mathscr{I}} z_i.$$

It is this quantity, the sum of initial buffer contents, that we use as our scaling parameter. For each $x = (\eta, z) \in \mathscr{X}$ with $|x| \neq 0$ and each $\omega \in \Omega$, we define fluid scaled processes $(\hat{D}^x, \hat{F}^x, \hat{T}^x, \hat{Z}^x)$ by setting

$$(6.37) \quad \left(\hat{D}^x(t, \omega), \hat{F}^x(t, \omega), \hat{T}^x(t, \omega), \hat{Z}^x(t, \omega) \right)$$

$$= \frac{1}{|x|} \left(D^x(|x|t, \omega), F^x(|x|t, \omega), T^x(|x|t, \omega), Z^x(|x|t, \omega) \right), \quad t \geq 0.$$

For each $x \in \mathscr{X}$ with $|x| \neq 0$ and each $\omega \in \Omega$, we observe that $\hat{T}^x(\cdot, \omega)$ belongs to the function space

$$\mathbb{C}(\mathbb{R}_+, \mathbb{R}^J) := \{ g : \mathbb{R}_+ \to \mathbb{R}^J, \ g \text{ is continuous} \}.$$

One can also check that $\hat{F}^x(\cdot, \omega)$ is an element of $\mathbb{D}(\mathbb{R}_+, \mathbb{R}^J)$, and $\hat{D}^x(\cdot, \omega)$ and $\hat{Z}^x(\cdot, \omega)$ are elements of $\mathbb{D}(\mathbb{R}_+, \mathbb{R}^I)$, where, for an integer $d > 0$,

$$\mathbb{D}(\mathbb{R}_+, \mathbb{R}^d) := \big\{ g : \mathbb{R}_+ \to \mathbb{R}^d, \ g \text{ is right continuous on}$$
$$\mathbb{R}_+ \text{ and has left limits in } (0, \infty) \big\}.$$

The following theorem shows that, for almost all ω, the family

$$\left(\hat{D}^x(\cdot, \omega), \hat{F}^x(\cdot, \omega), \hat{T}^x(\cdot, \omega), \hat{Z}^x(\cdot, \omega) \right),$$

indexed by $x \in \mathscr{X}$ with $|x| \neq 0$, has appropriate limits as $|x| \to \infty$. In its statement, a set $C \subset \mathscr{X}$ is said to be *unbounded* if $\{ |x| : x \in C \} \subset \mathbb{R}_+$ is unbounded in $\mathbb{R}$. Also, in the proof of the theorem we use the fact that, under our baseline stochastic assumptions (see Section 2.2), service times satisfy the following condition for each $j \in \mathscr{J}$ with probability one:

$$(6.38) \qquad \lim_{n \to \infty} \frac{1}{n} \max_{1 \leq \ell \leq n} v_j(\ell, \omega) = 0;$$

see Proposition B.8 for a proof.

Theorem 6.5. *Fix $\omega \in \Omega$ such that (2.14), (2.15), and (6.38) hold for the core stochastic elements. For any unbounded set $C \subset \mathcal{X}$ of initial states, there exists a sequence $\{x_n\} \subset C$ with $|x_n| \to \infty$ such that*

$$(6.39) \quad \left(\hat{D}^{x_n}(\cdot, \omega), \hat{F}^{x_n}(\cdot, \omega), \hat{T}^{x_n}(\cdot, \omega), \hat{Z}^{x_n}(\cdot, \omega) \right) \to (\hat{D}, \hat{F}, \hat{T}, \hat{Z})$$

as $n \to \infty$ for some functions $\hat{F}, \hat{T} \in \mathbb{C}(\mathbb{R}_+, \mathbb{R}^J)$, and $\hat{D}, \hat{Z} \in \mathbb{C}(\mathbb{R}_+, \mathbb{R}^I)$, where the convergence in (6.39) is uniform convergence on compact (u.o.c.) sets (see Definition A.6). Furthermore, the four-tuple $(\hat{D}, \hat{F}, \hat{T}, \hat{Z})$ satisfies the fluid equations (6.1) through (6.6) with $|\hat{Z}(0)| = 1$.

Definition 6.6. *A four-tuple $(\hat{D}, \hat{F}, \hat{T}, \hat{Z}) \in \mathbb{C}^I \times \mathbb{C}^J \times \mathbb{C}^I \times \mathbb{C}^J$ is said to be a fluid limit path of the SPN if there exists an $\omega \in \Omega$ and a sequence $\{x_n\} \subset \mathcal{X}$ with $|x_n| \to \infty$ such that (6.39) holds.*

The conclusion (6.39) involves convergence of four components. To prove Theorem 6.5, we now state two lemmas. The first lemma proves convergence of the third component, and the second lemma says that convergence of the third component implies convergence of the second one.

Lemma 6.7. *Fix $\omega \in \Omega$. For any unbounded set $C \subset \mathcal{X}$ of initial states, there exists a sequence $\{x_n\} \subset C$ with $|x_n| \to \infty$ such that*

$$(6.40) \quad \hat{T}^{x_n}(\cdot, \omega) \to \hat{T}(\cdot) \quad and \quad \hat{Z}^{x_n}(0, \omega) \to \hat{Z}(0) \quad as \quad n \to \infty$$

for some function $\hat{T} \in \mathbb{C}(\mathbb{R}_+, \mathbb{R}^J)$ and some $\hat{Z}(0) \in \mathbb{R}^I_+$.

Lemma 6.8. *Fix an $\omega \in \Omega$ such that (2.15) and (6.38) both hold. Consider an arbitrary sequence of initial states $\{x_n\} \subset \mathcal{X}$ and an arbitrary sequence $\{r_n\} \subset \mathbb{R}_+$ with $r_n \to \infty$. Then, for each $t \geq 0$,*

$$(6.41) \quad \lim_{n \to \infty} \frac{1}{r_n} F_j^{x_n}(r_n t, \omega)$$

exists if and only if

$$(6.42) \quad \lim_{n \to \infty} \frac{1}{r_n} T_j^{x_n}(r_n t, \omega)$$

exists. Furthermore, denoting by $\hat{F}_j(t)$ and $\hat{T}_j(t)$ the limits in (6.41) and (6.42) respectively, when they exist, we have

$$(6.43) \quad m_j \hat{F}_j(t) = \hat{T}_j(t).$$

Lemmas 6.7 and 6.8 will be proved at the end of this section. Assuming these two lemmas, we are now ready to prove Theorem 6.5.

Proof of Theorem 6.5. Fix an $\omega \in \Omega$ such that (2.14), (2.15), and (6.38) are all satisfied. By Lemma 6.7, there exists a sequence $\{x_n\} \subset C$ with $|x_n| \to \infty$ such that (6.40) holds. It follows from Lemma 6.8 that for each $t \geq 0$,

$$\hat{F}_j^{x_n}(t,\omega) \to \hat{F}_j(t) \quad \text{as } n \to \infty,$$

where $\hat{F}_j(t) := \mu_j \hat{T}_j(t)$. Because $\hat{F}_j(\cdot)$ is a continuous nondecreasing function, it follows from Lemma A.10 that

(6.44) $$\hat{F}_j^{x_n}(\cdot,\omega) \to \hat{F}_j(\cdot) \quad \text{as } n \to \infty$$

uniformly on compact sets. From the SLLN (2.14), we have

$$\hat{E}_i^{x_n}(t,\omega) \to \hat{E}_i(t),$$

where $\hat{E}_i(t) := \lambda_i t$ for each $t \geq 0$ and

$$\hat{E}_i^{x_n}(t,\omega) = \frac{1}{|x_n|} E_i(|x_n|t,\omega) \quad \text{for } t \in \mathbb{R}_+ \text{ and } n \geq 1.$$

Again, it follows from Lemma A.10 that

(6.45) $$\hat{E}_i^{x_n}(\cdot,\omega) \to \hat{E}_i(\cdot) \quad \text{u.o.c.}$$

From the basic system equation (2.10), we have that

$$\hat{Z}_i^{x_n}(t) = \hat{Z}_i^{x_n}(0) + \hat{E}_i^{x_n}(t)$$
$$+ \sum_{j \in \mathscr{J}} \frac{1}{|x_n|} \Phi_i^j(|x_n|\hat{F}_j^{x_n}(t)) - \sum_{j \in \mathscr{J}} B_{ij} \hat{F}_j^{x_n}(t), \quad t \geq 0.$$

Combining that last equation with (6.40), (6.44), (6.45), and the SLLN for output vectors in (2.15), one has that $\hat{Z}_i^{x_n}(\cdot) \to \hat{Z}_i(\cdot)$ u.o.c. as $n \to \infty$, where $\hat{Z}_i(t)$ is given by the right side of (6.1) for $t \geq 0$. $\qquad \square$

Proof of Lemma 6.7. Fix an $\omega \in \Omega$ and an initial state $x \in \mathscr{X}$ with $|x| \neq 0$. Each component of $\hat{T}^x(\cdot,\omega)$ is nondecreasing, and $\hat{T}^x(0,\omega) = 0$. Furthermore,

$$A\left(\hat{T}^x(t,\omega) - \hat{T}^x(u,\omega)\right) \leq (t-u)b$$

for each pair (u,t) with $0 \le u \le t$. We now observe that

$$(6.46) \qquad\qquad e'A \ge \tilde{e}',$$

where prime denotes transpose and e and $\tilde{e}$ are column vectors of ones of appropriate dimensions. Thus one has

$$\left| \hat{T}^x(t,\omega) - \hat{T}^x(u,\omega)) \right| = \sum_{j \in \mathscr{J}} \left(\hat{T}^x_j(t,\omega) - \hat{T}^x_j(u,\omega) \right) \le (t-u)|b|,$$

where $|b|$ is the 1-norm of b as usual. It follows that the family of functions

$$\left\{ \hat{T}^x(\cdot,\omega) : |x| \ne 0, \omega \in \Omega \right\}$$

is equi-Lipschitz as defined in (A.2). Also, $\{\hat{Z}^x(0), |x| \ne 0\}$ is a bounded set in $\mathbb{R}^I$. Then it follows from Corollary A.9 and Theorem A.5 that for each $\omega \in \Omega$ there exists a sequence $\{x_n\} \subset C$ with $|x_n| \to \infty$ such that (6.40) holds for some function $\hat{T} \in \mathbb{C}(\mathbb{R}_+, \mathbb{R}^J)$ and some $\hat{Z}(0) \in \mathbb{R}^I_+$. $\square$

The proof of Lemma 6.8 is similar to that of the SLLN for a standard renewal process, where event times are the successive positions of a random walk with positive steps sizes. (That is, the interevent times for a standard renewal process are i.i.d. positive random variables.) In proving the SLLN for a renewal process, one relies on (i) an SLLN for the corresponding random walk, and (ii) what we will call a "key relationship" that connects the renewal process and the random walk; see, for example, theorem 2.10 of Serfozo (2009).

For each $j \in \mathscr{J}$, $\omega \in \Omega$, and $x \in \mathscr{X}$, define $V^x_j(n,\omega)$ to be the sum of the first n type j service times, including services that are open at $t = 0$. That is,

$$(6.47)$$

$$V^x_j(n,\omega) = \begin{cases} \displaystyle\sum_{-N^x_j(0) < \ell \le 0} v^x_j(\ell,\omega) + \sum_{1 \le \ell \le n - N^x_j(0)} v_j(\ell,\omega) & \text{for } n \ge N^x_j(0), \\[2em] \displaystyle\sum_{-n < \ell \le 0} v^x_j(\ell,\omega) & \text{for } n < N^x_j(0). \end{cases}$$

For each $j \in \mathscr{J}$, one can describe V^x_j as a random walk that is "delayed" by initial processing variables, which may be different for different initial states $x \in \mathscr{X}$. To prove Lemma 6.8, we need to (i) prove an SLLN for this "delayed" random walk, and (ii) exploit a key relationship that connects the delayed random walk V_j^x and the service completion process F_j^x.

The following lemma is the required SLLN for V_j^x. Our service count bound (2.34) plays a critical role in this proof.

Lemma 6.9. *For each $\omega \in \Omega$ such that the SLLN (2.15) for service times holds, one has*

$$(6.48) \qquad \lim_{n \to \infty} \sup_{x \in \mathscr{X}} \left| \frac{1}{n} V_j^x(n, \omega) - m_j \right| = 0.$$

Proof. Let κ be the constant in the service count bound (2.34). From the definition of V^x and condition (6.36), one has the following for $n \geq \kappa \geq N_j^x(0)$, where ζ is the random variable defined immediately before (6.36):

$$(6.49) \qquad \sum_{\ell=1}^{n-\kappa} v_j(\ell, \omega) \leq V_j^x(n) \leq \kappa \zeta(\omega) + \sum_{\ell=1}^{n} v_j(\ell, \omega).$$

Both the upper and lower bounds in (6.49) are independent of the initial state x. It follows that

$$(6.50) \qquad \frac{1}{n} \sum_{\ell=1}^{n-\kappa} \left(v_j(\ell, \omega) - m_j \right) - \frac{1}{n} m_j \kappa \leq \frac{1}{n} V_j^x(n) - m_j$$

$$\leq \frac{1}{n} \kappa \zeta(\omega) + \frac{1}{n} \sum_{\ell=1}^{n} \left(v_j(\ell, \omega) - m_j \right).$$

The proof of (6.48) readily follows from (6.50) and the SLLN for service times in (2.15). $\qquad \square$

A key relationship. We now argue that, for each $x \in \mathscr{X}$, $\omega \in \Omega$, $j \in \mathscr{J}$, and $t \geq 0$, one has

$$(6.51) \qquad V_j^x(F_j^x(t)) - N_j^x(t) \max_{-N_j^x(0) < \ell \leq F_j^x(t) + N_j^x(t)} v_j(\ell) \leq T_j^x(t)$$

$$\leq V_j^x(F_j^x(t) + N_j^x(t)).$$

The second inequality in (6.51) says that the cumulative effort devoted to type j services by time t cannot be larger than the sum of the service times for the type j services that have been completed by t or are open at t. To understand the first inequality, first observe the following: if all type j services were completed in the order they were started, then the total effort devoted to type j services up to time t would be at least $V_j^x(F_j^x(t))$, and hence one has $V_j^x(F_j^x(t)) \leq T_j^x(t)$ in that scenario. The first inequality in (6.51) includes an adjustment to account for the $N_j^x(t)$ type j services that are open at time t and whose start times *might be earlier than* those of some completed type j services. In the special case where at most one type j service can be open at a time, (6.51) follows from the familiar relationship

$$(6.52) \quad F_j^x(t) = G_j^x(T_j^x(t)) \quad \text{and} \quad V_j^x(G_j^x(t)) \leq t < V_j^x(G_j^x(t)+1)$$

for $t \geq 0$, where G_j^x is the renewal process associated with the delayed random walk V_j^x. Thus, (6.51) is a generalization of (6.52) to allow for "service overtake," that is, services completed in an order different from the one in which they were started.

Proof of Lemma 6.8. Fix a $t \geq 0$ and an $\omega \in \Omega$ that satisfies (2.15) and (6.38). To simplify notation, we suppress the argument ω and the superscript x_n in this proof. It suffices to prove that

$$(6.53) \qquad m_j \limsup_{n \to \infty} \frac{F_j(r_n t)}{r_n} \leq \limsup_{n \to \infty} \frac{1}{r_n} T_j(r_n t),$$

$$(6.54) \qquad m_j \liminf_{n \to \infty} \frac{F_j(r_n t)}{r_n} \geq \liminf_{n \to \infty} \frac{1}{r_n} T_j(r_n t).$$

We first prove (6.53). If $\lim_{n \to \infty} F_j(r_n t) < \infty$, then

$$\lim_{n \to \infty} \frac{1}{r_n} F_j(r_n t) = 0$$

and (6.53) clearly holds. Assume that $\lim_{n \to \infty} F_j(r_n t) = \infty$. From the first inequality of the key relationship (6.51), one has

$$(6.55) \quad \frac{F_j(r_n t)}{r_n} \frac{1}{F_j(r_n t)} V_j(F_j(r_n t))$$

$$\leq \frac{1}{r_n} T_j(r_n t) + \kappa \frac{F_j(r_n t)}{r_n} \frac{1}{F_j(r_n t)} \max_{-N_j(0) < \ell \leq F_j(r_n t) + N_j(r_n t)} v_j(\ell)$$

for all $r > 0$. By (6.36),

$$\max_{-N_j(0) < \ell \leq F_j(r_n t) + N_j(r_n t)} v_j(\ell) \leq \zeta(\omega) + \max_{1 \leq \ell \leq F_j(r_n t) + \kappa} v_j(\ell, \omega).$$

It follows from (6.48), (6.38), and (6.35) that

$$(6.56) \qquad \lim_{n\to\infty} \frac{1}{F_j(r_nt)} V_j(F_j(r_nt)) = m_j > 0 \text{ and}$$

$$\lim_{n\to\infty} \frac{1}{F_j(r_nt)} \max_{-N_j(0)<\ell\leq F_j(r_nt)+N_j(r_nt)} v_j(\ell) = 0.$$

Assume further that $\limsup_{n\to\infty} F_j(r_nt)/r_n < \infty$. Then (6.53) follows from (6.55) by taking $\limsup$ on both sides. To complete the proof of (6.53), it remains to show that

$$(6.57) \qquad \limsup_{n\to\infty} \frac{F_j(r_nt)}{r_n} = \infty$$

implies

$$(6.58) \qquad \limsup_{n\to\infty} \frac{T_j(r_nt)}{r_n} = \infty.$$

To prove (6.58), it follows from (6.56) that there exists $n_0 > 0$ such that

$$\frac{1}{F_j(r_nt)} V_j(F_j(r_nt)) \geq m_j/2 \quad \text{and}$$

$$\kappa \frac{1}{F_j(r_nt)} \max_{-N_j(0)<\ell\leq F_j(r_nt)+N_j(r_nt)} v_j(\ell) < m_j/4$$

for $n \geq n_0$. Thus, (6.55) implies that

$$\frac{F_j(r_nt)}{r_n} (m_j/2 - m_j/4) \leq \frac{1}{r_n} T_j(r_nt)$$

for $n \geq n_0$, from which we have

$$(m_j/4) \limsup_{n\to\infty} \frac{F_j(r_nt)}{r_n} \leq \limsup_{n\to\infty} \frac{1}{r_n} T_j(r_nt),$$

proving (6.58) under condition (6.57).

We next prove (6.54). The right side of (6.51) becomes

$$\frac{1}{r_n} T_j(r_nt) \leq \frac{1}{r_n} V_j(F_j(r_nt) + N_j(r_nt)).$$

If $\lim_{n\to\infty} F(r_nt) < \infty$, then (6.54) follows from

$$\lim_{n\to\infty} \frac{1}{r_n} T_j(r_nt) \leq \lim_{n\to\infty} \frac{1}{r_n} V_j(F_j(r_nt) + N_j(r_nt)) = 0.$$

Now assume $\lim_{n\to\infty} F(r_n t) = \infty$. Then (6.54) follows from

$$\frac{1}{r_n} T_j(r_n t) \le \frac{F_j(r_n t) + N_j(r_n t)}{r_n} \frac{1}{F_j(r_n t) + N_j(r_n t)} V_j(F_j(r_n t) + N_j(r_n t)),$$

plus (6.35) and the first part of (6.56). □

6.5 Fluid Model Stability Implies SPN Stability

We are now positioned to complete the proof of Theorem 6.2.

Proof of Theorem 6.2.   By Lemma 3.7, it suffices to prove that stability of the fluid limit implies (3.5). To begin, the definition of fluid limit stability gives the following: there exists an $h > 0$ such that, for each fluid limit path $(\hat{T}, \hat{Z})$ with $|\hat{Z}(0)| = 1$, one has $\hat{Z}(h) = 0$. We now prove that

$$(6.59) \qquad \lim_{|x|\to\infty} \frac{1}{|x|} |Z^x(|x|h, \omega)| = 0$$

for each $\omega \in \Omega$ such that (2.14), (2.15), and (6.38) hold. Suppose there exists an ω such that (2.14) and (2.15) hold but (6.59) does not. Then there exists a sequence of initial states $\{x_n\} \subset \mathscr{X}$ satisfying $\lim_{n\to\infty} |x_n| = \infty$ such that

$$(6.60) \qquad \lim_{n\to\infty} \frac{1}{|x_n|} |Z^{x_n}(|x_n|h, \omega)| > 0.$$

By Theorem 6.5 in Section 6.4, there exists a subsequence $\{x_{n_k}\} \subset \{x_n\}$ such that

$$(6.61) \qquad \left(\hat{T}^{x_{n_k}}, \hat{Z}^{x_{n_k}}\right) \to (\hat{T}, \hat{Z}) \quad \text{as } k \to \infty$$

for some fluid limit path $(\hat{T}, \hat{Z})$, where, again, the fluid scaled processes are defined by

$$\left(\hat{T}^{x_{n_k}}(t), \hat{Z}^{x_{n_k}}(t)\right) = \frac{1}{|x_{n_k}|} \left(T^{x_{n_k}}(|x_{n_k}|t), Z^{x_{n_k}}(|x_{n_k}|t)\right) \quad t \ge 0.$$

Therefore, by (6.61),

$$\lim_{k\to\infty} \frac{1}{|x_{n_k}|} Z^{x_{n_k}}(|x_{n_k}|h) = \lim_{k\to\infty} \hat{Z}^{x_{n_k}}(h) = \hat{Z}(h) = 0,$$

contradicting (6.60). Thus we have proved (6.59).

From (6.59), Lemma 6.10, and Theorem B.2, one can check that condition (3.5) holds. Therefore, by Lemma 3.7, the ambient Markov chain X is positive recurrent, proving Theorem 6.2. $\qquad\square$

Lemma 6.10. *Fix an $h > 0$. The family of random variables*

$$\left\{ \frac{1}{|x|} \left| Z^x(\lfloor x \rfloor h) \right|, \quad x \in \mathscr{X} \text{ with } |x| \geq 1 \right\}$$

is uniformly integrable (see Definition B.1).

Proof. It follows from Proposition B.5 that the family of random variables

$$\left\{ \sum_{i \in \mathscr{I}} \frac{1}{|x|} E_i(\lfloor x \rfloor h), \quad x \in \mathscr{X} \text{ with } |x| \geq 1 \right\}$$

is uniformly integrable. For each $x \in \mathscr{X}$ with $|x| \neq 0$,

$$\frac{1}{|x|} |Z^x(\lfloor x \rfloor h)| \leq 1 + \sum_{i \in \mathscr{I}} \frac{1}{|x|} E_i(\lfloor x \rfloor h),$$

from which the lemma follows. $\qquad\square$

6.6 Sources and Literature

Dai (1995a) established Theorem 6.2 for queueing networks (see Section 2.6 for the meaning of that term). In that work, the interarrival and service time distributions are assumed to be general, so the ambient Markov process contains remaining interarrival and service times as components of the state description. Those components change continuously, and therefore the ambient Markov process does not have a discrete state space. Each fluid limit path in Dai (1995a) is shown to satisfy "delayed" fluid equations that are slightly different from the fluid equations in this chapter. Chen (1995) proved that the stability of the standard fluid model implies that of the delayed fluid model. Bramson (1998) and Bramson (2008) modified the definition of fluid limit in Dai (1995a) so that each fluid limit path satisfies the standard fluid equations. Under our baseline stochastic assumptions and Assumption 3.8,

(6.36) holds. Therefore, the complications in Dai (1995a), Chen (1995), and Bramson (1998) and Bramson (2008) do not apply.

Fluid models and fluid approximations of queueing networks have been extensively studied in the literature; see, for example, Newell (1982) and Chen and Mandelbaum (1991). The latter paper established a functional SLLN theorem whose content is the following: in a single-class queueing network, the fluid-scaled queue length process converges to a fluid limit that satisfies deterministic fluid equations. The fluid scaling there uses a generic large parameter in both time and space.

Rybko and Stolyar (1992) was the first paper to study fluid scaling, using the initial number of jobs in the system as the large parameter. Focusing on the Rybko–Stolyar network pictured in Figure 1.9, but assuming that both servers use an FCFS policy rather than the priority policy described in Section 6.2, that paper made the first connection in the literature between fluid limits and positive recurrence of the network's ambient Markov chain.

At about the same time, Dupuis and Williams (1994) made a connection between the positive recurrence of a multidimensional reflecting Brownian motion and the stability of a corresponding deterministic Skorohod problem (fluid model). Both Rybko and Stolyar (1992) and Dupuis and Williams (1994) inspired the framework developed in Dai (1995a). The work of Rybko and Stolyar (1992) was generalized in an independent, contemporaneous paper by Stolyar (1995) involving a discrete state space.

Theorem 6.2 provides a sufficient condition for positive recurrence of an ambient Markov chain. The converse of the theorem does not exist yet. Effort has been made by Meyn (1995), Dai (1996), and Pulhaskii and Rybko (2000) to provide a "partial converse" in the queueing network setting; see also Gamarnik and Hasenbein (2005).

Closely related to Dai (1995a) is Dai and Meyn (1995). That paper proved the following for queueing networks: assuming fluid limit stability and finite $(k + 1)$th moments for both interarrival and service time distributions, the network's steady-state total job count has a finite kth moment. For an SPN satisfying our baseline stochastic assumptions, which include Poisson arrival processes and phase-type service time distributions, one naturally expects that the steady-state total job count has finite moments of all orders if the fluid limit is stable, but that analysis has not been undertaken thus far.

Fluid limit paths are defined in the almost sure sense in Dai (1995a) and in this book. One can define a fluid limit as the weak convergence limit of a sequence of fluid-scaled stochastic processes as in Stolyar (1995). (See Billingsley, 1999, or Ethier and Kurtz, 1986, for the definition of weak convergence of stochastic processes.) Foss and Kovalevskii (1999) explored this notion of fluid limit and used it to study a polling model.

7

Fluid Equations That Characterize
Specific Policies

Theorem 6.2, stating that fluid limit stability implies SPN stability, is the centerpiece of this book. To make practical use of that result, one needs to know how fluid limit paths behave, and that analysis begins with Theorem 6.5, which justifies fluid equations (6.1) through (6.6) under *any* policy. Typically, however, the analysis of a particular control policy requires that additional policy-specific equations be identified. These additional equations must be justified through the same fluid limit procedure used in the proof of Theorem 6.5.

In Section 7.1, by carrying out the detailed fluid limit procedure, we justify the extra fluid equation (6.7) that was presented earlier for a queueing network operating under a nonidling policy. In Section 7.2, by carrying out a similar procedure, we rigorously justify an analogous fluid equation for a queueing network with nonpreemptive static buffer priorities. (This is Theorem 7.3.) The fluid equation developed there generalizes (6.15), which was stated without proof in our earlier discussion of the criss-cross network example. Section 7.3 concerns the first-come-first-served (FCFS) control policy for a queueing network; there we present the extra fluid equation (7.13) and its justification.

Section 7.4 is concerned with the generic PS network defined in Chapter 4. Specifically, we consider the equivalent head-of-line model for a PS network (see Section 4.4), with attention restricted to policies having a certain special structure, and we derive an additional fluid equation to characterize such a policy. It will be shown later (see Section 8.2) that one particular policy from the specified family is maximally stable. Finally, sources and literature are discussed briefly in Section 7.5.

7.1 Queueing Network with a Nonidling Policy

This section focuses on queueing networks, which were defined in Section 2.6 as a special class of SPNs. In such a network, there is one activity per buffer, and that activity consists of a server from one particular pool processing a job from the specified buffer. As in Section 5.1, we denote by $\mathscr{I}(k)$ the set of buffers that are processed by servers from pool k; the formal version of that definition is (2.43). A control policy for a queueing network is said to be *nonidling* if no server remains idle while there is a job waiting in any of the buffers that are processed by that server. A fluid equation that is satisfied by every nonidling policy is the following: for each pool k and each $t > 0$,

$$(7.1) \qquad \sum_{i \in \mathscr{I}(k)} \hat{Z}_i(t) > 0 \text{ implies that } \frac{d}{dt}\left(\sum_{i \in \mathscr{I}(k)} \hat{T}_i(t) \right) = b_k.$$

Remark 7.1. The statement in (7.1) says that if $\hat{Z}_j(t) > 0$ for some $j \in \mathscr{I}(k)$ and $t > 0$, then the sum $\sum_{i \in \mathscr{I}(k)} \hat{T}_i(t)$ is differentiable at time t, and moreover its derivative is equal to b_k. However, there is no guarantee that $\hat{T}_j(\cdot)$ is differentiable at t.

Theorem 7.2. *For a queueing network operating under a nonidling policy, each fluid limit path satisfies (7.1).*

Proof. Let $(\hat{D}, \hat{F}, \hat{T}, \hat{Z})$ be a fluid limit path with a corresponding sequence $\{x_n\} \subset \mathscr{X}$ and a corresponding $\omega \in \Omega$ such that (2.14), (2.15), (6.38), and (6.39) all hold. Fix this ω, a time $t > 0$, and a server pool $k \in \mathscr{K}$. Throughout the remainder of this proof, we suppress the argument ω. Assume that the buffers served by pool k contain a positive amount of fluid at time t, that is, $\sum_{i \in \mathscr{I}(k)} \hat{Z}_i(t) > 0$. By the continuity of $\hat{Z}$, there exists a $\delta \in (0, t)$ such that

$$(7.2) \qquad \epsilon := \min_{u \in [t-\delta, t+\delta]} \sum_{i \in \mathscr{I}(k)} \hat{Z}_i(u) > 0.$$

By (6.39), there exists an integer $L > 0$ such that for $n > L$ one has $|x_n|\epsilon/2 > b_k$ and

$$(7.3) \qquad \sup_{u \in [t-\delta, t+\delta]} |\hat{Z}^{x_n}(u) - \hat{Z}(u)| \leq \epsilon/2.$$

From (7.2) and (7.3), one has

$$\min_{u \in [t-\delta, t+\delta]} \sum_{i \in \mathscr{I}(k)} \hat{Z}_i^{x_n}(u) \geq \epsilon/2,$$

which is equivalent to

$$\sum_{i \in \mathscr{I}(k)} Z_i^{x_n}(u) \geq |x_n|\epsilon/2 > b_k$$

for $u \in \big(|x_n|(t-\delta), |x_n|(t+\delta)\big)$ and $n > L$. This means that for each $n > L$ and at each time $u \in \big(|x_n|(t-\delta), |x_n|(t+\delta)\big)$, there are always jobs waiting for pool k to process them. From the nonidling hypothesis, we have that

(7.4) $$\sum_{i \in \mathscr{I}(k)} \Big(T_i^{x_n}(u_2) - T_i^{x_n}(u_1)\Big) = (u_2 - u_1)b_k$$

for any $u_1, u_2 \in \big(|x_n|(t-\delta), |x_n|(t+\delta)\big)$ with $u_1 < u_2$. Therefore, for any $u_1, u_2 \in (t-\delta, t+\delta)$ with $u_1 < u_2$,

$$\sum_{i \in \mathscr{I}(k)} \Big(\hat{T}_i^{x_n}(u_2) - \hat{T}_i^{x_n}(u_1)\Big) = (u_2 - u_1)b_k \quad \text{for} \quad n > L.$$

Taking the limit as $n \to \infty$, we have

$$\sum_{i \in \mathscr{I}(k)} \Big(\hat{T}_i(u_2) - (\hat{T}_i(u_1))\Big) = (u_2 - u_1)b_k$$

for any $u_1, u_2 \in (t-\delta, t+\delta)$ with $u_1 < u_2$, from which (7.1) follows, proving the theorem. $\qquad\square$

7.2 Queueing Network with Nonpreemptive Static Buffer Priorities

In this section, we focus again on queueing networks, considering a special class of non-idling policies called *static buffer priority* (SBP) policies. To define this policy class, let us first recall that $\mathscr{I} := \{1,\ldots,I\}$ is the set of all buffers, while $\mathscr{K} := \{1,\ldots,K\}$ is the set of all server pools, and $\mathscr{I}(k)$ is defined via (2.43) to be the set of buffers that are processed by server pool k. Now for each $i \in \mathscr{I}$, let $p(i)$ to be the server pool that processes jobs from buffer i, that is, the unique $k \in \mathscr{K}$ such that $i \in \mathscr{I}(k)$.

An SBP policy is defined by a permutation $\sigma : \mathscr{I} \to \mathscr{I}$. Specifically, if two buffers $i,j \in \mathscr{I}$ have $p(i) = p(j) = k$ (that is, buffers i and j are both served by pool k), then servers in pool k give priority to class i over class j if and only if $\sigma(i) < \sigma(j)$. Under this SBP policy, a newly freed server in pool k will choose for its next service a job from whatever nonempty buffer $i \in \mathscr{I}(k)$ has the lowest value of the priority index $\sigma(i)$; specifically, it chooses the *oldest* job in that buffer, that is, the one whose most recent arrival into the buffer occurred first. If all buffers $i \in \mathscr{I}(k)$ are empty when the server is freed, then it simply waits for a job of some class $i \in \mathscr{I}(k)$ to arrive. (Here as elsewhere, the word "arrival" is used to include both external and internal arrivals, the latter being created via class transitions following service completions.)

We shall consider the nonpreemptive version of the SBP policy, which means the following: if all servers in pool k are busy and a job arrives into some buffer $i \in \mathscr{I}(k)$, that job simply waits in its buffer, even though some of the jobs being served by pool k may have lower priority than the new arrival. This control policy fits within the framework of our basic SPN model, in which services are carried through to completion at full intensity (that is, with full service effort) once they have been started. Moreover, it is a simply structured control policy, as that term was defined in Section 2.3.

In contrast, one may consider the preemptive version of the static SBP policy, which means the following. If all servers in pool k are busy when a job arrives into some buffer $i \in \mathscr{I}(k)$, and if one or more of the jobs currently being served by pool k has lower priority than the new arrival, then service is suspended for one of the jobs currently being served, and the server freed by that suspension begins processing the new arrival. The priority ranking σ is used in the obvious way to determine the class of the job whose service is suspended, and if there is more than one job of that class being served at the moment of the new arrival, then the youngest such job (that is, the one that arrived most recently into its current buffer) is selected for suspension. When the suspended service is eventually resumed (using the stated criteria for commencement of new services, first by priority ranking and then by age of job), it requires only as much additional service effort as needed to complete its original service time. This is an example of a relaxed control policy. More specifically, it is one that interrupts services under some circumstances but never splits the capacity of a server between two or more jobs.

For a buffer $j \in \mathscr{I}$, let

(7.5) $\mathscr{H}(j) := \{i \in \mathscr{I} : p(i) = p(j) \text{ with } \sigma(i) \leq \sigma(j)\}.$

In words, $\mathscr{H}(j)$ is the set of same-pool buffers whose priorities are at least as high as that of buffer j. The fluid equation corresponding to the nonpreemptive SBP policy is the following: for each buffer $j \in \mathscr{I}$ and each $t > 0$,

(7.6) $\sum\limits_{i \in \mathscr{H}(j)} \hat{Z}_i(t) > 0$ implies that $\dfrac{d}{dt}\left(\sum\limits_{i \in \mathscr{H}(j)} \hat{T}_i(t)\right) = b_k,$

where $k = p(j)$ is the pool serving class j jobs.

Theorem 7.3.   *For a queueing network operating under a nonpreemptive SBP policy, each fluid limit path satisfies (7.6).*

Remark 7.4.   This theorem also holds under the preemptive version of the SBP policy. In fact, the proof is significantly easier in that case, because contrary to what is said in the third sentence of the following proof, (7.4) *does* hold under preemption if $\mathscr{I}(k)$ is replaced by $\mathscr{H}(j)$.

Proof.   Let $(\hat{D}, \hat{F}, \hat{T}, \hat{Z})$ be a fluid limit path with a corresponding sequence $\{x_n\} \subset \mathscr{X}$ and a corresponding $\omega \in \Omega$ such that (2.14), (2.15), (6.38), and (6.39) all hold. Fix this ω and a time $t > 0$. The proof mimics that of Theorem 7.2, with $\mathscr{H}(j)$ replacing $\mathscr{I}(k)$ everywhere, except that the $\mathscr{H}(j)$ version of (7.4) does not hold under the non-preemption assumption. Indeed, (7.4) is replaced by the following:

(7.7)
$$(u_2 - u_1)b_k - R^{x_n}(n, t, \omega) \leq \sum_{i \in \mathscr{H}(j)} \left(T_i^{x_n}(u_2, \omega) - (T_i^{x_n}(u_1, \omega)\right) \leq (u_2 - u_1)b_k$$

for any $u_1, u_2 \in \left(|x_n|(t-\delta), |x_n|(t+\delta)\right)$ with $u_1 < u_2$, where $R^{x_n}(n, t, \omega)$ is the total remaining processing time for all jobs that are being processed at time $|x_n|(t-\delta)$ by servers from pool k. The first inequality in (7.7) holds because of the priority policy and the fact that

$$\sum_{i \in \mathscr{H}(j)} Z_i^{x_n}(u, \omega) > b_k \text{ for all } u \in \left(|x_n|(t-\delta), |x_n|(t+\delta)\right).$$

To prove (7.6), following the rest of the proof of Theorem 7.2, it suffices to show that

$$(7.8) \qquad \lim_{n \to \infty} \frac{1}{r_n} R^{X_n}(n, t, \omega) = 0,$$

where $r_n := |x_n|$. To bound the total remaining service time $R^{X_n}(n, t)$, suppose there is one class j job in service at time $r_n(t - \delta)$. The remaining service time of this job is bounded by

$$\max_{-N_j^{X_n}(0) < \ell < F_j^{X_n}(r_n(t-\delta), \omega) + N_j^{X_n}(r_n(t-\delta), \omega)} v_j(\ell, \omega),$$

which is further bounded by

$$\max \left(\zeta(\omega), \max_{j \in \mathscr{J}} \max_{1 < \ell < F_j^{X_n}(r_n(t-\delta), \omega) + \kappa} v_j(\ell, \omega) \right),$$

where ζ is defined immediately before (6.36) and κ is the constant appearing in (2.34). Because each class has at most κ jobs in service at any given time, one has

$$R^{X_n}(n, t, \omega) \le \kappa J \left(\xi(\omega) + \sum_{j \in \mathscr{J}} \max_{1 < \ell < F_j^{X_n}(r_n(t-\delta), \omega) + \kappa} v_j(\ell, \omega) \right),$$

where, again, $J = I$ is the number of classes in the network. We claim that for each $j \in \mathscr{J}$,

$$(7.9) \qquad \lim_{n \to \infty} \frac{1}{r_n} \max_{1 < \ell < F_j^{X_n}(r_n(t-\delta), \omega) + \kappa} v_j(\ell, \omega) = 0,$$

from which (7.8) immediately follows. If $\lim_{n \to \infty} F_j^{X_n}(r_n(t - \delta), \omega) < \infty$, then (7.9) clearly holds. In the opposite case, the left side of (7.9) is equal to

$$\lim_{n \to \infty} \frac{F_j^{X_n}(r_n(t - \delta), \omega)}{r_n} \frac{1}{F_j^{X_n}(r_n(t - \delta), \omega)} \max_{1 < \ell < F_j^{X_n}(r_n(t-\delta), \omega) + \kappa} v_j(\ell, \omega)$$

$$= \hat{F}_j(t - \delta) \lim_{n \to \infty} \frac{1}{F_j^{X_n}(r_n(t - \delta), \omega)} \max_{1 < \ell < F_j^{X_n}(r_n(t-\delta), \omega) + \kappa} v_j(\ell, \omega)$$

$$= 0,$$

where the first equality follows from (6.39), and the second equality follows from the choice of ω to satisfy (6.38). □

7.3 Queueing Network with FCFS Control

Continuing to focus on queueing networks, let us now consider the FCFS control policy that was introduced in Section 2.6. FCFS is a nonidling policy, and unlike all other control policies studied in this book, it is *not* simply structured, as that term was defined in Section 2.3. To minimize technicalities in our derivation of its characteristic fluid equation, we restrict attention to the case of single-server pools, that is, $b_k = 1$ for each $k \in \mathcal{K}$. Under this assumption, the FCFS policy can also be called first-in-first-out (FIFO) processing, as noted in Section 2.6.

As in Section 2.6, let λ_i denote the external arrival rate (possibly zero) for jobs of class $i \in \mathcal{I}$, let m_i denote the mean service time for such jobs, and let P_{ij} denote the probability that a class i job next becomes a class j job after its service is completed. Also, we denote by $\mathcal{I}(k)$ the set of job classes that are processed by server $k \in \mathcal{K}$, and we use the term "station k" to mean server k plus the buffers containing jobs from classes $i \in \mathcal{I}(k)$. Other notation and terminology will be carried forward from Chapter 2 without comment.

Equations (2.8), (2.9), and (2.10) govern the evolution of a queueing network under *any* control policy, and we now develop an additional equation that is specific to the FCFS policy. For that purpose, let

$$(7.10) \qquad G_i(t) := E_i(t) + \sum_{j \in \mathcal{I}} \Phi_i^j(D_j(t)),$$

interpreting this as the cumulative number of arrivals (both external and internal) into class $i \in \mathcal{I}$ over the interval $(0, t]$. We denote by $W_k(t)$ the *immediate workload* for server k at time t, which means the following: $W_k(t)$ is the amount of time required for server k to complete the processing of all jobs of classes $i \in \mathcal{I}(k)$ that are waiting or in service at time t, assuming that no further arrivals into those classes (either external or internal) are allowed after time t. To express this mathematically, let $V_i(n)$ denote the sum of the first n class i service times, including possibly that of a job in service initially. That is,

$$V_i(n) = \begin{cases} v_i(0) + \sum_{\ell=1}^{n-1} v_i(\ell) & \text{if } Z_i(0) \geq 1, \\ \sum_{\ell=1}^{n} v_i(\ell) & \text{if } Z_i(0) = 0. \end{cases}$$

This definition is consistent with (6.47). Also, given a vector $\xi = (\xi_1, \ldots, \xi_I) \in \mathbb{Z}_+^I$, let $V(\xi) = (V_1(\xi_1), \ldots, V_I(\xi_I))'$. The immediate workload vector $W(t) = (W_1(t),\dots,W_K(t))'$ is then given by the following expression:

$$(7.11) \qquad\qquad W(t) = AV\big(G(t)+Z(0)\big) - AT(t),$$

where T is the cumulative service effort process defined via (2.22), and $A = (A_{ki})$ is the capacity consumption matrix as usual, that is, $A_{ki} = 1$ if $i \in \mathscr{I}(k)$ and $A_{ki} = 0$ otherwise. In words, (7.11) says the following: for each $k \in \mathscr{K}$, the immediate workload $W_k(t)$ equals the total work for server k that has arrived up to time t, minus the total time that server k has been busy up to t.

Let us consider the set of all jobs residing in buffers $i \in \mathscr{I}(k)$ at time t. Under the FCFS control policy, server k will complete the processing of those jobs precisely at time $t + W_k(t)$, because jobs arriving at station k after time t will have lower priority than the ones already present at time t. This observation leads to the following identity, which is the key to further analysis:

$$(7.12) \quad D_i(t + W_k(t)) = Z_i(0) + G_i(t), \quad \text{for } t \ge 0, \ i \in \mathscr{I}(k), \ k \in \mathscr{K}.$$

In words, (7.12) says that, for each $i \in \mathscr{I}(k)$, the number of departures from class i up to time $t + W_k(t)$ equals the number of arrivals into that class up to time t.

Theorem 7.5. *For a queueing network operating under the FCFS control policy, each fluid limit path $(\hat{D},\hat{F},\hat{T},\hat{Z})$ satisfies (6.1) through (6.6) and*

$$(7.13) \qquad \hat{D}_i(t + \hat{W}_k(t)) = \hat{G}_i(t), \quad \text{for } t \ge 0, \ i \in \mathscr{I}(k), \ k \in \mathscr{K},$$

where

$$(7.14) \qquad\qquad \hat{G}_i(t) = \lambda_i t + \sum_{j=1}^{J} P_{ji}\hat{D}_j(t), \quad i \in \mathscr{I},$$

$$(7.15) \qquad\qquad \hat{W}_k(t) = \sum_{j \in \mathscr{I}(k)} m_j \hat{Z}_j(t), \quad k \in \mathscr{K}.$$

Proof. Let $(\hat{D},\hat{F},\hat{T},\hat{Z})$ be a fluid limit path with a corresponding sequence of initial states $\{x_n\} \subset \mathscr{X}$ and a corresponding $\omega \in \Omega$ such that (2.14), (2.15), (6.38), and (6.39) all hold. Fix this ω. Define

$$\hat{G}^{X_n}(t) := \frac{1}{|x_n|} G^{X_n}(|x_n|t, \omega), \quad \hat{W}^{X_n}(t) := \frac{1}{|x_n|} W^{X_n}(|x_n|t, \omega).$$

It follows from (7.10), (2.14), and (2.15) that

(7.16) $$\hat{G}^{X_n} \to \hat{G}$$

u.o.c. as $n \to \infty$, where $\hat{G}$ is given by (7.14). It follows from (7.11), (2.15), and (6.39) that

(7.17) $$\hat{W}^{X_n} \to \hat{W}$$

u.o.c. as $n \to \infty$, where

$$\begin{aligned}
\hat{W}(t) &= AM\big(\hat{G}(t) + \hat{Z}(0)\big) - A\hat{T}(t) \\
&= AM\big(\hat{G}(t) + \hat{Z}(0) - \hat{D}(t)\big) \\
&= AM\hat{Z}(t), \quad t \geq 0.
\end{aligned}$$

From (7.16), (7.17), (6.39), and (7.12), the fluid model equation (7.13) follows. □

7.4 Unitary Network with Specially Structured Control

We consider here a unitary network (see Section 2.6 for the meaning of that term) operating under a relaxed control policy with a certain special structure. This analysis prepares the way for our later study (see Section 10.4) of unitary networks with "proportionally fair" resource allocation. Recall that single-server pools are assumed in our relaxed SPN formulation, and b is a K-vector of server capacities, expressed in units like megabytes per second. Also, in a unitary network we have a one-to-one correspondence between job classes and service types, so both are indexed by $i \in \mathcal{I}$.

In the relaxed control setting of Section 2.4, let t be a decision time and let $\hat{z}$ be the updated job count vector at t, as defined in Section 2.3. We consider relaxed control policies under which the I-vector β of service rates going forward from t has the form

(7.18) $$\beta = h(\hat{z}),$$

where $h : \mathbb{R}_+^I \to \mathbb{R}_+^I$ is a policy function that meets the capacity constraint

(7.19) $$Ah(\cdot) \leq b.$$

Thus it is required that the service rate vector β be determined entirely by the vector $\hat{z}$ of updated job counts, whereas in our general relaxed formulation β can depend on both $\hat{n}$ and $\hat{z}$. To obtain the fluid model equation corresponding to (7.18), we make the following assumptions.

Assumption 7.6. The policy function $h(\cdot)$ in (7.18) is *continuous away from* 0, by which we mean the following: for each $i \in \mathscr{I}$ and each $z \in \mathbb{R}_+^I$ with $z_i > 0$, $h_i(\cdot)$ is continuous at z. Moreover,

(7.20) $$h(cz) = h(z) \quad \text{for all } c > 0 \text{ and } z \in \mathbb{R}_+^I.$$

Remark 7.7. Property (7.20) is expressed verbally by saying that $h(\cdot)$ is *homogeneous of degree zero*. Together, (7.18) and (7.20) say that β depends on $\hat{z}$ only through the *relative* magnitudes of its components.

Theorem 7.8. *Consider a unitary network operating under the relaxed control policy defined by (7.18), where $h : \mathbb{R}_+^I \to \mathbb{R}_+^I$ satisfies (7.19) and Assumption 7.6. Each fluid limit path $(\hat{D}, \hat{F}, \hat{T}, \hat{Z})$ satisfies the following: for each $i \in \mathscr{I}$ and $t > 0$,*

(7.21) $$\hat{Z}_i(t) > 0 \quad \text{implies that} \quad \frac{d}{dt}\hat{T}_i(t) = h_i(\hat{Z}(t)).$$

Remark 7.9. In addition to (7.21), it is tempting to write, for each $i \in \mathscr{I}$,

(7.22) $$\hat{Z}_i(t) = 0 \quad \text{implies that} \quad \frac{d}{dt}\hat{T}_i(t) = 0,$$

because (7.22) is analogous to the following property of the unitary network: when buffer i is empty, no service effort is allocated to that job class. However, (7.22) need not be true. To be specific, $\hat{T}_i$ may be nondifferentiable at t when $\hat{Z}_i(t) = 0$, and even if it is differentiable, its derivative need not be zero. See, for example, (10.70) in Chapter 10. A corrected version of (7.22) will be stated later as Lemma 8.9.

Proof of Theorem 7.8. Let $(\hat{D}, \hat{F}, \hat{T}, \hat{Z})$ be a fluid limit path with corresponding $\omega \in \Omega$ and a sequence $\{x_n\} \subset \mathscr{X}$ such that (6.39) holds. It follows that

(7.23) $$\hat{Z}^{x_n}(\cdot, \omega) \to \hat{Z}(\cdot) \quad \text{u.o.c.} \quad \text{as } n \to \infty.$$

Fix a time $t > 0$ and a buffer $i \in \mathscr{I}$. Assume that $\hat{Z}_i(t) > 0$. We need to prove that $\hat{T}_i$ is differentiable at t and its derivative is given by the right side of (7.21).

By the continuity of $\hat{Z}$, there exists a $\delta \in (0, t)$ such that

$$(7.24) \qquad \epsilon := \min_{u \in [t-\delta, t+\delta]} \hat{Z}_i(u) > 0.$$

Thus, by (7.23), there exists $L_0 > 0$ such that

$$Z_i^{x_n}(u) \geq |x_n| \epsilon / 2 \geq 1$$

for $u \in \big(|x_n|(t-\delta), |x_n|(t+\delta) \big)$ and $n > L_0$. It follows from (2.28) and (7.18), where a superscript x was omitted from (2.28) in the interest of visual simplicity, that

$$T_i^{x_n}(|x_n|u_2) - T_i^{x_n}(|x_n|u_1) = \int_{|x_n|u_1}^{|x_n|u_2} h_i\big(Z^{x_n}(u)\big) du$$

$$= \int_{u_1}^{u_2} |x_n| h_i\big(Z^{x_n}(|x_n|u)\big) du$$

for any $u_1, u_2 \in (t-\delta, t+\delta)$, which implies that

$$(7.25) \quad \hat{T}_i^{x_n}(u_2) - \hat{T}_i^{x_n}(u_1) = \int_{u_1}^{u_2} h_i\big(|x_n|\hat{Z}^{x_n}(u)\big) du = \int_{u_1}^{u_2} h_i\big(\hat{Z}^{x_n}(u)\big) du$$

for any $u_1, u_2 \in (t-\delta, t+\delta)$; in obtaining the last equality of (7.25), we have used the fact that h is homogeneous of degree zero, which is part of Assumption 7.6.

Condition (7.19) ensures that the policy function h is bounded, and Assumption 7.6 ensures that it is continuous away from 0. Combining those properties with (7.23) and the dominated convergence theorem (Corollary B.3), we take limits on both sides of (7.25) as $n \to \infty$, concluding that

$$\hat{T}_i(u_2) - \hat{T}_i(u_1) = \int_{u_1}^{u_2} h_i\big(\hat{Z}(u)\big) du$$

for any $u_1, u_2 \in (t-\delta, t+\delta)$, from which one has that the derivative of $\hat{T}_i$ exists at t and the derivative is given by the right side of (7.21). $\qquad \square$

7.5 Sources and Literature

A variant of Theorem 7.2 appeared as theorem 4.1 of Dai (1995a). The model considered there was a queueing network with *general* service time distributions and with single-server pools; in its analysis, the nonidling fluid equation (7.1) appeared in an alternative, integrated form involving cumulative server idle time and complementarity relationships.

A restrictive version of Theorem 7.3 appeared as theorem 7.1 of Dai (1995a). That result concerns the special kind of queueing network called a reentrant line (see Figure 1.11 for an example), operating under a specific SBP policy, namely, the *preemptive* version of first-buffer-first-served. Again the key fluid equation, which appears as (7.6) in this chapter, was expressed in an alternative, integrated form.

For a queueing network with single-server stations, the key to analysis of the FCFS control policy is (7.12). One early work that used this characterization of FCFS, but in a slightly different form, was Harrison and Viên Nguyen (1993); we do not know of earlier references, but they very likely exist. The FIFO fluid model equation (7.13) first appeared as equation (2.6) of Bramson (1996a). An alternative version of that FIFO fluid model equation, expressed in terms of the inverse of $t + \hat{W}_k(t)$, appeared as equation (29) of Chen and Zhang (1997). Bramson (1994) showed by example that a queueing network operating under the FCFS control policy can be unstable even though it satisfies the standard load condition $\rho < e$ that was discussed in Section 2.6. In a later paper, that same author proved a contrasting positive result for a queueing network *of Kelly type* with single-server pools; the italicized phrase means that all classes processed by any given server have a common mean service time. To be specific, Bramson (1996a) proved that the FCFS fluid model is stable for such a queueing network if the standard load condition is satisfied. The proof is lengthy, involving an entropy Lyapunov function, and it will not be included in this book.

Theorem 7.8 covers a wide a variety of control policies for queueing networks, including the HLSPS and HLPPS policies that were introduced in Section 4.6.

8

Proving Fluid Model Stability Using
Lyapunov Functions

To make use of Theorem 6.2, one must prove that a fluid model corresponding to a given SPN is stable. Section 8.1 presents a key lemma on fluid model stability, establishing a sufficient condition that involves Lyapunov functions. That condition, known as the *drift condition for fluid models*, is typically easier to apply than the analogous drift condition for Markov chains, as we show by example in Section 8.2. In Sections 8.3 and 8.4, we construct Lyapunov functions for two fluid models that were derived earlier in Chapter 7, demonstrating the effectiveness of the fluid model methodology.

Notational change for fluid solutions. In this chapter, we use (D, F, T, Z), *without hats over the components*, to denote a fluid model solution. (See Section 6.2 for the meaning of that term.) Dropping the hats in this way simplifies notation, and it should cause no confusion, because here we study only fluid models, not the SPNs from which they are derived.

8.1 Fluid Model Calculus and Lyapunov Functions

In the following definition, d and m are positive integers.

Definition 8.1. A function $g : \mathbb{R}^d \to \mathbb{R}^m$ is said to be *Lipschitz continuous* (or just Lipschitz) if for any bounded set $B \subset \mathbb{R}^d$ there exists a constant $\kappa(B) > 0$ such that

$$|g(x) - g(y)| \leq \kappa(B)|x - y| \quad \text{for any } x, y \in B.$$

It is said to be *globally Lipschitz continuous* (or just globally Lipschitz) if the constant $\kappa(B)$ is independent of B.

160

It is clear that a function $g = (g_1, \ldots, g_m)$ is Lipschitz if and only if each component of g is Lipschitz. The same definition applies when $\mathbb{R}^d$ is replaced by $\mathbb{R}_+^d$.

Lemma 8.2. *Assume that $g : \mathbb{R}^m \to \mathbb{R}$ and $h : \mathbb{R}_+ \to \mathbb{R}^m$ are both Lipschitz. Define $f(t) := g(h(t))$ for each $t \in \mathbb{R}_+$. Then $f : \mathbb{R}_+ \to \mathbb{R}$ is Lipschitz.*

Proof. The proof follows immediately from the fact that the continuous function h is bounded on any bounded set of $\mathbb{R}_+$. □

Lemma 8.3. *Any solution (D, F, T, Z) of (6.1) through (6.6) is globally Lipschitz.*

Proof. Using the fact that $\tilde{e}' \le e'A$, where e and $\tilde{e}$ are vectors of ones of appropriate dimensions, constraint (6.6) implies that

$$|T(t) - T(s)| \le (t - s)|b| \quad \text{for any } 0 \le s < t,$$

proving that T is globally Lipschitz. Then (6.4) implies that F is globally Lipschitz, (6.3) implies that D is globally Lipschitz, and (6.1) implies that Z is globally Lipschitz. □

The following definition involves *Lebesgue measure*, which extends the notion of "length" from intervals to general subsets of $\mathbb{R}$; see, for example, Royden and Fitzpatrick (2010). This concept is not used elsewhere in this book.

Definition 8.4. A collection {property$(t) : t \in [a, b]$} is said to hold for *almost every* $t \in [a, b]$ if the Lebesgue measure of the set $\{t \in [a, b] : \text{property}(t) \text{ does not hold}\}$ is zero.

Lyapunov functions. A function H that satisfies all the hypotheses of Lemma 8.5 is said to be a *Lyapunov function* for the fluid model. All stability results for the fluid models in this book can be proved either by applying this result with an appropriate Lyapunov function, or else by applying a generalization of this result (Lemma 8.11) that appears at the end of the current section. Another result appearing later in this section, Lemma 8.9, often plays an important role in applying Lemma 8.5. Recall that, for a function $f : \mathbb{R}_+ \to \mathbb{R}$, we use $\dot{f}(t)$ to denote the derivative of f at t. As in Section 6.2, whenever $\dot{f}(t)$ is used, it is assumed that the derivative of f at t exists.

Lemma 8.5.   *Let $H : \mathbb{R}^I_+ \to \mathbb{R}_+$ be Lipschitz continuous. Assume that $H(0) = 0$ and $H(x) \neq 0$ when $x \neq 0$. Consider a fluid model consisting of (6.1) through (6.6) plus possibly other equations. For a fluid model solution (D, F, T, Z), let*

$$(8.1) \qquad f(t) := H(Z(t)) \quad for \ t \geq 0.$$

Suppose there exists an $\epsilon > 0$ such that

$$(8.2) \qquad \dot{f}(t) \leq -\epsilon \quad for \ almost \ all \ points \ t \ with \ Z(t) \neq 0.$$

Then $Z(t) = 0$ for $t \geq t_0 = f(0)/\epsilon = H(Z(0))/\epsilon$.

Proof.   We first prove that f is a nonincreasing function. By Lemmas 8.2 and 8.3, the function f defined in (8.1) is Lipschitz continuous. Therefore, by Lemma A.2, it is absolutely continuous. It follows from Lemma A.3 that f has a derivative at almost all points $t \in \mathbb{R}_+$. Fix a $t > 0$. Assume $Z(t) = 0$. Then $f(t) = H(0) = 0$. Therefore, f attains its minimum at t. If f has a derivative at t, then $\dot{f}(t) = 0$. This fact, together with condition (8.2), proves that $\dot{f}(t) \leq 0$ for almost all $t \in \mathbb{R}_+$. Because f is absolutely continuous, it follows from Lemma A.3 and (A.1) that

$$(8.3) \qquad f(b) - f(a) = \int_a^b \dot{f}(t)dt \leq 0$$

for any $0 \leq a < b$, proving that f is nonincreasing.

Next we prove $f(t_0) = 0$. Suppose on the contrary that $f(t_0) > 0$. By the continuity of f, there exists a $\delta > 0$ such that $f(t_0 + \delta) > 0$. Now the monotonicity of f implies that $f(t) > 0$ for $t \in [0, t_0 + \delta]$. It follows that $Z(t) \neq 0$ for $t \in [0, t_0 + \delta]$. Then (8.2) gives $\dot{f}(t) \leq -\epsilon$ for almost all $t \in [0, t_0 + \delta]$. By (A.1),

$$f(t_0 + \delta) - f(0) = \int_0^{t_0 + \delta} \dot{f}(t)dt \leq -\epsilon(t_0 + \delta)$$

$$= -f(0) - \epsilon\delta,$$

which implies

$$0 \leq f(t_0 + \delta) \leq -\epsilon\delta,$$

which is a contradiction.   □

When the function H in (8.1) is quadratic, the following lemma is often useful.

Lemma 8.6. *In the setting of Lemma 8.5, let us again define f via (8.1), but assume the following in place of (8.2): there exists an $\epsilon > 0$ such that*

$$(8.4) \qquad \dot{f}(t) \leq -\epsilon \sqrt{f(t)} \quad \text{for almost all points } t \text{ with } Z(t) \neq 0.$$

Then $Z(t) = 0$ for $t \geq t_0 = \sqrt{f(0)}/\epsilon$.

Proof. Because f is absolutely continuous and $\dot{f}(t) \leq 0$ for almost all $t \in \mathbb{R}_+$, (8.3) implies that f is nonincreasing on $\mathbb{R}_+$. Define $t_1 = \inf\{t \in \mathbb{R}_+ : f(t) = 0\}$. (Recall that by convention, $\inf(\emptyset) = \infty$.) It follows that $f(t) > 0$ for $t \in [0, t_1)$. Then by the monotonicity of f we have that $f(t) = 0$ for $t \geq t_1$ when $t_1 < \infty$. It suffices to prove that

$$(8.5) \qquad\qquad t_1 \leq t_0.$$

If $t_1 = 0$, then (8.5) clearly holds. Otherwise, define $g(t) = \sqrt{f(t)}$. For any interval $[a, b] \subset (0, t_1)$ and any $s, t \in [a, b]$,

$$|g(t) - g(s)| \leq \frac{1}{\sqrt{f(b))}} |f(t) - f(s)|,$$

where $f(b) > 0$. Thus g is (locally) Lipschitz continuous, hence absolutely continuous in the open interval $(0, t_1)$, and the following holds:

$$\dot{g}(t) = \frac{\dot{f}(t)}{\sqrt{f(t)}} \leq -\epsilon \quad \text{for almost all } t \text{ in } (0, t_1).$$

Therefore, for any $t \in (0, t_1)$,

$$g(t) - g(0) = \int_0^t \dot{g}(s) ds \leq -t\epsilon.$$

Because $g(t) \geq 0$, the preceding inequality implies that $t\epsilon \leq g(0)$, and because t can be arbitrarily close to t_1, one has $t_1 \leq g(0)/\epsilon = t_0$, proving (8.5). $\qquad\square$

Definition 8.7. A point $t > 0$ is said to be a *regular point* for a fluid model solution (D, F, T, Z) if all of its components are differentiable at t.

Remark 8.8.   Because (6.4) defines F in terms of T, (6.3) defines D in terms of F, and (6.1) defines Z in terms of F and D, we see that t is a regular point for a fluid model solution (D, F, T, Z) if and only if $T(\cdot)$ is differentiable at t.

Because each fluid model solution is Lipschitz continuous (Lemma 8.3) and hence absolutely continuous (Lemma A.2), the set of nonregular points has Lebesgue measure zero. Thus nonregular points can be excluded in verifying the drift condition (8.2). A first benefit of restricting attention to regular points is expressed by the following lemma.

Lemma 8.9.   *Assume $t > 0$ is a regular point of a fluid model solution (D, F, T, Z). Then $Z_i(t) = 0$ implies that $\dot{Z}_i(t) = 0$.*

Proof.   Assume $Z_i(t) = 0$. Then $Z_i(\cdot)$ achieves its minimum at t. Because $Z_i(\cdot)$ is assumed to be differentiable at t, we have $\dot{Z}_i(t) = 0$.   □

When $Z_i(t) = 0$, this lemma shows that the fluid departure rate from buffer i is equal to the fluid arrival rate into that buffer. The actual rate can often be determined through other means. In particular, it is not necessarily true that the fluid departure rate is zero at time t. The following lemma is particularly useful when the function H in (8.1) is piecewise linear.

Lemma 8.10.   *For an integer $d > 0$, let $f_j : \mathbb{R}_+ \to \mathbb{R}$ be given functions on $\mathbb{R}_+$ for $j = 1, \ldots, d$. For each $t \in \mathbb{R}_+$ define*

$$f(t) = \max_{j=1,\ldots,d} f_j(t).$$

Fix a $t > 0$ and an $i \in \{1, \ldots, d\}$ such that $f_i(t) = f(t)$. Assume that f_i and f are both differentiable at t. Then

$$\dot{f}(t) = \dot{f}_i(t).$$

Proof.   For each $\delta > 0$,

$$f(t \pm \delta) - f(t) \geq f_i(t \pm \delta) - f(t)$$
$$= f_i(t \pm \delta) - f_i(t).$$

Thus

$$\dot{f}(t) = \lim_{\delta \downarrow 0} \frac{f(t+\delta) - f(t)}{\delta} \geq \lim_{\delta \downarrow 0} \frac{f_i(t+\delta) - f_i(t)}{\delta} = \dot{f}_i(t),$$

$$\dot{f}(t) = \lim_{\delta \downarrow 0} \frac{f(t-\delta) - f(t)}{-\delta} \leq \lim_{\delta \downarrow 0} \frac{f_i(t-\delta) - f_i(t)}{-\delta} = \dot{f}_i(t),$$

proving the lemma. $\qquad\square$

In Lemma 8.11, f is *not* assumed to be Lipschitz continuous, a condition that is sometimes difficult to check. This result generalizes Lemma 8.5. Here we denote by $D^+f(t)$ the Dini derivative of a function f at t, which is defined via (A.9).

Lemma 8.11. *Assume that $f : \mathbb{R}_+ \to \mathbb{R}_+$ satisfies the following three conditions. (a) f is continuous on $(0,\infty)$. (b) For each interval $[a,b) \subset \mathbb{R}_+$ there exists a constant $M > 0$ such that*

$$(8.6) \qquad D^+f(t) \leq M \quad \text{for each } t \in [a,b).$$

(c) There exists an $\epsilon > 0$ such that

$$(8.7) \qquad D^+f(t) \leq -\epsilon \quad \text{for almost all } t \in \mathbb{R}_+ \text{ with } f(t) > 0.$$

Then $f(t) = 0$ for $t \geq f(0)/\epsilon$.

Proof. It will be shown that

$$(8.8) \qquad f(b) - f(a) \leq \int_a^b D^+f(u)\,du$$

for any interval $[a,b) \subset \mathbb{R}_+$. Once (8.8) is established, the proof of the current lemma is identical to that of Lemma 8.5. Incidentally, without condition (8.6) the inequality (8.8) is generally not true; see Massoulié (2007) for a counterexample.

For the proof of (8.8), fix an interval $[a,b) \in \mathbb{R}_+$ with $a < b$. Conditions (a) and (b) then give us

$$(8.9) \qquad f(t) - f(s) \leq (t-s)M \quad \text{for } s,t \in [a,b) \text{ with } s < t;$$

see, for example, theorem 3.4.5 of Kannan and Krueger (1996). Next, define $g(u) := f(u) - Mu$ for $u \in [a,b)$. Inequality (8.9) implies that g is nonincreasing in $[a,b)$, and thus $\dot{g}(u)$ exists for almost every $u \in [a,b)$.

It follows from (A.9) that $D^+ f(u) = \dot{f}(u)$ when $\dot{f}(u)$ exists. Also, we have from the definition of g that $\dot{f}(u) = \dot{g}(u) + M$ exists when $\dot{g}(u)$ exists. Thus $\dot{f}(u)$ exists for almost every $u \in [a, b)$, and (8.8) is equivalent to the following:

$$(8.10) \qquad\qquad f(b) - f(a) \leq \int_a^b \dot{f}(u)\,du.$$

Finally, to prove (8.10), for each integer n we partition $[a, b)$ into 2^n subintervals of equal length. For $k = 1, \ldots, 2^n$ and

$$u \in \left[\frac{k-1}{2^n}(b-a) + a, \frac{k}{2^n}(b-a) + a \right),$$

define

$$\lfloor u \rfloor_n := \frac{k-1}{2^n}(b-a) + a \quad \text{and} \quad \lceil u \rceil_n := \frac{k}{2^n}(b-a) + a.$$

Thus, for each n, $\lfloor u \rfloor_n$ and $\lceil u \rceil_n$ are well defined and piecewise constant on $[a, b)$.

Now one can write

$$f(b) - f(a) = \sum_{k=1}^{2^n} \left(f\big(k(b-a)2^{-n} + a\big) - f\big((k-1)(b-a)2^{-n} + a\big) \right)$$

$$= \int_a^b \frac{f(\lceil u \rceil_n) - f(\lfloor u \rfloor_n)}{\lceil u \rceil_n - \lfloor u \rfloor_n}\,du \quad \text{for each positive integer } n,$$

and then by taking $n \to \infty$ one has

$$f(b) - f(a) = \lim_{n \to \infty} \int_a^b \frac{f(\lceil u \rceil_n) - f(\lfloor u \rfloor_n)}{\lceil u \rceil_n - \lfloor u \rfloor_n}\,du$$

$$\leq \int_a^b \limsup_{n \to \infty} \frac{f(\lceil u \rceil_n) - f(\lfloor u \rfloor_n)}{\lceil u \rceil_n - \lfloor u \rfloor_n}\,du$$

$$= \int_a^b \dot{f}(u)\,du;$$

here the inequality follows from Fatou's lemma (Lemma B.4), because the integrand is bounded above by the constant M in (8.9), and the last equality is due to the assumed existence of $\dot{f}(u)$ for almost all $u \in [a, b]$. This establishes (8.8), completing the proof. $\qquad\square$

8.2 Advantage of Fluid Models over Markov Chains

Condition (8.2) for fluid models resembles the drift condition (D.36) that is used for proving positive recurrence of continuous-time Markov chains (CTMCs). Let us consider the application of the latter criterion to an SPN model that can be embedded in a CTMC (see Section 3.1 for the meaning of that phrase). The verification of the Markov chain drift condition (D.36) is typically quite easy for states in which all buffers have positive contents. It is for states in which some buffers are empty that the verification may become difficult. In that regard, however, there are important differences between the fluid model drift condition and the Markov chain drift condition. For a Markov chain, (D.36) requires the drift to be negative in *all* "large" states. For a fluid model, (8.2) allows us to ignore some times t at which the buffer contents process $Z(\cdot)$ may be large but is not differentiable. Also, at times t when a buffer is empty and $Z(\cdot)$ *is* differentiable, the fluid input rate to the empty buffer is necessarily equal to the fluid output rate from it (see Lemma 8.9). Most of the advantage of fluid model methodology over direct analysis of Markov chains comes from these two observations.

Tandem queueing network. As a simple example to illustrate these subtle but important points, let us consider the tandem queueing network depicted in Figure 1.1, operating under the nonidling first-come-first-served control policy. Assuming that the external input process is Poisson and the service time distribution at each station is exponential, the two-dimensional buffer contents process Z is a CTMC; this is the "original CTMC" referred to in the paragraphs that follow. When we refer to "the corresponding fluid model," that means the fluid model defined by (6.1) through (6.6) plus the non-idling condition (6.7).

Theorem 8.12. *Assuming that the standard load condition* (1.1) *holds, the fluid model corresponding to the tandem queueing system in Figure 1.1 is stable.*

This is a special case of Theorem 8.14, which will be proved in Section 8.3 using a piecewise linear Lyapunov function. Here we construct a *linear* Lyapunov function to prove Theorem 8.12, and then show the following: when that same Lyapunov function is applied to the original CTMC, the drift condition (D.36) fails to hold.

Proof of Theorem 8.12. Let (D, F, T, Z) be a solution of the fluid model equations (6.1) through (6.7). Eliminating D and F from those equations, as one can always do (see comments following (6.6) in Section 6.1), and then substituting the special structure of the tandem queueing system, the fluid model equations reduce to

(8.11) $Z_1(t) = Z_1(0) + \lambda_1 t - \mu_1 T_1(t),$

(8.12) $Z_2(t) = Z_2(0) + \mu_1 T_1(t) - \mu_2 T_2(t),$

(8.13) $Z_i(t) \geq 0, \quad i = 1, 2,$

(8.14) $T_i(0) = 0$ and $0 \leq T_i(t) - T_i(s) \leq t - s$ for $0 \leq s < t, \quad i = 1, 2,$

(8.15) $Z_i(t) > 0$ implies $\dot{T}_i(t) = 1, \quad i = 1, 2.$

For each $t \geq 0$, define

$$f(t) := Z_1(t) + Z_2(t),$$

interpreting $f(t)$ as the total amount of fluid in the system at time t. This function f has the form $f(t) = H(Z(t))$, as in (8.1), where

(8.16) $H : (z_1, z_2) \in \mathbb{R}_+^2 \rightarrow H(z_1, z_2) = z_1 + z_2 \in \mathbb{R}_+.$

Because H is a linear function, it is Lipschitz continuous. Now we verify that condition (8.2) in Lemma 8.5 is satisfied.

Using (8.11) and (8.12), one has

$$f(t) = f(0) + \lambda_1 t - \mu_2 T_2(t).$$

Thus

$$\dot{f}(t) = \lambda_1 - \mu_2 \dot{T}_2(t),$$

which is the difference between the arrival rate into buffer 1 and the departure rate from buffer 2. Assume $f(t) > 0$ and t is a regular point of (D, F, T, Z). We now verify that (8.2) is satisfied. (a) If $Z_2(t) > 0$, then the nonidling condition (8.15) for station 2 implies that $\dot{T}_2(t) = 1$. Thus

$$\dot{f}(t) = \lambda_1 - \mu_2.$$

(b) If $Z_2(t) = 0$, then it must be true that $Z_1(t) > 0$, because $f(t)$ is assumed to be positive. Lemma 8.9 then gives $\dot{Z}_2(t) = 0$. Thus

(8.17) $\dot{f}(t) = \dot{Z}_1(t) = \lambda_1 - \mu_1 \dot{T}_1(t) = \lambda_1 - \mu_1,$

where the last equality holds because of the nonidling condition (8.15) for station 1. Under the load condition (1.1), condition (8.2) is satisfied with

$$\epsilon = \min(\mu_1 - \lambda_1, \mu_2 - \lambda_1) > 0. \qquad \square$$

CTMC drift condition. Now we show that for our original Markov chain Z the *same* linear Lyapunov function H in (8.16) does *not* yield negative drift for all "large" states. The off-diagonal entries of the generator matrix for Z (see Appendix D.1) have the following form:

$$\begin{aligned}
\Lambda\big((z_1, z_2), (z_1 + 1, z_2)\big) &= \lambda_1, \\
\Lambda\big((z_1, z_2), (z_1 - 1, z_2 + 1)\big) &= \mu_1 \text{ when } z_1 > 0, \\
\Lambda\big((z_1, z_2), (z_1, z_2 - 1)\big) &= \mu_2 \text{ when } z_2 > 0
\end{aligned}$$

for each $z = (z_1, z_2) \in \mathbb{Z}_+^2$. Thus the CTMC drift condition (D.36) requires that

$$(8.18) \qquad \sum_{z' \in \mathbb{Z}_+^2} \Lambda(z, z') H(z') \leq -c_1 + c_2 1_{\{|z| \geq c_3\}} \quad \text{for } z \in \mathbb{Z}_+^2$$

for some constants $c_i > 0$ $(i = 1, 2, 3)$. For the function H in (8.16), the drift in state $z = (z_1, z_2)' \in \mathbb{Z}_+^2$ is

$$\sum_{z' \in \mathbb{Z}_+^2} \Lambda(z, z') H(z') = \begin{cases} \lambda_1 - \mu_2 & \text{if } z_2 > 0, \\ \lambda_1 & \text{if } z_2 = 0. \end{cases}$$

Thus the drift is always positive when $z_2 = 0$, no matter how large z_1 is. This conclusion derives from the fact that server 2 idles when buffer 2 is empty ($z_2 = 0$), and hence the departure rate from buffer 2 is zero. In contrast, when buffer 2 is empty at a regular time point in the fluid model, the departure rate from buffer 2 is equal to the departure rate from buffer 1, never zero; see (8.17).

Continuing discussion of the tandem queueing system, the simplest known Lyapunov function for the original CTMC is

$$H(z_1, z_2) = z_1^2 + a(z_1 + z_2)^2,$$

where $a > 0$ is a constant to be specified. Setting $H_1(z_1, z_2) = z_1^2$ and $H_2(z_1, z_2) = (z_1 + z_2)^2$, readers can easily verify that

$$\Lambda H_1(z_1,z_2) = \begin{cases} 2(\lambda_1 - \mu_1)z_1 + (\lambda_1 + \mu_1) & \text{if } z_1 > 0, \\ \lambda_1 & \text{if } z_1 = 0, \end{cases} \quad \text{and}$$

$$\Lambda H_2(z_1,z_2) = \begin{cases} 2(\lambda_1 - \mu_2)(z_1 + z_2) + (\lambda_1 + \mu_2) & \text{if } z_2 > 0, \\ 2\lambda_1 z_1 + \lambda_1 & \text{if } z_2 = 0. \end{cases}$$

From the load condition (1.1), one therefore has

$$\Lambda H(z_1,z_2) \le \lambda_1 + \mu_1 + a(\lambda_1 + \mu_2) \quad \text{for each } (z_1,z_2) \in \mathbb{Z}_+^2.$$

By choosing $a \in (0, \mu_1/\lambda_1 - 1)$, we ensure that

$$\lambda_1 - \mu_1 + a\lambda_1 < 0,$$

from which one can verify that

(8.19)
$$\Lambda H(z_1,z_2) \le -1$$

for each $(z_1,z_2) \in \mathbb{Z}_+^2$ with $z_1 + z_2$ large enough. Therefore, the proposed function H produces a negative drift for *all* large states of the original CTMC. This illustrates a phenomenon often observed in which the fluid model corresponding to a particular SPN admits a Lyapunov function that is simpler (in our case, a linear function) than the simplest known Lyapunov function for the original CTMC (in our case, a quadratic function).

More significantly, there are several important families of SPNs for which (a) fluid model stability has been proved using Lyapunov functions, but (b) Lyapunov functions have yet to be found for the original Markov chains. One such family consists of subcritical queueing networks operating under the HLPPS control policy, which we define in Section 4.6. In Sections 10.5 and 10.6 (specifically, see Corollary 10.18 and the discussion that follows it), entropy Lyapunov functions will be used to prove stability of the corresponding fluid models, but no Lyapunov function is known for the original Markov chain.

Exceptional set   For the tandem fluid model, assume $\mu_1 > \mu_2$. The fluid model solution Z starting from a typical initial state $Z(0)$ is piecewise linear (see Figure 8.1). There are two points t_1 and t_2 at which (D,F,T,Z) are nondifferentiable. The first point t_1 is the time of first emptiness for buffer 1, namely,

$$t_1 := \frac{Z_1(0)}{\mu_1 - \lambda_1}.$$

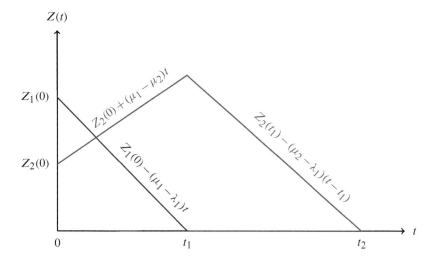

Figure 8.1 Dynamics of the tandem fluid model.

The second point $t_2 > t_1$ is the time of first emptiness for buffer 2, namely,

$$t_2 := \frac{Z_1(0) + Z_2(0)}{\mu_2 - \lambda_1}.$$

In checking the fluid drift condition (8.2), the time point t_1 can be excluded from the analysis. (At t_1, the departure rate from buffer 1 is *undefined*, changing from μ_1 to λ_1.)

8.3 Feedforward Queueing Network (Piecewise Linear Lyapunov Function)

The following formalizes a definition that was stated verbally in Section 1.3. Here we denote by $P = (P_{ij})$ the routing matrix for a queueing network model, as in Section 2.6, and by $p(i)$ the server pool (or station) that serves buffer i, as in Section 7.2. Also, $\mathscr{I}(k)$ is the set of buffers served by pool k, as defined in (2.43).

Definition 8.13. A queueing network is said to be *feedforward* if its stations can be numbered in such a way that $p(i) > p(j)$ implies $P_{ij} = 0$ for $i, j \in \mathscr{I}$.

In this definition, feedback to the same station is allowed. That is, P_{ij} can be positive when $p(i) = p(j)$. In this section, we consider the fluid model for a feedforward queueing network operating under a nonidling policy. We denote by (D, F, T, Z) a solution of the fluid model equations (6.1) through (6.7). Eliminating D and F as in the previous section, and using the special structure of a queueing network, we have the following reduced set of fluid model equations: for each $i \in \mathscr{I}$,

$$(8.20) \quad Z_i(t) = Z_i(0) + \lambda_i t + \sum_{\ell \in \mathscr{I}} P_{\ell i} \mu_\ell T_\ell(t) - \mu_i T_i(t) \quad \text{for each } t \geq 0,$$

$(8.21) \quad Z_i(t) \geq 0 \quad$ for each $t \geq 0$,

$(8.22) \quad T_i(0) = 0$ and $T_i(\cdot)$ is nondecreasing,

$$(8.23) \quad \sum_{\ell \in \mathscr{I}(k)} (T_\ell(t) - T_\ell(s)) \leq (t - s) b_k \quad \text{for each } k \in \mathscr{K} \text{ and } t > s \geq 0.$$

Theorem 8.14. *Consider a feedforward queueing network. If the standard load condition (5.1) holds, then the fluid model under any nonidling policy is stable, and thus by Theorem 6.2 the feedforward queueing network is also stable under any nonidling policy.*

Before proving Theorem 8.14, we present a lemma that holds for the fluid model of a *general* queueing network, without the feedforward restriction. For its statement, we define a linear function $W : \mathbb{R}_+^I \to \mathbb{R}_+^K$ as follows:

$$(8.24) \qquad W(z) := AM(I - P')^{-1} z \quad \text{for } z \in \mathbb{R}_+^I.$$

For an interpretation, let $z \in \mathbb{Z}_+^I$ be an arbitrary buffer contents vector. Arguing as in the derivation of formula (2.40) for the load vector ρ, one sees that $W_k(z)$ represents the *total expected workload* for servers in pool k that is embodied in the buffer contents vector z. That is, if external arrivals were turned off after time t, then $W_k(Z(t))$ represents the expected total busy time (or service effort) required from the b_k servers in pool k before all buffers are empty. This workload may currently reside *anywhere* in the network, but eventually it will come to station k.

In the fluid model, $W(z)$ has the same interpretation, except that the word "expected" is deleted: the total service effort required from pool k to drain all buffers to emptiness is *exactly* $W_k(z)$ when starting with buffer contents vector z. For future reference, note that the definition

(2.40) of the load vector $\rho = (\rho_1, \ldots, \rho_K)'$ can be expressed in terms of the workload operator W as

$$(8.25) \qquad \rho = W(\lambda) = AM(I - P')^{-1}\lambda.$$

Lemma 8.15. *Consider a queueing network operating under a nonidling policy. If a fluid model solution (D, F, T, Z) is differentiable at $t > 0$, then for each station k the following holds:*

$$(8.26) \qquad \sum_{i \in \mathscr{I}(k)} Z_i(t) > 0 \quad \text{implies that} \quad \frac{d}{dt} W_k(Z(t)) = \rho_k - b_k.$$

Proof. From (8.20) through (8.23), one has

$$(8.27) \qquad \begin{aligned} W(Z(t)) &= W(Z(0)) + AM(I - P')^{-1}\lambda t - AT(t) \\ &= W(Z(0)) + \rho t - AT(t). \end{aligned}$$

Equation (8.26) then follows from (8.27) and the nonidling condition (6.7). $\qquad\square$

Proof of Theorem 8.14. Let (T, Z) be a solution of the reduced fluid equations (8.20) through (8.23). Define $\delta_1 = 1$. We claim that under the feedforward assumption there exists a constant $\delta_k > 0$ such that, for all $z \in \mathbb{R}_+^I$ with $\sum_{i \in \mathscr{I}(k)} z_i = 0$,

$$(8.28) \qquad \delta_k W_k(z) \le \max_{\ell \le k-1} \delta_\ell W_\ell(z) \text{ for } k = 2, \ldots, K.$$

Assuming (8.28), we now complete the proof of the theorem. For $z \in \mathbb{R}_+^I$, let

$$H(z) := \max_{k \in \mathscr{K}} \delta_k W_k(z).$$

Thus H is a piecewise linear function of z. Clearly, $H(z) = 0$ if and only if $z = 0$. We now prove that $f(t) = H(Z(t))$ satisfies condition (8.2), from which the theorem is immediate. Because $Z(t)$ is Lipschitz continuous in t, so is $W_k(Z(t))$ for each $k \in \mathscr{K}$. Therefore, f is also Lipschitz continuous. Thus, for almost all $t > 0$, f and (T, Z) are differentiable at t. Fix such a $t > 0$. Assume $Z(t) \ne 0$. We now verify that (8.2) is satisfied for

$$\epsilon := \min_{k \in \mathscr{K}} \delta_k(b_k - \rho_k),$$

which is positive under condition (5.1). To verify (8.2), let $k \in \mathcal{K}$ be the smallest index such that

$$\delta_k W_k(Z(t)) = f(t).$$

Condition (8.28) implies that $\sum_{i \in \mathcal{I}(k)} Z_i(t) > 0$, because otherwise k would not be the smallest index. Condition (8.2) follows from

$$\dot{f}(t) = \delta_k \frac{d}{dt} W_k(Z(t))$$

$$= \delta_k(\rho_k - b_k) \le -\epsilon,$$

where the first equality follows from Lemma 8.10 and the second one follows from Lemma 8.15.

It remains to prove (8.28). The $I \times I$ matrix P can be decomposed into the following block matrix: for $k, \ell \in \mathcal{K}$, elements of the block $P^{(k\ell)}$ are

$$P_{ij}, \quad i \in \mathcal{I}(k), j \in \mathcal{I}(\ell).$$

By the feedforward assumption and the convention for numbering stations, the routing matrix P is "block upper-triangular," meaning that $P^{(k\ell)} = 0$ if $k > \ell$. Also, $P^{(kk)}$ is a square matrix for each $k \in \mathcal{K}$. It follows from the Neumann expansion (2.37) that $Q = (I - P')^{-1}$ has a lower triangular block structure, and that all of its diagonal elements are strictly positive. For each $z \in \mathbb{R}_+^I$ and $k \in \mathcal{K}$, let

$$z^{(k)} := (z_i)_{i \in \mathcal{I}(k)} \quad \text{and} \quad M^{(k)} := \operatorname{diag}(m_i, i \in \mathcal{I}(k)).$$

Then,

$$W_k(z) = e' M^{(k)} \sum_{\ell=1}^{k} Q^{(k\ell)} z^{(\ell)}, \quad k \in \mathcal{K},$$

where e is the vector of ones of appropriate dimension. Fix a station $k \ge 2$. For $\ell = 1, \ldots, k-1$,

$$W_\ell(z) = e' M^{(\ell)} \sum_{\ell'=1}^{\ell} Q^{(\ell, \ell')} z^{(\ell')}$$

$$\ge e' M^{(\ell)} Q^{(\ell, \ell)} z^{(\ell)} \ge \sum_{i \in \mathcal{I}(\ell)} m_i Q_{ii} z_i$$

$$\ge \left(\min_{i \in \mathcal{I}(\ell)} m_i Q_{ii} \right) \sum_{i \in \mathcal{I}(\ell)} z_i.$$

On the other hand, when $z^{(k)} = 0$,

$$W_k(z) = e' M^{(k)} \sum_{\ell=1}^{k-1} Q^{(k\ell)} z^{(\ell')}$$

$$= \sum_{j \in \mathscr{I}(k)} \sum_{\ell=1}^{k-1} \sum_{i \in \mathscr{I}(\ell)} m_j Q_{ji}^{(k\ell)} z_i$$

$$\leq \left(\sum_{j \in \mathscr{I}(k)} m_j \max_{\ell \leq k-1, i \in \mathscr{I}(\ell)} Q_{ji}^{(k\ell)} \right) \sum_{\ell=1}^{k-1} \sum_{i \in \mathscr{I}(\ell)} z_i.$$

By choosing $\delta_k > 0$ small enough such that

$$(k-1)\delta_k \left(\sum_{j \in \mathscr{I}(k)} m_j \max_{\ell \leq k-1, i \in \mathscr{I}(\ell)} Q_{ji}^{(k\ell)} \right) \leq \min_{\ell \leq k-1} \delta_\ell \left(\min_{i \in \mathscr{I}(\ell)} m_i Q_{ii} \right) > 0,$$

one has

$$\delta_k W_k(z) \leq \frac{1}{k-1} \sum_{\ell=1}^{k-1} \delta_\ell W_\ell(z) \leq \max_{\ell \leq k-1} \delta_\ell W_\ell(z),$$

proving (8.28). $\square$

Remark 8.16.   Instead of constructing the Lyapunov function, one could use induction on $k = 1, 2, \ldots, K$ and Lemma 8.9 to prove that there exists a $t_k > 0$ such that

(8.29) $W_1(t) = 0, \ldots, W_k(t) = 0$ for $t \geq t_k |Z(0)|$.

8.4 Queueing Network with HLSPS Control
(Linear Lyapunov Function)

A queueing network with HLSPS control (see Section 4.6) is a unitary network having the special structure specified in Section 7.4. Specifically, the policy function h appearing in (7.18) is given by (4.31) for the HLSPS control policy, and readers can easily verify that this h satisfies Assumption 7.6, regardless of how the proportion vector γ may be chosen. Thus, by Theorem 7.8, each fluid limit path under HLSPS control satisfies the additional fluid equation (7.21).

Definition 8.17. The *fluid model corresponding to the HLSPS control policy* consists of (6.1) through (6.6) and (7.21), where h is defined via (4.31).

For a queueing network, recall that the vector α of total arrival rates is given by formula (2.38), which is the unique solution of the traffic equations (2.36). For each station $k \in \mathcal{K}$, let

$$(8.30) \qquad \gamma_i := \frac{\alpha_i m_i}{\rho_k}, \quad i \in \mathcal{I}(k),$$

where the station k load ρ_k is defined by (2.42).

Theorem 8.18. *Consider a queueing network subject to relaxed control, and let γ be the I-vector defined by (8.30). If the standard load condition (5.1) is satisfied, then the fluid model corresponding to the HLSPS control policy with proportion vector γ is stable, and thus by Theorem 6.2 the queueing network is also stable under that HLSPS control policy.*

Before this theorem is proved, consider the special case of a *generalized Jackson network*, which is a queueing network having a one-to-one correspondence between job classes and server pools (that is, $I = J = K$, and each server pool processes jobs from just one buffer). When the HLSPS policy is specialized to such a network, we have $\gamma_i = 1$ for each buffer $i \in \mathcal{I}$, which means that each server pool allocates all of its capacity to the one buffer for which it is responsible, processing those jobs in head-of-line fashion (that is, in the order of their arrival). Equivalently stated, HLSPS reduces to the nonidling FCFS policy for generalized Jackson networks, and hence we have the following.

Corollary 8.19. *For a generalized Jackson network, if the standard load condition (5.1) is satisfied, then the fluid model is stable under the nonidling FCFS policy.*

To prove Theorem 8.18, we first state a lemma whose proof will follow that of the theorem. The proof of this lemma constitutes the bulk of this section.

Lemma 8.20. *Fix an $\epsilon > 0$. Assume that (D, F, T, Z) is a fluid model solution satisfying (6.1) through (6.6) and*

$$(8.31) \qquad Z_j(t) > 0 \text{ implies that } \dot{D}_j(t) \geq \alpha_j + \epsilon.$$

Then $Z(t) = 0$ for $t \geq |(I - P')^{-1} Z(0)| / \epsilon$.

Proof of Theorem 8.18. Let (D, F, T, Z) be a solution of the fluid model equations (6.1) through (6.6) and (7.21), with the policy function h defined via (4.31). Specializing (6.3) and (6.4) to the queueing network setting, one has

(8.32) $\qquad D_i(t) = \mu_i T_i(t)$ for each $i \in \mathscr{I}$ and $t \in \mathbb{R}_+$.

Fix a buffer $j \in \mathscr{I}$ with $Z_j(t) > 0$. Then

$$\dot{D}_j(t) = \mu_j \dot{T}_j(t) = \mu_j b_k \gamma_i = (b_k / \rho_k) \alpha_j$$
$$= \alpha_j + (b_k / \rho_k - 1) \alpha_j,$$

where the second equality follows from (7.21). Therefore, condition (8.31) is satisfied with

$$\epsilon := \min_{k \in \mathscr{K}, j \in \mathscr{I}(k)} (b_k / \rho_k - 1) \alpha_j.$$

Now $\epsilon > 0$ by the standard load condition (5.1), so the theorem follows from Lemma 8.20. $\qquad \square$

Proof of Lemma 8.20. Let

$$f(t) := e'(I - P')^{-1} Z(t).$$

In the queueing network, the jth component of $(I - P')^{-1} Z(t)$ is the expected number of class j services required by jobs that are located *anywhere* in the network at time t. Thus $f(t)$ is the expected total number of services required to complete processing of jobs located anywhere in the network at time t.

From (6.1) through (6.6), we have

$$f(t) = f(0) + e' \alpha t - e' D(t).$$

Thus, for $Z(t) \neq 0$,

(8.33) $\qquad \dot{f}(t) = \sum_{j=1}^{J} \left(\alpha_j - \dot{D}_j(t) \right)$

$$= \sum_{j: Z_j(t) \neq 0} \left(\alpha_j - \dot{D}_j(t) \right) + \sum_{j: Z_j(t) = 0} \left(\alpha_j - \dot{D}_j(t) \right)$$

$$\leq -\epsilon |\{j : Z_j(t) \neq 0\}| + \sum_{j: Z_j(t) = 0} \left(\alpha_j - \dot{D}_j(t) \right)$$

(8.34) $\qquad \leq -\epsilon,$

where $|\mathscr{S}|$ is the cardinality of a finite set $\mathscr{S}$. The inequality (8.33) follows from (8.31), and (8.34) is a consequence of $Z(t) \neq 0$ and the following claim:

$$(8.35) \qquad \dot{D}_j(t) \geq \alpha_j \quad \text{when } Z_j(t) = 0.$$

Assuming (8.35) for the moment, f is a Lyapunov function satisfying the conditions in Lemma 8.5. Thus $Z(t) = 0$ for $t \geq f(0)/\epsilon = |(I - P')^{-1} Z(0)|/\epsilon$.

To prove (8.35), we introduce the following convention. For a vector $x \in \mathbb{R}^J$ and a set $\mathbf{j} \subset \mathscr{J}$, we let $x_{\mathbf{j}}$ denote the subvector $(x_j, j \in \mathbf{j})$. For a $J \times J$ matrix M, we use $M_{\mathbf{jj'}}$ to denote the submatrix $(M_{j\ell})$ with $j \in \mathbf{j}$ and $\ell \in \mathbf{j'}$. Now let $\mathbf{j} = \{j : Z_j(t) = 0\}$. Because t is a fixed regular point, we omit the dependency on t in the definition of $\mathbf{j}$. Also, let $\mathbf{j}^c$ denote the complement of $\mathbf{j}$. Because

$$\dot{Z}(t) = \lambda + (P' - I)\dot{D}(t),$$

we have

$$\dot{Z}_{\mathbf{j}}(t) = \lambda_{\mathbf{j}} + ((P')_{\mathbf{jj}} - I)\dot{D}_{\mathbf{j}}(t) + (P')_{\mathbf{jj}^c}\dot{D}_{\mathbf{j}^c}(t).$$

By Lemma 8.9, $\dot{Z}_{\mathbf{j}}(t) = 0$, and thus

$$\begin{aligned}
\dot{D}_{\mathbf{j}}(t) &= (I - (P')_{\mathbf{jj}})^{-1}\lambda_{\mathbf{j}} + (I - (P')_{\mathbf{jj}})^{-1}(P')_{\mathbf{jj}^c}\dot{D}_{\mathbf{j}^c}(t) \\
&\geq (I - (P')_{\mathbf{jj}})^{-1}\lambda_{\mathbf{j}} + (I - (P')_{\mathbf{jj}})^{-1}(P')_{\mathbf{jj}^c}\alpha_{\mathbf{j}^c} \\
&= \alpha_{\mathbf{j}}.
\end{aligned}$$

The last equality in this display uses the definition (2.36), and the inequality depends on the following two facts: (a) each entry of $(I - (P')_{\mathbf{jj}})^{-1}(P')_{\mathbf{jj}^c}$ is nonnegative, and (b) $\dot{D}_{\mathbf{j}^c}(t) \geq \alpha_{\mathbf{j}^c}$. □

8.5 Assembly Operation with Complementary Side Business

Consider the SPN pictured in Figure 5.5. Using the proof techniques in Chapter 7, one can prove that each fluid limit path satisfies the following additional fluid model equations:

$$(8.36) \qquad \dot{Z}_1(t) - \dot{Z}_2(t) = \lambda_1 - \lambda_2 \quad \text{if } Z_1(t) > Z_2(t),$$

$$(8.37) \qquad \dot{Z}_1(t) - \dot{Z}_2(t) = \lambda_1 + \mu_2 - \lambda_2 \quad \text{if } Z_1(t) < Z_2(t),$$

$$(8.38) \qquad \dot{Z}_1(t) = \lambda_1 - \mu_1 \quad \text{if } Z_1(t) > 0 \text{ and } Z_2(t) > 0,$$

$$(8.39) \qquad \dot{Z}_1(t) \leq \lambda_1.$$

Theorem 8.21. *Assume that the three inequalities in (5.33) are all satisfied. Then the fluid model is stable.*

Proof. Let Z be a fluid model solution with $|Z(0)| = 1$. It suffices to prove that $Z(t) = 0$ for $t \geq \delta$, where $\delta > 0$ is some constant independent of Z. Choose an $\alpha > 0$ such that

$$(8.40) \qquad \alpha\lambda_1 < \lambda_2 - \lambda_1,$$
$$(8.41) \qquad \alpha\lambda_1 < \lambda_1 + \mu_2 - \lambda_2.$$

Define the piecewise linear Lyapunov function

$$f(t) := |Z_1(t) - Z_2(t)| + \alpha Z_1(t).$$

Clearly, $f(t) = 0$ if and only if $Z(t) = 0$. Let t be a regular point of Z and $|Z_1 - Z_2|$. Computing $\dot{f}(t)$, we have

$$\dot{f}(t) \leq \begin{cases} \lambda_1 - \lambda_2 + \alpha\lambda_1 & \text{if } Z_1(t) > Z_2(t), \\ \lambda_2 - (\lambda_1 + \mu_2) + \alpha\lambda_1 & \text{if } Z_1(t) < Z_2(t), \\ \alpha(\lambda_1 - \mu_1) & \text{if } Z_1(t) = Z_2(t) > 0. \end{cases}$$

Let

$$\epsilon := \min\left\{\lambda_2 - \lambda_1 - \alpha\lambda_1, (\lambda_1 + \mu_2) - \lambda_2 - \alpha\lambda_1, \alpha(\mu_1 - \lambda_1)\right\}.$$

Then $\epsilon > 0$ and $\dot{f}(t) \leq -\epsilon$ for each t with $f(t) > 0$. By Lemma 8.5, $f(t) = 0$ for $t \geq f(0)/\epsilon$, proving the theorem. □

8.6 Global Stability of Ring Networks

In this section and the next one, we consider queueing network models, as defined in Section 2.6, further restricting attention to the case of single-server stations (that is, $b_k = 1$ for each $k \in \mathcal{K}$). For such a model, Proposition 5.1 states that subcriticality is equivalent to the standard load condition $\rho < e$ (the K-vector of ones). Our focus here is on the following property.

Definition 8.22. A queueing network is said to be *globally stable* if it is stable in the sense of Definition 3.6 (that is, its ambient Markov chain X is positive recurrent) under every simply structured, nonidling control policy.

By Theorem 5.4, subcriticality is a *necessary* condition for stability, and hence for global stability as well. It follows from Theorem 8.14 and

Proposition 5.1 that a subcritical *feedforward* queueing network (see Definition 8.13) is globally stable, and it was widely conjectured up until the early 1990s that the same was true for all queueing networks, not just the feedforward family. However, we have seen in Section 1.6 that global stability *need not hold* for the Rybko–Stolyar example pictured in Figure 1.9, even in the subcritical case, and the same is true for the Dai–Wang reentrant line example pictured in Figure 1.11. To be more specific, for each of those examples we have exhibited parameter values and a nonidling control policy such that (a) the system is subcritical, but (b) buffer contents grow without bound when the system's dynamic evolution is simulated. In this section and the next one, we analyze other specially structured families of queueing network models, striving to provide insight as to what conditions, beyond the standard load condition, are required for global stability.

The fluid model of a general SPN is defined through (6.1) through (6.6), which are equivalent to (8.20) through (8.23) for a queueing network. Under a nonidling policy (*any* nonidling policy), the queueing network's fluid model further includes (6.7), which we reproduce here (specialized to the case $b = e$) for ease of reference: for each station $k \in \mathcal{K}$,

$$(8.42) \qquad \sum_{i \in \mathcal{I}(k)} Z_i(t) > 0 \quad \text{implies} \quad \frac{d}{dt} \left(\sum_{i \in \mathcal{I}(k)} T_i(t) \right) = 1.$$

Given our restriction to the case $b = e$, the data for a queueing network's fluid model are (λ, m, A, P), where λ is the I-vector of external arrival rates, m is the I-vector of mean service times, A is the $K \times I$ capacity consumption matrix, and P is the $I \times I$ routing matrix. As usual, we define the service rates $\mu_i := 1/m_i$ for $i \in \mathcal{I}$.

Definition 8.23. The fluid model of a queueing network is said to be *globally stable* if there exists a constant $\gamma > 0$ such that, for every fluid model solution (T, Z) that satisfies (8.20) through (8.23) and (8.42), one has $Z(t) = 0$ for all $t \geq \gamma |\hat{Z}(0)|$.

It follows from Theorem 6.2 that if the fluid model of a queueing network is globally stable, then the queueing network itself is globally stable. In this section, we identify another family of queueing networks, in addition to the feedforward family, that exhibits global stability in the subcritical case. This analysis uses a piecewise linear Lyapunov function applied to the network's fluid model, and it illustrates once again the power of fluid model methodology for stability analysis.

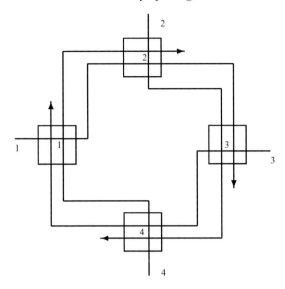

Figure 8.2 A symmetric four-station ring network.

Unidirectional ring. A unidirectional ring network with L types of customers and K single-server stations has the following structure. Customers of any given type ℓ enter the network at some station k, then follow a deterministic route with visitation sequence $k, k+1, \ldots, k+n_\ell$ for some $n_\ell \geq 1$; here we use the convention that whenever $k+j > K$, station $k+j$ is understood to be station $k+j-K$. An example of a symmetric four-station ring network is pictured in Figure 8.2. We assume that the mean service times of all classes served at station k are the same, denoting that common value as β_k. Let λ_ℓ be the arrival rate of type ℓ customers. One can easily map this network description into the standard queueing network description in Section 2.6, where customer class (ℓ, j) denotes those customers from type ℓ at stage j of their route.

Theorem 8.24. *If the standard load condition $\rho < e$ holds, then the fluid model of the unidirectional ring network is globally stable.*

Proof. Let $Z_{\ell j}(t)$ be the level of type ℓ fluid at stage j of its route, and define

$$Z_{\ell j}^+(t) := \sum_{i=1}^{j} Z_{\ell i}(t) \quad \text{for } j = 1, \ldots, n_\ell \quad \text{and} \quad \ell = 1, \ldots, L.$$

Also, let $p(\ell,j)$ be the station at which type ℓ customers are served at stage j of their route, and define

$$\alpha_k := \sum_{(\ell,j):p(\ell,j)=k} \lambda_\ell,$$

and

$$B_k(t) := \sum_{(\ell,j):p(\ell,j)=k} T_{(\ell,j)}(t).$$

One interprets α_k as the total arrival rate to station k, and $B_k(t)$ as the cumulative amount of time that server k is busy in $(0,t]$. Next, let

$$(8.43) \qquad G_k(t) := \sum_{(\ell,j):p(\ell,j)=k} Z_{\ell j}^+(t) = G_k(0) + \alpha_k t - 1/\beta_k B_k(t)$$

for each station $k \in \mathcal{K}$, invoking (8.20) to justify the second equality. Finally, completing preparations for the main argument, we define

$$H_k(t) := \sum_{(\ell,j):p(\ell,j)=k} Z_{\ell j}(t),$$

for each $k \in \mathcal{K}$, interpreting this as the total amount of fluid at station k. It then follows from (8.42) that

$$(8.44) \qquad H_k(t) > 0 \quad \text{implies} \quad \dot{G}_k(t) = -(1/\beta_k - \alpha_k) < 0,$$

where the final inequality is due to the standard load condition $\rho < e$ and the fact that $\rho_k = \alpha_k \beta_k$.

Our piecewise linear Lyapunov function is

$$(8.45) \qquad h(t) := \max\{G_1(t),\ldots,G_K(t)\}.$$

It is globally Lipschitz continuous, so it has a derivative almost everywhere. Fix a $t > 0$. Assume that $h(t) > 0$ and that t is a regular point for $h(\cdot)$, for $G_1(\cdot),\ldots$, and for $G_K(\cdot)$. We would like to prove that

$$(8.46) \qquad \dot{h}(t) \le -\epsilon,$$

where

$$\epsilon := \min_{k \in \mathcal{K}} (1/\beta_k - \alpha_k).$$

Once (8.46) is proved, it follows from Lemma 8.5 that $h(t) = 0$ for $t \geq h(0)/\epsilon$, and thus that the fluid model is globally stable. To prove (8.46), let $k_1,\ldots,k_q$ be the stations such that

$$G_{k_1}(t) = G_{k_2}(t) = \ldots = G_{k_q}(t) = h(t)$$

and $G_k(t) < h(t)$ for $k \notin \{k_1,\ldots,k_q\}$; here q is an integer with $1 \leq q \leq K$. From Lemma 8.10 we have that

$$\dot{h}(t) = \dot{G}_{k_1}(t) = \ldots = \dot{G}_{k_q}(t).$$

First suppose that $\{k_1,\ldots,k_q\} = \mathscr{K}$, or equivalently, that $G_k(t) = h(t)$ for all $k \in \mathscr{K}$. Then there exists at least one station $k \in \mathscr{K}$ such that $H_k(t) > 0$. For that k, we have $\dot{h}(t) = \dot{G}_k(t) \leq -\epsilon$, using (8.44) to justify the inequality, which establishes (8.46).

Now suppose on the other hand that there exists a $k \in \mathscr{K}$ for which $G_k(t) < h(t)$. In that case, we can choose a station k such that $k \notin \{k_1,\ldots,k_q\}$ but $k+1 \in \{k_1,\ldots,k_q\}$. (Recall that by convention one interprets station $k+1$ as station 1 when $k = K$.) Now the following argument shows that $H_{k+1}(t) > 0$: if it were true that $H_{k+1}(t) = 0$, then we would have

$$G_{k+1}(t) = \sum_{(\ell,j):p(\ell,j)=k+1} Z_{\ell j}^+(t)$$

$$= \sum_{(\ell,j):p(\ell,j)=k+1} Z_{\ell(j-1)}^+(t) \leq \sum_{(\ell,j):p(\ell,j)=k} Z_{\ell j}^+(t) < h(t),$$

which contradicts $k+1 \in \{k_1,\ldots,k_q\}$. Thus,

$$\dot{h}(t) = \dot{G}_{k+1}(t) \leq -\epsilon,$$

again proving (8.46), where the inequality follows from (8.44).   □

8.7 Global Stability of Reentrant Lines

In the previous section, we have seen that, because of its distinctive routing structure, a unidirectional ring network is globally stable if it is subcritical, without any further restrictions on the vector λ of external arrival rates and the vector m of mean service times. For a general queueing network, however, the situation is more complicated. Fixing the routing matrix P and capacity consumption matrix A, and maintaining our assumption of single-server stations ($b = e$), let us consider the set of all (λ, m) pairs with which the network's fluid model is globally stable, calling this the fluid model's *global stability region*. In Section 2.6, the load vector ρ for a unitary network was defined in terms of λ, m, A, and P by (2.40). Of course, the global stability region can only include (λ, m) pairs such that $\rho < e$, and except in very special cases (including unidirectional ring networks) it is strictly smaller than that.

What are the additional constraints, beyond the standard load condition, that define the global stability region? In general, that remains an open question, but in this section we answer it for two examples of reentrant lines, a special class of queueing networks introduced previously in Section 1.6, having one input stream and a single deterministic route. In that context, it will be seen that two new types of load condition appear as requirements for global stability.

Theorem 8.25. *Consider the two-station, five-class reentrant queueing network depicted in Figure 8.3. Its fluid model is globally stable if and only if*

(8.47) $$\lambda_1 (m_1 + m_3 + m_5) < 1,$$

(8.48) $$\lambda_1 (m_2 + m_4) < 1,$$

(8.49) $$\lambda_1 (m_2 + m_5) < 1.$$

Conditions (8.47) and (8.48) together are equivalent to the standard load condition $\rho < e$, whereas (8.49) is a separate type of load condition analogous to (1.9) for the Rybko–Stolyar example pictured in Figure 1.9. Dai and Vande Vate (2000) described the added requirements (8.49) and (1.9) as *virtual station conditions* for their respective queueing

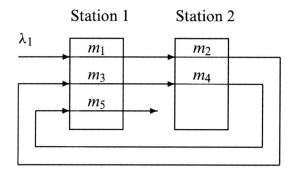

Figure 8.3 A two-station, five-class reentrant queueing network.

networks. Readers are reminded that, in our previous discussions of the Rybko–Stolyar example, a specific static buffer priority (SBP) policy was assumed. One can show that priority scheme to be an *extreme* control policy for the Rybko–Stolyar network, which means that if the network is stable under the specified policy, then it is globally stable.

As explained in Section 1.6, the virtual station condition (1.9) is necessary in the Rybko–Stolyar example to avoid a *mutual blocking* phenomenon under the extreme SBP policy, and that same statement applies to the added condition (8.49) for the reentrant line currently under discussion. Specifically, it can be shown that an extreme control policy for the current example is any SBP policy that gives highest priority to class 5 at station 1, and also gives highest priority to class 2 at station 2.

To prove that (8.49) is necessary for global stability of the current example, one constructs an unstable solution for the associated fluid model under an extreme SBP policy; the unstable fluid model solution is one that cycles when (8.49) is violated. This construction is analogous to the one in Figure 6.1 for the Rybko–Stolyar example, so it is omitted.

Proof of sufficiency. Assume that (8.47) through (8.49) are satisfied. Let (T, Z) be a fluid model solution satisfying (8.20) through (8.23) and (8.42). We shall construct a piecewise linear Lyapunov function to prove global stability. It is similar to the one in (8.45), but with an important difference: here the Lyapunov function is parameterized by a vector $x = (x_1, x_2, x_3, x_4, x_5) > 0$. Specifically, let

$$(8.50) \qquad G_1(t) := x_1 Z_1^+(t) + x_3 Z_3^+(t) + x_5 Z_5^+(t),$$
$$(8.51) \qquad G_2(t) := x_2 Z_2^+(t) + x_4 Z_4^+(t).$$

This can be restated in vector form as

$$(8.52) \qquad G(t) := A \operatorname{diag}(x)(I - P')^{-1} Z(t).$$

When $x = (1, 1, 1, 1, 1)$, (8.52) is consistent with our earlier definition of $G(\cdot)$ in (8.43). Our piecewise linear Lyapunov function for the current example is the following, where $G(t)$ is defined via (8.52):

$$(8.53) \qquad h(t) := \max\big(G_1(t), G_2(t)\big),$$

Fix a $t > 0$. Assume that $h(t) > 0$ and that t is a regular point for both $T(\cdot)$ and $h(\cdot)$. We would like to prove that there exists a five-vector $x > 0$ for use in (8.50) and (8.51), and a scalar $\epsilon > 0$, such that

(8.54) $$\dot{h}(t) \le -\epsilon.$$

By Lemma 8.5, this will establish stability of the fluid model and thus conclude the proof of sufficiency. Let

$$H_1(t) := Z_1(t) + Z_3(t) + Z_5(t) \quad \text{and} \quad H_2(t) := Z_2(t) + Z_4(t).$$

To prove (8.54), following the proof of Theorem 8.24, it will suffice to show that

(8.55) $H_1(t) = 0$ implies $G_1(t) \le G_2(t),$

(8.56) $H_2(t) = 0$ implies $G_2(t) \le G_1(t),$

and

(8.57) $H_1(t) > 0$ implies $\dot{G}_1(t) \le -\epsilon,$

(8.58) $H_2(t) > 0$ implies $\dot{G}_2(t) \le -\epsilon.$

Thus it remains to prove the following: if (8.47) through (8.49) hold, then there exists a five-vector $x > 0$ for use in (8.50) and (8.51), and a scalar $\epsilon > 0$, such that (8.55) through (8.58) are satisfied. That conclusion follows directly from Lemmas 8.26 and 8.27.   □

Lemma 8.26.   *(a) If $x_2, x_4, \epsilon > 0$ satisfy the following, then (8.58) holds:*

(8.59) $$\lambda_1(x_2 + x_4) + \epsilon \le \mu_2 x_2,$$

(8.60) $$\lambda_1(x_2 + x_4) + \epsilon \le \mu_4 x_4.$$

(b) If $x_1, x_3, x_5, \epsilon > 0$ satisfy the following, then (8.57) holds:

(8.61) $$\lambda_1(x_1 + x_3 + x_5) + \epsilon \le \mu_1 x_1,$$

(8.62) $$\lambda_1(x_1 + x_3 + x_5) + \epsilon \le \mu_3 x_3,$$

(8.63) $$\lambda_1(x_1 + x_3 + x_5) + \epsilon \le \mu_5 x_5.$$

(c) If $x > 0$ satisfies the following, then (8.56) holds:

(8.64) $$x_2 + x_4 \le x_1 + x_3 + x_5,$$

(8.65) $$x_4 \le x_3 + x_5.$$

(d) If $x > 0$ satisfies the following, then (8.55) holds:

(8.66) $$x_3 + x_5 \le x_2 + x_4,$$

(8.67) $$x_5 \le x_4.$$

Proof. (a) Assume $H_2(t) > 0$. One has from (8.51) and (8.20) that

$$(8.68) \qquad \dot{G}_2(t) = \lambda_1(x_2 + x_4) - x_2\mu_2\dot{T}_2(t) - x_4\mu_4\dot{T}_4(t).$$

Now $\dot{T}_2(t) + \dot{T}_4(t) = 1$ by (8.42), and making that substitution in (8.68) gives

$$(8.69) \qquad \begin{aligned} \dot{G}_2(t) &= \big(\lambda_1(x_2 + x_4) - x_2\mu_2\big)\dot{T}_2(t) \\ &\quad + \big(\lambda_1(x_2 + x_4) - x_4\mu_4\big)\dot{T}_4(t) \le -\epsilon, \end{aligned}$$

where the inequality follows from (8.59) and (8.60). This proves (8.58), as desired, and the proof of part (b) is similar. (c) When $H_2(t) = 0$, one has

$$\begin{aligned} G_2(t) &= x_2 Z_1(t) + x_4\big(Z_1(t) + Z_3(t)\big) = (x_2 + x_4)Z_1(t) + x_4 Z_3(t), \\ G_1(t) &= x_1 Z_1(t) + x_3\big(Z_1(t) + Z_3(t)\big) + x_5\big(Z_1(t) + Z_3(t) + Z_5(t)\big) \\ &= (x_1 + x_3 + x_5)Z_1(t) + (x_3 + x_5)Z_3(t) + x_5 Z_5(t). \end{aligned}$$

Thus (8.56) holds, as desired, and the proof of part (d) is similar. $\square$

Lemma 8.27. *Assume conditions (8.47) through (8.49) are satisfied. Then there exist a five-vector $x > 0$ and a scalar $\epsilon > 0$ such that (8.59) through (8.67) hold.*

Proof. It is enough to prove that there exists a five-vector $x > 0$ satisfying

$$(8.70) \qquad \lambda_1(x_1 + x_3 + x_5) < \mu_1 x_1,$$
$$(8.71) \qquad \lambda_1(x_1 + x_3 + x_5) < \mu_3 x_3,$$
$$(8.72) \qquad \lambda_1(x_1 + x_3 + x_5) < \mu_5 x_5,$$
$$(8.73) \qquad \lambda_1(x_2 + x_4) < \mu_2 x_2,$$
$$(8.74) \qquad \lambda_1(x_2 + x_4) < \mu_4 x_4,$$
$$(8.75) \qquad x_3 + x_5 \le x_2 + x_4,$$
$$(8.76) \qquad x_5 \le x_4,$$
$$(8.77) \qquad x_2 + x_4 \le x_1 + x_3 + x_5,$$
$$(8.78) \qquad x_4 \le x_3 + x_5.$$

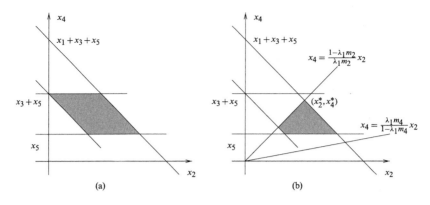

Figure 8.4 (a) The set of (x_2, x_4) pairs constrained by (8.80) and (8.81). (b) The region in (a) intersected with the region (8.79).

Assume that (8.47) through (8.49) hold. Note that (8.73) and (8.74) are equivalent to

$$(8.79) \qquad x_2 \frac{\lambda_1 m_4}{1 - \lambda_1 m_4} < x_4 < \frac{1 - \lambda_1 m_2}{\lambda_1 m_2} x_2,$$

and for given values $(x_1, x_3, x_5) > 0$, requiring that (x_2, x_4) satisfy (8.75) through (8.78) is equivalent to requiring that (x_2, x_4) belongs to the following parallelogram:

$$(8.80) \qquad x_5 \leq x_4 \leq x_3 + x_5,$$

$$(8.81) \qquad x_3 + x_5 \leq x_2 + x_4 \leq x_1 + x_3 + x_5;$$

see the shaded region in Figure 8.4a. With (x_1, x_3, x_5) fixed, note that the line

$$x_4 = \frac{(1 - \lambda_1 m_2)}{\lambda_1 m_2} x_2$$

intersects the line

$$x_2 + x_4 = x_1 + x_3 + x_5,$$

at

$$(x_2^*, x_4^*) = \Big(\lambda_1 m_2 (x_1 + x_3 + x_5), \ (1 - \lambda_1 m_2)(x_1 + x_3 + x_5) \Big).$$

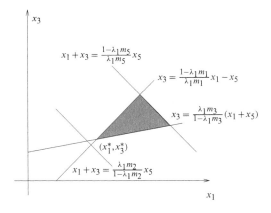

Figure 8.5 The set of (x_1, x_3) pairs constrained by (8.70) through (8.72) and (8.82).

Therefore, the region (8.79) and the parallelogram have a nonempty intersection (the shaded region in Figure 8.4b), if and only if $x_4^* > x_5$, or equivalently,

$$(8.82) \qquad (1 - \lambda_1 m_2)(x_1 + x_3 + x_5) > x_5.$$

It remains to show that there exist $x_1, x_3, x_5 > 0$ satisfying (8.70) through (8.72) and (8.82). Let $x_5 > 0$ be fixed. Then (8.70) and (8.71) are equivalent to

$$(8.83) \qquad \frac{\lambda_1 m_3}{1 - \lambda_1 m_3}(x_1 + x_5) < x_3 < \frac{1 - \lambda_1 m_1}{\lambda_1 m_1} x_1 - x_5.$$

The set of pairs $(x_1, x_3) > 0$ satisfying (8.83) is nonempty, and that region's two boundaries intersect at

$$(x_1^*, x_3^*) = \left(\frac{\lambda_1 m_1}{1 - \lambda_1 (m_1 + m_3)} x_5, \ \frac{\lambda_1 m_3}{1 - \lambda_1 (m_1 + m_3)} x_5 \right);$$

see Figure 8.5. From (8.47),

$$x_1^* + x_3^* = \frac{\lambda_1 (m_1 + m_3)}{1 - \lambda_1 (m_1 + m_3)} x_5 < \frac{1 - \lambda_1 m_5}{\lambda_1 m_5} x_5.$$

Therefore, for any fixed $x_5 > 0$, the set of (x_1, x_3) pairs constrained by (8.70) through (8.72) is nonempty; see the shaded region in Figure 8.5.

Because (8.49) holds, this region has a nonempty intersection with the region constrained by (8.82).   ☐

To recapitulate, the fluid model for our two-station, five-class reentrant line is globally stable if and only if its parameters satisfy, in addition to the standard load conditions (8.47) and (8.48), the virtual station condition (8.49). Dai and Vande Vate (2000) made a general study of global stability for fluid models of two-station reentrant lines, characterizing precisely the parameter combinations that yield global stability. Their characterization involves three types of parameter constraints, namely, standard load conditions, virtual station conditions, and what they called *push start conditions*; the last group is empty for our two-station, five-class example.

To explain the nature of push start conditions, let us consider the two-station, six-class reentrant queueing network depicted in Figure 8.6, for which Dai and Vande Vate (2000) found the following: the fluid model for this network is globally stable if and only if

(8.84) $$\lambda_1(m_1 + m_3 + m_5) < 1,$$

(8.85) $$\lambda_1(m_2 + m_4 + x_6) < 1,$$

(8.86) $$\lambda_1(m_2 + m_5) < 1,$$

(8.87) $$\lambda_1\left(\frac{m_3}{1 - \lambda_1 m_1} + m_6\right) < 1.$$

Of course, (8.84) and (8.85) are the standard load conditions for this example. Also, (8.86) is identical to the virtual station condition (8.49)

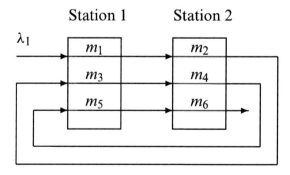

Figure 8.6 A two-station, six-class reentrant queueing network.

in our two-station, five-class example. Its necessity in the six-class network becomes clear by considering a policy that gives lowest priority to class 6 at station 2; classes 1 through 5 then form a subnetwork that is identical to the two-station, five-class example, and therefore (8.86) is necessary for global stability of the two-station, six-class fluid model.

Finally, condition (8.87) is of a new type, with which we associate the term "push start" for the following reason. Assume that (8.84), (8.85), and (8.86) are all satisfied, and furthermore that the network is operated under a control policy that gives highest priority to class 1 at station 1. Given condition (8.84), buffer 1 will empty after a finite amount of time and remain empty ever afterward. Once buffer 1 has become empty, fluid arriving from the outside world will reach buffer 2 instantaneously, its arrival rate there being the constant λ_1. The fraction of time that server 1 spends processing class 1, thereby keeping buffer 1 empty, is $\lambda_1 m_1$, so server 1 has a fraction $1 - \lambda_1 m_1$ of its time left over to work on classes 3 and 5; this will be referred to as *residual capacity* in the following discussion.

One may say that classes 2 through 6 form a two-station, five-class *reduced fluid network* with input rate λ_1. When server 1 devotes all of its residual capacity to serving class $i = 3$ or 5, services of that class are completed at a rate of

$$(8.88) \qquad (1 - \lambda_1 m_1)\mu_i.$$

In the reduced network consisting of classes 2 through 6, server 1 sees only classes 3 and 5, and the mean duration of a service for class $i = 3$ or 5 is the reciprocal of the service rate in (8.88). In particular, what plays the role of the mean service time for class 3 fluid in the reduced network is

$$\hat{m}_3 = \frac{m_3}{1 - \lambda_1 m_1}.$$

Applying the same logic that produced the virtual station condition (8.49) for our earlier two-station, five-class example (the one pictured in Figure 8.3), one obtains condition (8.87) for the reduced fluid network in which classes 2 through 6 are processed.

This derivation of (8.87) is based on a scenario where class 1 is given highest priority, a control policy that pushes fluid at the maximum possible rate through the first class or buffer on the reentrant line's one route, regardless of fluid levels in other buffers that are dependent on the associated server for their processing. This led Dai and Vande

Vate (2000) to call (8.87) a *push start* condition. In general, push start conditions involve a compounding of two effects, namely, a virtual station condition arising in a reduced network of classes occurring later in the reentrant line's one route, and the magnifying influence of highest priority being granted to a class occurring early in the route.

8.8 Sources and Literature

A slightly more general version of Lemma 8.5 was stated and used in Dai (1995a), and again in Dai and Weiss (1996). Lemma 8.9 first appeared as part (b) of proposition 4.2 in Dai and Weiss (1996). Lemma 8.10 was first stated and proved as lemma 3.2 of that same paper, in which the term "regular point" was also introduced. Our statement of Lemma 8.11 is patterned after that of lemma 6 in Massoulié (2007), but its proof follows that of lemma 2 in Bramson et al. (2017).

Theorem 8.14 in Section 8.3 was stated as corollary 6.2 of Dai (1995a), whose proof was based on induction. The current proof, using a piecewise linear Lyapunov function, appears to be new. Lemma 8.20 in Section 8.4 is essentially the same as theorem 4 of Bramson (1998). Its proof employs a linear Lyapunov function. Such a Lyapunov was also used in Dai (1995a) to prove the fluid model stability of a generalized Jackson network, and in Dai and Weiss (1996) for the fluid model stability of a reentrant line under the last-buffer–first-served policy. Linear Lyapunov functions were systematically explored in Chen and Zhang (2000) to obtain sufficient conditions for the fluid model stability of queueing networks under SBP policies.

The concept of global stability (see Section 8.6) was first introduced in Dai (1995b). This same concept is called *universal stability* in adversarial queueing models for packet routing. That class of queueing models was introduced in pioneering work by Borodin et al. (2001), and was studied by Gamarnik (2000), Andrews et al. (2001), Goel (2001), and Gamarnik (2003).

Theorem 8.24 and its proof are taken from Dai and Weiss (1996). Tassiulas and Georgiadis (1994) proved an analogous theorem when interarrival times and service times are deterministically constrained as in Cruz (1991). The piecewise linear Lyapunov function used in the proof of Theorem 8.24 generalizes the ones used by Botvich and Zamyatin (1994). Later work by Down and Meyn (1994) and by Dai and Vande Vate (2000) systematically explored that class of Lyapunov functions.

The findings reported in Section 8.7 for two-station reentrant lines are special cases of general results developed by Dai and Vande Vate (2000); the terms "virtual station condition" and "push start condition" first appeared in their paper. In earlier work, Bertsimas et al. (1996) proved that a linear program characterizes the global stability region of a two-station fluid model. Dai et al. (1999) studied global stability of a three-station fluid model, showing that the global stability region is *not* necessarily monotone in mean service times. The possible non-monotonicity of a global stability region was first demonstrated by Humes (1994) for deterministic networks and later by Bramson (1998) for stochastic networks. Using quadratic Lyapunov functions, Chen (1995) provided a sufficient condition for global stability of a general queueing network. Hasenbein (1997) established necessary conditions for global stability of a general queueing network. Characterizing the global stability region for the fluid model of a general queueing network remains an open problem.

9

Max-Weight and Back-Pressure Control

We consider in this chapter an important special case of the general SPN model formulated in Chapter 2. Specifically, attention is restricted to what we call *Leontief networks*, and we introduce in that context the so-called *back-pressure (BP)* control policy, which is called *max-weight* (MW) control in the case of single-hop networks (see Sections 9.2 and 9.4 for elaboration). Actually, two versions of BP control will be considered, one tailored to our basic SPN formulation (Section 2.3) and the other to our relaxed formulation (Section 2.4). Both are simply structured policies, which means, among other things, that they only make decisions at times $t > 0$ when an external arrival or a service completion occurs.

The BP policy, or BP algorithm, was originally developed for use in digital communication networks, and in that context it provides a comprehensive approach to routing and scheduling. That is, BP logic can be used to address both the question of *where* to send packets next on their journey through a physical network, and the *order* in which packets competing for capacity of a switch or link are to be processed. As we shall see, the BP policy is maximally stable for a very broad class of networks, and as such it is commonly invoked as a "gold standard" against which other policies are compared.

For an explanation of the basic idea underlying BP control, first recall that the letter β was used in Section 2.4 to denote the J-vector of total service rates (one component for each activity or service type) chosen by the system manager at a given decision time. Also, for ease of reference we recall the definition (5.3) of the input–output matrix R:

$$(9.1) \qquad\qquad R := (B - \Gamma)M^{-1}.$$

Here B and Γ are nonnegative $I \times J$ matrices and $M = \text{diag}(m_1, \ldots, m_J)$; one interprets R_{ij} as the average decrease in buffer i contents per unit of effort devoted to activity j. Let us now define the bilinear function

$$(9.2) \qquad p(\beta, z) := z \cdot R\beta \quad \text{for } \beta \in \mathbb{R}_+^J \text{ and } z \in \mathbb{R}_+^I,$$

where $\cdot$ denotes inner product as usual. At each decision time t, given the updated state descriptors $\hat{n}$ and $\hat{z}$ defined by (2.17) and (2.18), the back-pressure principle is to choose a vector β that maximizes $p(\beta, \hat{z})$, subject to capacity constraints and material availability constraints. However, as we shall see shortly, the implementation of that principle is somewhat different in the basic and relaxed model contexts.

Section 9.1 describes the special structure of Leontief networks, and back-pressure control policies are defined and characterized in Sections 9.2 through 9.4. Stability properties of the BP policies are studied in Sections 9.6 and 9.7. Specifically, we prove that the relaxed BP policy is maximally stable for Leontief networks, as is the basic BP policy for Leontief networks where all activities involve a single server.

9.1 Leontief Networks

The special network structure that we study in this chapter involves two assumptions, which are identified as Assumptions 9.1 and 9.2.

Assumption 9.1. Each column of the input–output matrix R contains exactly one positive element.

Hereafter we shall denote by $i(j)$ the unique buffer $i \in \mathscr{I}$ such that $R_{ij} > 0$ $(j \in \mathscr{J})$. In applications, SPN models that satisfy Assumption 9.1 typically have the following two structural features. First, each column j of the material requirements matrix B contains a single positive element, which appears in row $i(j)$, and the rest zeros. That is, activity j requires as its input one or more items from buffer $i(j)$, and no other items. Moreover, it is almost always a single item from buffer $i(j)$ that serves as input to a type j service. Hereafter, we shall say that activity j *serves* buffer $i(j)$.

The second structural feature referred to in the preceding paragraph is that $\Gamma_{ij} < B_{ij}$ for all $j \in \mathscr{J}$ and $i = i(j)$. That is, the expected number of items returned to buffer $i(j)$ as output from a type j service is less than the number of such items removed from that buffer by the service. In most models of interest, the former number is actually zero.

The unitary networks defined in Section 2.6, which include queueing networks, satisfy Assumption 9.1, but it is also satisfied by model families in which jobs of a given class can be processed using any one of several different activities or service types. The following is an essential complement to Assumption 9.1, as we shall explain in the remainder of this section.

Assumption 9.2. There exists a vector $x \geq 0$ such that $Rx > 0$.

To explain the significance of Assumption 9.2, we shall use the term *basis* to mean a set of I activities or service types (among the $J \geq I$ activities available to the system manager) such that exactly one of them serves each of the I different buffers. The elements of such a set will be called *basic activities*. The corresponding *basis matrix* is the square submatrix $\hat{R}$ that is formed from R by selecting columns that correspond to the basic activities. (Our use of these terms is slightly narrower than their standard meaning in linear programming theory.) Given a choice of basis, it follows from (9.1) that one can renumber activities so that the corresponding basis matrix has the form

$$(9.3) \qquad\qquad \hat{R} = (I - Q)\Delta^{-1},$$

where Q is a nonnegative square matrix and Δ is a diagonal matrix with positive diagonal elements. Proposition 9.4 provides an insightful equivalent statement of Assumption 9.2 in terms of basis matrices, and the following will be used in its proof.

Lemma 9.3. *Given that Assumption 9.1 holds, let $y > 0$ be such that $Rx = y$ for at least one vector $x \in \mathbb{R}_+^J$. Then there exists a basis matrix $\hat{R}$ and a vector $\hat{x} \in \mathbb{R}_+^I$ such that*

$$(9.4) \qquad\qquad \hat{R}\hat{x} = y.$$

Proof. Let y be as in the lemma's statement and consider the following linear program: choose $x \in \mathbb{R}_+^J$ to minimize $e'x$ subject to $Rx = y$, where e is the J-vector of ones. By hypothesis, this problem is feasible (that is, there exists at least one feasible solution x), and its objective value is bounded below by zero. Thus, basic linear programming theory ensures the existence of an optimal solution x^* having no more than I positive components; see, for example, section 2.6 of Bertsimas and Tsitsiklis (1997). Let $\hat{x} \in \mathbb{R}_+^I$ be an I-vector obtained from x^* by deleting $J - I$ components with $x_j^* = 0$, and let $\hat{R}$ be the corresponding $I \times I$ submatrix of R (that is, the submatrix formed by deleting from R the same columns j with $x_j^* = 0$). Then $\hat{x}$ and $\hat{R}$ jointly satisfy (9.4).

It remains to show that $\hat{R}$ is a basis matrix, or equivalently, that exactly one of its I positive elements occurs in each row. If there were a row i of $\hat{R}$ with all elements non-positive, then the ith component of the vector y in (9.4) would be nonpositive as well, which is a contradiction. On the other hand, if there were a row with at least two positive elements, there would have to be a row with no positive element, because $\hat{R}$ has exactly I positive elements by Assumption 9.1. This again leads to a contradiction. □

The following is a standard result, for which references will be provided in Section 9.8, but we include a proof for completeness.

Proposition 9.4. *Given that Assumption 9.1 holds, Assumption 9.2 is satisfied if and only if there exists a choice of basis such that the matrix Q in (9.3) has spectral radius < 1.*

Proof. Suppose that Assumption 9.1 holds and there is a choice of basis for which $\rho(Q) < 1$, where $\rho(\cdot)$ denotes spectral radius. From Theorem F.1, it follows that $Q^n \to 0$ as $n \to \infty$, that $\hat{R}$ is nonsingular, and that $\hat{R}^{-1}$ has the following Neumann expansion:

$$(9.5) \qquad \hat{R}^{-1} = \Delta^{-1}(I - Q)^{-1} = \Delta^{-1}\left(I + Q + Q^2 + \cdots\right) \geq 0.$$

Fix a $y \in \mathbb{R}^I$ with $y > 0$. Define $\hat{x} := \hat{R}^{-1}y$. It follows from (9.5) that $\hat{x} \geq 0$ and $\hat{R}\hat{x} = y$. Define $x = (x_j) \in \mathbb{R}^J$ via

$$x_j := \begin{cases} \hat{x}_j & \text{if activity } j \text{ is in the basis,} \\ 0 & \text{otherwise.} \end{cases}$$

Then $x \geq 0$ and $Rx = \hat{R}\hat{x} = y$. Thus, R satisfies Assumption 9.2.

Conversely, suppose that Assumptions 9.1 and 9.2 both hold. Then there exists an $x \geq 0$ such that $y := Rx > 0$, and hence by Lemma 9.3 there exist a basis matrix $\hat{R}$ and a vector $\hat{x} \in \mathbb{R}_+^I$ such that (9.4) holds. Renumbering components if necessary, we have the representation (9.3) for $\hat{R}$, and hence (9.4) can be rewritten as

$$\hat{x} = \Delta^{-1}y + Q\hat{x}.$$

Iterating that relationship gives

$$\hat{x} = \left(I + Q + Q^2 + \cdots + Q^n\right)\Delta^{-1}y + Q^{n+1}\hat{x}$$

for all $n = 1, 2, \ldots$. Using the fact that $Q \geq 0$ and $\hat{x} > 0$, one has

$$\left(I + Q + Q^2 + \cdots + Q^n\right)\Delta^{-1}y \leq \hat{x}$$

for all $n \geq 1$. Because $\hat{x} < \infty$ and the diagonal elements of Δ are positive, this implies that $Q^n \to 0$ as $n \to \infty$, which is equivalent to $\rho(Q) < 1$ by Theorem F.1. $\qquad\qquad\square$

To see why Assumption 9.2 is indispensable for SPN modeling, consider a system manager confronting a vector $z > 0$ of initial buffer contents, and suppose that all external input processes are turned off. Also, to keep things simple, let us restrict attention to the usual case where each column of B contains a single one and the rest zeros. In this setting, the following hold for any choice of basis. (a) Each basic activity requires as input a single item from a different one of the system's I buffers. (b) The matrix Q in (9.3) is a square submatrix of Γ. (c) The diagonal matrix Δ in (9.3) is the principal submatrix of M^{-1} corresponding to basic activities; that is, the diagonal elements of Δ are mean service rates for basic activities.

Assume that the manager chooses a fixed set of basic activities with which to drain the buffers of their initial contents, and denote by $\hat{R} = (I - Q)\Delta^{-1}$ the corresponding basis matrix, as in (9.3). Defining u_i as the expected number of type i services that will be needed before all buffers are empty, and $u = (u_1, \ldots, u_I)$, one can generalize the reasoning used in the final paragraph of Appendix F to express u in terms of z and powers of Q. Specifically, in the current context one has the formula

$$(9.6) \qquad\qquad u = \left(I + Q + Q^2 + \cdots\right)z.$$

Suppose that Assumption 9.2 is satisfied. Combining Proposition 9.4, Theorem F.1, and formula (9.6), we then have the following conclusion: there exists a choice of basis such that all components of the vector u are finite. If Assumption 9.2 is *not* satisfied, those same three results imply that no choice of a fixed basis will allow us to empty the system with a finite expected number of services. Furthermore, it is not difficult to show in the latter case that every other strategy for processing the initial buffer contents is doomed to fail in the same sense, regardless of how it may switch dynamically among the various activities that serve various buffers. Thus, given that Assumption 9.1 holds, Assumption 9.2 is necessary for the SPN model to be meaningful.

Definition 9.5. An SPN model whose data satisfy Assumptions 9.1 and 9.2 will hereafter be called a *Leontief network*.

This name derives from standard terminology in applied mathematics, where most authors (but unfortunately not all) call a matrix R *Leontief* if it satisfies Assumptions 9.1 and 9.2. Such matrices have a prominent role in the sector-level modeling of economic production that was propounded in the midtwentieth century by economist Wassily Leontief. See Section 9.8 for pointers to the associated literature.

9.2 Basic Back-Pressure Policy

In our basic SPN model (Section 2.3), the following equivalences hold at each decision time t: $\beta = \tilde{n} = \hat{n} + u$. That is, the vector β of total service rates immediately after t equals the vector $\tilde{n}$ of service counts going forward from t, which is equal in turn to the updated service count vector $\hat{n}$ plus the vector u of new service starts that is chosen at t. Thus, choosing β to maximize $p(\beta, \hat{z})$, given the current values of $\hat{n}$ and $\hat{z}$, amounts to choosing u to maximize $p(\hat{n} + u, \hat{z}) = p(\hat{n}, \hat{z}) + p(u, \hat{z})$. Because $\hat{n}$ is fixed at the time of decision making, we therefore solve the following optimization problem:

(9.7)   maximize $p(u, \hat{z})$

(9.8)   subject to $u \in \mathbb{Z}_+^J$, $A(\hat{n} + u) \le b$ and $B(\hat{n} + u) \le \hat{z}$.

Recall that, in our basic SPN model, a simply structured policy chooses $u = g(\hat{n}, \hat{z})$ at each decision time, where g is a fixed policy function. Denoting by u^* the optimal solution of (9.7) and (9.8), the basic BP policy takes $g(\hat{n}, \hat{z}) = u^*$. For each activity $j \in \mathscr{J}$, the policy initiates u_j^* new type j services, each of which proceeds at full speed until the next decision time, as do each of the $\hat{n}_j$ services of type j that are open and still incomplete.

Single-hop example. Consider the N-model pictured in Figure 9.1, which is a modified version of Figure 1.7. This is a parallel-server system (see Section 4.7 for the meaning of that term) having $I = 2$ buffers or job classes, $K = 2$ single-server pools, and $J = 3$ activities or

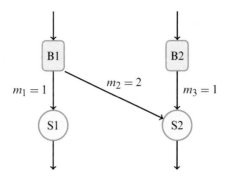

Figure 9.1 N-model (based on Figure 1.7).

service types that were specified in Section 1.6. Given the mean service times m_j shown in the figure, we have $\mu_1 = \mu_3 = 1$ and $\mu_2 = 1/2$, so the matrices R, A, and B for this example are as follows:

$$R = \begin{pmatrix} 1 & 1/2 & 0 \\ 0 & 0 & 1 \end{pmatrix}, \quad A = \begin{pmatrix} 1 & 0 & 0 \\ 0 & 1 & 1 \end{pmatrix} \quad \text{and} \quad B = \begin{pmatrix} 1 & 1 & 0 \\ 0 & 0 & 1 \end{pmatrix}.$$

Thus, at any given decision time, the optimization problem (9.7) and (9.8) amounts to the following. Choose $u \in \mathbb{Z}_+^3$ to

$$(9.9) \qquad \qquad \text{maximize} \quad \hat{z}_1 u_1 + \frac{1}{2}\hat{z}_1 u_2 + \hat{z}_2 u_3$$

subject to the following four constraints:

$$(9.10) \qquad u_1 \leq 1 - \hat{n}_1 \quad \text{and} \quad u_2 + u_3 \leq 1 - (\hat{n}_2 + \hat{n}_3),$$

$$(9.11) \qquad u_1 + u_2 \leq \hat{z}_1 - (\hat{n}_1 + \hat{n}_2) \quad \text{and} \quad u_3 \leq \hat{z}_3 - \hat{n}_3.$$

The right-hand sides of the two constraints in (9.10) are, respectively, the number of idle or uncommitted servers in pool 1 at the decision time (a binary quantity), and the number of uncommitted servers in pool 2 at the decision time (also binary). Similarly, the right-hand sides in (9.11) are the numbers of uncommitted items or jobs in buffer 1 and buffer 2, respectively, at the decision time.

   Let us consider a decision time at which server 2 is not committed (that is, $\hat{n}_2 = \hat{n}_3 = 0$), further assuming that both buffers contain

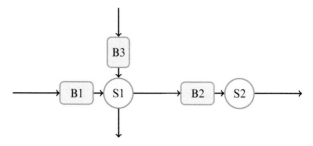

Figure 9.2 Criss-cross network (based on Figure 1.2).

uncommitted jobs (that is, $\hat{z}_1 > \hat{n}_1$ and $\hat{z}_2 > 0$). Then the objective (9.7) dictates that server 2

$$\text{initiate a new type 2 service if } \frac{1}{2}\hat{z}_1 > \hat{z}_2,$$

$$\text{initiate a new type 3 service if } \frac{1}{2}\hat{z}_1 < \hat{z}_2,$$

$$\text{do either one if } \frac{1}{2}\hat{z}_1 = \hat{z}_2.$$

As stated in the preamble to this chapter, we use the term "max-weight control" to describe the policy determined by (9.9) through (9.11), because the underlying model is single-hop. See Section 9.4 for an explanation of that terminology.

Multihop example. The criss-cross network was pictured earlier in Figure 1.2, and for ease of reference that diagram is reproduced here as Figure 9.2. This SPN has $I = 3$ buffers or job classes, $K = 2$ single-server pools, and $J = 3$ service types that are defined in the obvious way. The matrices R, A, and B for the criss-cross example are as follows, where μ_i denotes as usual the mean service rate for class i jobs:

$$R = \begin{pmatrix} \mu_1 & 0 & 0 \\ -\mu_1 & \mu_2 & 0 \\ 0 & 0 & \mu_3 \end{pmatrix}, \quad A = \begin{pmatrix} 1 & 0 & 1 \\ 0 & 1 & 0 \end{pmatrix} \quad \text{and} \quad B = \begin{pmatrix} 1 & 0 & 0 \\ 0 & 1 & 0 \\ 0 & 0 & 1 \end{pmatrix}.$$

Thus, at any given decision time, the optimization problem (9.7) and (9.8) is the following. Choose $u \in \mathbb{Z}_+^3$ to

(9.12) $$\text{maximize} \quad \mu_1(\hat{z}_1 - \hat{z}_2)u_1 + \mu_2\hat{z}_2 u_2 + \mu_3\hat{z}_3 u_3$$

subject to five constraints:

(9.13) $u_1 + u_3 \leq 1 - (\hat{n}_1 + \hat{n}_3)$ and $u_2 \leq 1 - \hat{n}_2$,

(9.14) $u_1 \leq \hat{z}_1 - \hat{n}_1$, $u_2 \leq \hat{z}_2 - \hat{n}_2$,   and $u_3 \leq \hat{z}_3 - \hat{n}_3$.

The right-hand sides of the two constraints in (9.13) are binary indicators of server availability, as in (9.10), and the right-hand sides in (9.14) are the numbers of uncommitted jobs in the three buffers.

Consider a decision time at which server 1 is uncommitted (that is, $\hat{n}_1 = \hat{n}_3 = 0$), further assuming that buffers 1 and 3 both contain uncommitted jobs (that is, $\hat{z}_1 > \hat{n}_1$ and $\hat{z}_3 > \hat{n}_2$). Then the objective (9.12) dictates that server 1

initiate a new type 1 service if $\mu_1(\hat{z}_1 - \hat{z}_2) > \mu_3 \hat{z}_3$,

initiate a new type 3 service if $\mu_1(\hat{z}_1 - \hat{z}_2) < \mu_3 \hat{z}_3$ and $\hat{z}_3 > 0$,

initiate a new service of either type 1 or type 3 if $\mu_1(\hat{z}_1 - \hat{z}_2) = \mu_3 \hat{z}_3 > 0$,

remain idle (do nothing) if $\hat{z}_1 < \hat{z}_2$ and $\hat{z}_3 = 0$,

remain idle (do nothing) or initiate a new service of type 1
     if $\hat{z}_1 = \hat{z}_2 > 0$ and $\hat{z}_3 = 0$.

Thus we see that increasing the content of buffer 2 provides a *dis*incentive for serving jobs from buffer 1, and in particular, server 1 will refuse to serve buffer 1 if $\hat{z}_1 < \hat{z}_2$.

Basis for the name "back-pressure."   More generally, in a network setting where some jobs follow multihop routes, a build-up of congestion in any given buffer will cause the system manager to deemphasize service of jobs that are currently waiting in another buffer if jobs in it will move on to the congested one after completing their next service. That is, congestion in a "downstream buffer" produces a "back-pressure" against service of jobs in the corresponding "upstream buffer." In a single-hop network, however, there *are* no "downstream buffers," and thus the name "back-pressure" is inappropriate. In a single-hop setting, we therefore use the alternative name "max-weight," the basis for which will be explained in Section 9.4.

9.3 Relaxed Back-Pressure Policy

To define a relaxed BP policy, we first review some salient features of the relaxed SPN model formulation in Section 2.4. Attention is restricted there to the case of single-server pools, and for each $k \in \mathcal{K}$

we interpret $b_k > 0$ as the capacity of server k. Also, no more than one service of each type $j \in \mathcal{J}$ is allowed to be open at any given time. At each decision time t, the system manager determines the updated state descriptors $\hat{n}$ and $\hat{z}$, then chooses a J-vector β of service rates, or service effort levels, subject to capacity constraints and material availability constraints expressed by (9.16) and (9.17). To implement the back-pressure principle in this setting, we take $\beta = h(\hat{n}, \hat{z})$ as the solution of the following optimization problem:

(9.15) $$\text{maximize } p(\beta, \hat{z})$$

subject to

(9.16) $$\beta \in \mathbb{R}_+^J, \ A\beta \leq b \text{ and } B(\hat{n} + u) \leq \hat{z},$$

where $u = (u_1, \ldots, u_J)'$ is determined from β via

(9.17) $$u_j = \begin{cases} 1 & \text{if } \beta_j > 0 \text{ and } \hat{n}_j = 0, \\ 0 & \text{otherwise.} \end{cases}$$

As in the previous section, one interprets u as a *service initiation vector*, meaning that a service of type j is to be initiated at the decision time in question if and only if $u_j = 1$. In the proof of Proposition 9.6, we derive an equivalent problem statement that is in some ways simpler and more natural, with the system manager first choosing u from a finite set of possibilities and then choosing β given u.

We now develop a characterization of the back-pressure optimization problem (9.15) through (9.17) that simplifies the fluid model analysis to follow. First, recalling that A denotes the $K \times J$ capacity consumption matrix for an SPN (see Section 2.1), let us define the polytope

$$\mathcal{A} := \{\beta \in \mathbb{R}_+^J : A\beta \leq b\},$$

and denote by $\mathcal{E}$ the finite set of extreme points of $\mathcal{A}$. Elements of $\mathcal{E}$ will be called *extreme allocations*.

Proposition 9.6. *For each system state* $(\hat{n}, \hat{z}) \in \mathbb{Z}_+^J \times \mathbb{Z}_+^I$ *there exists an optimal solution for the back-pressure optimization problem (9.15) through (9.17) that belongs to* $\mathcal{E}$.

Remark 9.7. Hereafter we follow the convention that solutions of the back-pressure optimization problem (9.15) through (9.17) are always chosen from $\mathcal{E}$.

Proof of Proposition 9.6.   Fix a state $(\hat{n}, \hat{z})$. In the following argument, e denotes the J-vector of ones, and a vector is said to be binary if each of its components is either zero or one.

Let $\mathscr{U}$ be the finite set of all binary vectors u such that $\hat{n} + u \leq e$ and $B(\hat{n} + u) \leq \hat{z}$. In words, this is the set of all service initiation vectors that are possible at a decision time with updated state descriptors $\hat{n}$ and $\hat{z}$. For each $u \in \mathscr{U}$, let $\mathscr{N}(u)$ be the set of $j \in \mathscr{J}$ such that $\hat{n}_j + u_j = 0$, and define

$$\mathscr{A}(u) := \{\beta \in \mathscr{A} : \beta_j = 0 \text{ for all } j \in \mathscr{N}(u)\}.$$

Thus $\mathscr{A}(u)$ is the set of all service rate vectors β that are achievable going forward from the decision time, given that the service initiation vector u has been chosen. Obviously, $\mathscr{A}(u)$ is a polytope for each $u \in \mathscr{U}$, and we denote by $\mathscr{E}(u)$ the finite set of its extreme points. It is easy to verify that $\mathscr{E}(u) \subset \mathscr{E}$ for each $u \in \mathscr{U}$.

For any given $u \in \mathscr{U}$, consider the optimization problem $\hat{\beta}(u) = \text{argmax}\{p(\beta, \hat{z}) : \beta \in \mathscr{A}(u)\}$. This is a linear program with a bounded feasible region and at least one feasible solution (namely, $\beta = 0$), so we know from LP theory that the optimum is achieved at an extreme point of the constraint set. That is, we can assume without loss of generality that there exists an optimal solution $\hat{\beta}(u) \in \mathscr{E}(u) \subset \mathscr{E}$. Moreover, defining

(9.18) $$\hat{\beta} := \text{argmax}\{p(\hat{\beta}(u), \hat{z}) : u \in \mathscr{U}\},$$

it is easy to verify that $\hat{\beta}$ is an optimal solution of (9.15) through (9.17). That is, our back-pressure optimization problem (9.15) through (9.17) is equivalent to maximizing $p(\cdot, \hat{z})$ over the union of the polytopes $\{\mathscr{A}(u), u \in \mathscr{U}\}$. Because $\hat{\beta}(u) \in \mathscr{E}(u) \subset \mathscr{E}$ by construction for each $u \in \mathscr{U}$, the optimal solution $\hat{\beta}$ specified in (9.18) is also an element of $\mathscr{E}$, which completes the proof.   □

9.4 More about Max-Weight and Back-Pressure Policies

Rather than focusing directly on the bilinear objective function that defines a max-weight or back-pressure control policy, Section 9.6 develops an alternative interpretation of these as policies that myopically *maximize the expected rate of decrease for a quadratic Lyapunov function.* Because there are real computations involved in each successive control decision, many authors speak in terms of the max-weight or back-pressure *algorithm* for network control, using the term "network algorithm" as a synonym for what we have called a "control policy."

Basis for the name "max-weight." The control policies that we consider in this chapter encourage removal of items from high-content buffers, and in symmetric fashion, discourage the creation or transfer of items into such buffers. To substantiate that claim, we observe that the biliner function $p(\cdot, \cdot)$ can be expressed in the form

$$(9.19) \qquad p(\beta, \hat{z}) = \sum_{j \in \mathcal{J}} \mu_j w_j(\hat{z}) \beta_j,$$

where the "weight" associated with completion of a type j service is

$$(9.20) \qquad w_j(\hat{z}) := \sum_{i \in \mathcal{I}} \hat{z}_i(B_{ij} - \Gamma_{ij}) \quad \text{for} j \in \mathcal{J}.$$

To animate the objective function (9.19) and (9.20), one may imagine that a reward of ℓ will be received for every item removed from a buffer containing ℓ items, and in symmetric fashion, a penalty of ℓ will be incurred for every item *added to* a buffer containing ℓ items. Given that the updated buffer contents vector $\hat{z}$ is observed at a decision time t, one may interpret $w_j(\hat{z})$ as the expected net reward (in units like dollars) generated by a type j service completion. It follows that (9.19) is the expected net reward *rate* (in units like dollars per hour) generated by a vector β of total service rates. The back-pressure control principle is to choose service rates at each decision time to maximize that expected net reward rate.

The name "max-weight," which can be viewed as shorthand for "maximum weighted sum," is widely used to describe such control policies, and we have chosen to maintain that terminology in the case of single-hop networks. For reasons noted earlier in Section 9.2, we use the more vivid term "back-pressure control" in multihop settings.

Generalized back-pressure policies. For a more general version of relaxed back-pressure control, let $h_1, \ldots, h_I > 0$ be fixed constants, and then liberalize the definition (9.20) of the weights $w_j(\hat{z})$ as follows:

$$(9.21) \qquad w_j(\hat{z}) := \sum_{i \in \mathcal{I}} h_i \hat{z}_i(B_{ij} - \Gamma_{ij}), \quad j \in \mathcal{J}.$$

Readers will see later that all results developed in this chapter remain valid in the more general setting. In this book, however, the terms "max-weight" and "back-pressure," without any modifiers, refer to the original case $h_1 = \cdots = h_I = 1$.

Negative features of back-pressure control. As readers will see later in this chapter, max-weight and back-pressure policies are attractive as means of achieving system stability. There are, however, some negative considerations to be weighed against that notable virtue. One of these is the potential complexity of the optimization problem (9.7) and (9.8), or (9.15) through (9.17), which must be solved at every decision time.

Also, in our earlier discussion of the criss-cross example (Figure 9.2) the following feature was revealed: *in certain circumstances* (specifically, when $\hat{z}_3 = 0$ and $\hat{z}_2 > \hat{z}_1 > 0$ in that example), *the back-pressure policy may direct a server to remain idle (do nothing), even though there is work for that server to do.* This suggests that the policy's performance (in terms of long-run average buffer contents, for example) can probably be improved, even though it ensures long-run stability.

Another drawback of the back-pressure policy concerns its information requirements. Consider, for example, a network of data switches, whose formulation as a stochastic processing network will be discussed in Chapter 12 of this book. In that context, it is switches that play the role of servers, and each switch processes data packets (the units of flow) from a collection of associated input buffers. Moreover, a packet will typically pass through other switches and their associated input buffers before reaching its final destination. To implement back-pressure control in that environment requires more real-time information on system status than is realistically obtainable. For elaboration, see Section 12.6, where we also discuss an alternative approach that is less data hungry.

9.5 The Characteristic Fluid Equation

In this section, we identify a fluid equation that characterizes the relaxed back-pressure policy, given that Assumption 9.1 holds. (Assumption 9.2 is not needed for this result.) It will be shown later (see Section 9.7) that fluid limits under the basic BP policy satisfy this same equation if each activity or service type is conducted by one server acting alone.

Theorem 9.8. *Consider an SPN that satisfies Assumption 9.1 and operates under the relaxed back-pressure control policy. Each fluid limit path $(\hat{D}, \hat{F}, \hat{T}, \hat{Z})$ satisfies*

$$(9.22) \quad R\dot{\hat{T}}(t) \cdot \hat{Z}(t) = \max_{\beta \in \mathscr{A}} p(\beta, \hat{Z}(t)) \quad \text{at each regular point } t \in \mathbb{R}_+.$$

The rest of this section is devoted to the proof of Theorem 9.8, and for that we need two preparatory lemmas. For each $z \in \mathbb{R}_+^I$ let

$$(9.23) \qquad \begin{aligned} \mathscr{J}(z,0) &:= \{j \in \mathscr{J} : z_{i(j)} = 0\} \quad \text{and} \\ \mathscr{J}(z,+) &:= \{j \in \mathscr{J} : z_{i(j)} > 0\}. \end{aligned}$$

Thinking of z as a vector of current buffer contents, one can describe the two sets in (9.23) as follows: $\mathscr{J}(z,0)$ is the set of activities that serve buffers that are currently empty, while $\mathscr{J}(z,+)$ is the set of activities that serve buffers that currently have positive contents. When $z \in \mathbb{Z}_+^I$ represents current buffer levels in an actual SPN, no activity in $\mathscr{J}(z,0)$ can be employed, because those activities lack the necessary input materials. However, if $z \in \mathbb{R}_+^I$ represents current buffer levels in the corresponding *fluid model*, an activity $j \in \mathscr{J}(z,0)$ may still be conducted at a positive level, processing jobs in buffer $i(j)$ even though $z_{i(j)} = 0$. That is, the total service rate $\beta_j = \dot{\hat{T}}_j$ may still be positive in the fluid model when $z_{i(j)} = 0$. Consider, for example, the tandem fluid model pictured in Figure 8.1. For $t \in (t_1, t_2)$, one has $Z_1(t) = 0$, but to keep buffer 1 empty one must process the fluid that is arriving at rate λ_1, which requires a service rate $\dot{T}_1(t) = \lambda_1/\mu_1 > 0$. That is, over an interval of positive length, buffer 1 remains empty but server 1 continues to process fluid from that buffer at a positive rate.

Definition 9.9. Fix $z \in \mathbb{R}_+^I$. An allocation $\beta \in \mathscr{A}$ is said to be *z-maximal* if

$$(9.24) \qquad\qquad p(\beta, z) = \max_{\alpha \in \mathscr{A}} p(\alpha, z).$$

The following lemma says that, even in the fluid setting, for each $z \in \mathbb{R}_+^I$ one can find a z-maximal allocation that assigns zero service rate to any activity whose corresponding buffer is empty.

Lemma 9.10. *Consider an SPN whose data satisfy Assumption 9.1. For any $z \in \mathbb{R}_+^I$ there exists a z-maximal extreme allocation $\beta(z) \in \mathscr{E}$ such that $\beta_j(z) = 0$ for $j \in \mathscr{J}(z,0)$.*

Proof. From linear programming theory, we know that $p(\cdot,z)$ achieves its maximum at an extreme point of the polytope $\mathscr{A}$. Let α be one such z-maximal extreme allocation, and then define a corresponding vector $\tilde{\alpha}$ as follows:

$$\tilde{\alpha}_j := \begin{cases} 0, & j \in \mathscr{J}(z,0), \\ \alpha_j, & j \in \mathscr{J}(z,+). \end{cases}$$

Because $\alpha \in \mathscr{A}$ and $\tilde{\alpha} \le \alpha$, it follows that $\tilde{\alpha} \in \mathscr{A}$. Note that

$$p(\alpha,z) = \sum_i \sum_j z_i R_{ij}\alpha_j = \sum_i \sum_{j\in \mathscr{J}(z,0)} z_i R_{ij}\alpha_j + \sum_i \sum_{j\in \mathscr{J}(z,+)} z_i R_{ij}\alpha_j$$

$$= \sum_i \sum_{j\in \mathscr{J}(z,0)} z_i R_{ij}\alpha_j + z' R\tilde{\alpha}$$

$$= \sum_i \sum_{j\in \mathscr{J}(z,0)} z_i R_{ij}\alpha_j + p(\tilde{\alpha},z)$$

$$\le p(\tilde{\alpha},z),$$

where the inequality follows from the facts that $R_{ij} \le 0$ for $i \neq i(j)$ and

$$\sum_i \sum_{j\in \mathscr{J}(z,0)} z_i R_{ij}\alpha_j = \sum_{j\in \mathscr{J}(z,0)} \sum_{i\neq i(j)} z_i R_{ij}\alpha_j \le 0.$$

Thus, $\tilde{\alpha}$ is also a z-maximal allocation. If $\tilde{\alpha}$ is an extreme allocation, then the proof is complete by choosing $\beta(z) = \tilde{\alpha}$. Otherwise, it is a linear combination of two or more z-maximal extreme allocations. Choose any one of those extreme allocations as $\beta(z)$, proving the lemma. $\square$

For each allocation $\beta \in \mathscr{E}$, let $Y^\beta(t)$ be the cumulative amount of time that allocation β has been employed over $[0,t]$ in an SPN. By Proposition 9.6 and Remark 9.7,

$$(9.25) \qquad T_j(t) = \sum_{\beta\in\mathscr{E}} \beta_j Y^\beta(t) \quad \text{for each } t \ge 0 \text{ and } j \in \mathscr{J}.$$

Clearly,

$$(9.26) \qquad Y^\beta(\cdot) \text{ is nondecreasing for each allocation } \beta \in \mathscr{E},$$

and

$$(9.27) \qquad \sum_{\beta\in\mathscr{E}} Y^\beta(t) = t \text{ for each } t \ge 0.$$

Setting $Y = (Y^\beta : \beta \in \mathscr{E})$, we now consider the *augmented SPN process*

$$(D, F, T, Z, Y).$$

In addition to satisfying (2.9), (2.10), and (2.23), which hold under any control policy, the augmented process (D, F, T, Z, Y) also satisfies (9.25) through (9.27). Properties (9.26) and (9.27) imply that Y^β is Lipschitz continuous for each $\beta \in \mathscr{E}$. As a result, the fluid limit of (D, F, T, Z, Y) can be defined as in Theorem 6.5 in Section 6.4.

Lemma 9.11. *Given that Assumption 9.1 holds, each fluid limit path* $(\hat{D}, \hat{F}, \hat{T}, \hat{Z}, \hat{Y})$ *under the relaxed back-pressure control policy satisfies the basic fluid equations (6.1) through (6.6) plus the following:*

$$(9.28) \qquad \hat{T}_j(t) = \sum_{\beta \in \mathscr{E}} \beta_j \hat{Y}^\beta(t) \quad \text{for each } t \geq 0 \text{ and } j \in \mathscr{J},$$

$$(9.29) \qquad \hat{Y}^\beta(\cdot) \text{ is nondecreasing for each allocation } \beta \in \mathscr{E},$$

$$(9.30) \qquad \sum_{\beta \in \mathscr{E}} \hat{Y}^\beta(t) = t \quad \text{for each } t \geq 0,$$

and for each allocation $\beta \in \mathscr{E}$,

$$(9.31) \qquad p(\beta, \hat{Z}(t)) < \max_{\alpha \in \mathscr{E}} p(\alpha, \hat{Z}(t)) \quad \text{implies that} \quad \dot{\hat{Y}}^\beta(t) = 0.$$

Proof. Consider a fluid limit path $(\hat{D}, \hat{F}, \hat{T}, \hat{Z}, \hat{Y})$, as defined in Theorem 6.5 (Section 6.4). There is a corresponding sequence $\{x_n\} \subset \mathscr{X}$ and a corresponding $\omega \in \Omega$ such that (2.14), (2.15), (6.38), and the following limit all hold as $n \to \infty$:

$$(9.32)$$
$$\left(\hat{D}^{x_n}(\cdot, \omega), \hat{F}^{x_n}(\cdot, \omega), \hat{T}^{x_n}(\cdot, \omega), \hat{Z}^{x_n}(\cdot, \omega), \hat{Y}^{x_n}(\cdot, \omega) \right) \to (\hat{D}, \hat{F}, \hat{T}, \hat{Z}, \hat{Y}),$$

where the convergence in (9.32) is uniform convergence on compact (u.o.c.) sets (see Definition A.6). It is easy to verify that this fluid limit path satisfies (9.28) through (9.30). It remains to prove that $\hat{Y}$ satisfies (9.31).

Fix $t > 0$ and fix an allocation $\beta \in \mathscr{E}$ that satisfies the inequality in (9.31). Let α be a $\hat{Z}(t)$-maximal extreme allocation (guaranteed to exist by Lemma 9.10). Then $p(\beta, \hat{Z}(t)) < p(\alpha, \hat{Z}(t))$. Let $\mathscr{I}(+) := \{i \in \mathscr{I} : \hat{Z}_i(t) > 0\}$ be the set of buffers with positive fluid at time t. Because

$\min_{i \in \mathscr{I}(+)} \hat{Z}_i(t) > 0$ and $\hat{Z}(\cdot)$ is continuous, there exist an $\epsilon > 0$ and a $\delta > 0$ such that, for each $u \in [t - \epsilon, t + \epsilon]$ and each $i \in \mathscr{I}(+)$,

$$p\big(\beta, \hat{Z}(u)\big) + \delta \le p\big(\alpha, \hat{Z}(u)\big) \quad \text{and} \quad \hat{Z}_i(u) \ge \delta.$$

Thus, when n is sufficiently large, $p\big(\beta, Z^{x_n}(|x_n|u)\big) + |x_n|\delta/2 \le p\big(\alpha, Z^{x_n}$ $(|x_n|u)\big)$ and $Z_i^{x_n}(|x_n|u) \ge |x_n|\delta/2$ for each $i \in \mathscr{I}(+)$ and each $u \in [t - \epsilon, t + \epsilon]$. Because $|x_n| \to \infty$ as $n \to \infty$, one can choose n large enough that $|x_n|\delta/2 \ge 1$. Therefore, for each time $u \in \big[|x_n|(t - \epsilon), |x_n|(t + \epsilon)\big]$ we have

$$(9.33) \qquad\qquad p\big(\beta, Z^{x_n}(u)\big) < p\big(\alpha, Z^{x_n}(u)\big)$$

and

$$(9.34) \qquad\qquad Z_i^{x_n}(u) \ge 1 \quad \text{for each } i \in \mathscr{I}(+).$$

Because α is a $\hat{Z}(t)$-maximal extreme allocation, we have that $i(j) \in \mathscr{I}(+)$ for each activity $j \in \mathscr{J}$ with $\alpha_j > 0$. Condition (9.34) implies that, for each activity $j \in \mathscr{J}$ with $\alpha_j > 0$,

$$(9.35) \qquad\qquad Z_{i(j)}^{x_n}(u) \ge 1 \text{ for each } u \in \big[|x_n|(t - \epsilon), |x_n|(t + \epsilon)\big].$$

Condition (9.35) ensures that buffer contents are adequate at any time $u \in [|x_n|(t - \epsilon), |x_n|(t + \epsilon)]$ for the allocation α to be feasible. By (9.33) and the definition of the relaxed back-pressure policy, the allocation β will not be employed during the time interval $[|x_n|(t - \epsilon), |x_n|(t + \epsilon)]$. Therefore,

$$(9.36) \qquad\qquad Y^\beta(|x_n|(t + \epsilon)) - Y^\beta(|x_n|(t - \epsilon)) = 0,$$

which implies $\hat{Y}^\beta(t + \epsilon) - \hat{Y}^\beta(t - \epsilon) = 0$, and hence $\dot{\hat{Y}}^\beta(t) = 0$. $\qquad\square$

Proof of Theorem 9.8. Let $(\hat{D}, \hat{F}, \hat{T}, \hat{Z}, \hat{Y})$ be a fluid limit path. We would like to prove that $(\hat{Z}, \hat{T})$ satisfies the fluid equation (9.22). By Lemma 9.11 and the fact that

$$\sum_{\beta \in \mathscr{E}} \dot{\hat{Y}}^\beta(t) = 1,$$

we have

$$(9.37)$$
$$\sum_{\beta \in \mathscr{E}} \dot{\hat{Y}}^\beta(t) p(\beta, \hat{Z}(t)) = \sum_{\beta \in \mathscr{E}} \dot{\hat{Y}}^\beta(t) \max_{\alpha \in \mathscr{A}} p(\alpha, \hat{Z}(t)) = \max_{\alpha \in \mathscr{A}} p(\alpha, \hat{Z}(t)).$$

Therefore,

$$
\begin{aligned}
R\dot{\hat{T}}(t) \cdot \hat{Z}(t) &= \sum_{i \in \mathcal{I}} \hat{Z}_i(t) \sum_{j \in \mathcal{J}} R_{ij} \dot{\hat{T}}_j(t) = \sum_{i \in \mathcal{I}} \hat{Z}_i(t) \sum_{j \in \mathcal{J}} R_{ij} \sum_{\beta \in \mathcal{E}} \beta_j \dot{\hat{Y}}^\beta(t) \\
&= \sum_{\beta \in \mathcal{E}} \dot{\hat{Y}}^\beta(t) p(\beta, \hat{Z}(t)) \\
&= \max_{\alpha \in \mathcal{A}} p(\alpha, \hat{Z}(t)),
\end{aligned}
$$

which proves (9.22). $\qquad\square$

9.6 Maximal Stability of Relaxed BP (Quadratic Lyapunov Function)

In this section, we consider a subcritical Leontief network operating under the relaxed back-pressure control policy. Using a quadratic Lyapunov function, we show that its fluid limit model is stable, which implies stability of the SPN itself by Theorem 6.2. Combining that result with Corollary 5.6, we conclude that the relaxed BP policy is maximally stable for Leontief networks. As in Section 9.5, we define $\mathcal{A} = \{\beta \in \mathbb{R}_+^J : A\beta \leq b\}$, where A is the $K \times J$ capacity consumption matrix for an SPN. For ease of exposition, vectors $\beta \in \mathbb{R}_+^J$ will be called *allocations* (of server capacities to processing activities) throughout the remainder of this chapter.

Theorem 9.12. *Consider a Leontief network that operates under the relaxed back-pressure control policy. Assume the static planning problem (5.5) through (5.8) has optimal objective value $\gamma^* < 1$. Then the corresponding fluid limit is stable, and as a consequence, the network's ambient Markov chain is positive recurrent.*

Proof. Let $(\hat{T}, \hat{Z})$ be a fluid model solution satisfying (6.1) through (6.6) and (9.22). Define the quadratic Lyapunov function

$$
(9.38) \qquad f(t) := \sum_{i \in \mathcal{I}} \left(\hat{Z}_i(t) \right)^2 \quad \text{for } t \in \mathbb{R}_+.
$$

Then

$$
(9.39) \qquad \dot{f}(t) = 2\hat{Z}(t) \cdot \dot{\hat{Z}}(t).
$$

Suppose that $(\tilde{x}, \gamma)$ is a feasible solution to (5.5) through (5.8). By Assumption 9.2, there exists an $\hat{x} \geq 0$ such that $R\hat{x} > 0$. Clearly, $\hat{x}$ can be scaled so that $\sum_j A_{kj}\hat{x}_j \leq (1 - \gamma)b$ for each server pool k. Let $x^* = \tilde{x} + \hat{x}$. One can check that $Ax^* \leq b$ (and hence $x^* \in \mathcal{A}$) and that $Rx^* = R\tilde{x} + R\hat{x} = \lambda + R\hat{x} \geq \lambda + \delta e$, where $\delta = \min_i \sum_j R_{ij}\hat{x}_j > 0$. By (9.22) and the fact that $x^* \in \mathcal{A}$,

$$R\dot{\hat{T}}(t) \cdot \hat{Z}(t) \geq Rx^* \cdot \hat{Z}(t) = (\lambda + R\hat{x}) \cdot \hat{Z}(t)$$

$$\geq \lambda \cdot \hat{Z}(t) + \delta \sum_i \hat{Z}_i(t) \geq \lambda \cdot \hat{Z}(t) + \delta\sqrt{f(t)}.$$

It follows from (6.1) through (6.6) that $\dot{\hat{Z}}(t) = \lambda - R\dot{\hat{T}}(t)$. Therefore,

$$\dot{f}(t) = 2\dot{\hat{Z}}(t) \cdot \hat{Z}(t) = 2(\lambda - R\dot{\hat{T}}(t)) \cdot \hat{Z}(t) \leq -2\delta\sqrt{f(t)}.$$

It follows from Lemma 8.6 that $\hat{Z}(t) = 0$ for $t \geq \sqrt{f(0)}/(2\delta)$. $\square$

9.7 Maximal Stability of Basic BP with Single-Server Activities

Does Theorem 9.12 remain true if we substitute the basic BP policy for the relaxed BP policy? In general, the answer is negative. Consider, for example, the bandwidth sharing model pictured in Figure 5.3 and discussed immediately thereafter. If server sharing is disallowed in that example, as is the case in our basic SPN formulation, there are parameter values for which (a) the system is subcritical, but (b) no stable control policy exists. Here we further narrow the class of SPN models considered, imposing two additional restrictions: first, $b_k = 1$ for each $k \in \mathcal{K}$, which means that the processing network contains exactly K servers, each of which constitutes a server pool by itself; and second, each column of the capacity consumption matrix A contains a single one and the rest zeros, which means that each activity or service type involves exactly one of the network's K servers. With these added restrictions, we obtain the following precise analog of Theorem 9.12.

Theorem 9.13. *Consider a Leontief network operating under the basic back-pressure control policy. Assume that $b_k = 1$ for each $k \in \mathcal{K}$, and that each column of A contains a single one and the rest zeros. If the static planning problem (5.5) through (5.8) has optimal objective value $\gamma^* < 1$, then the corresponding fluid limit is stable, and hence the network's ambient Markov chain is positive recurrent.*

Given the proof of Theorem 9.12, it will suffice to show that, under the stated hypotheses, the fluid equation (9.22) remains valid under the basic BP policy. In preparation for that task, let us first define the following partition of $\mathscr{J}$:

$$(9.40) \qquad \mathscr{J}(k) := \{j \in \mathscr{J} : A_{kj} = 1\} \quad \text{for each } k \in \mathscr{K}.$$

In words, $\mathscr{J}(k)$ is the set of activities for which server k is responsible, or equivalently, the set of service types that are conducted by server k. Next, for each buffer contents vector $z \in \mathbb{R}_+^I$ let

$$(9.41) \qquad q(j,z) := \begin{cases} \sum_{i \in \mathscr{I}} R_{ij} z_i = \mu_j w_j(z) & \text{for each } j \in \mathscr{J}, \\ 0 & \text{for } j = 0, \end{cases}$$

where $w_j(z)$ is defined via (9.20); an interpretation of this quantity is given in the paragraph following (9.20).

Continuing preparations for the proof of Theorem 9.13, let

$$\mathscr{N} := \{n \in \mathbb{Z}_+^J : An \le e\},$$

where e is again the vector of ones. It follows from (9.19) that for each allocation $\beta \in \mathscr{N}$,

$$(9.42) \qquad p(\beta,z) = \sum_{k \in \mathscr{K}} \sum_{j \in \mathscr{J}(k)} \beta_j q(j,z).$$

Now fix an allocation vector $\beta \in \mathscr{N}$. For each $k \in \mathscr{K}$, there is at most one activity $j \in \mathscr{J}(k)$ with $\beta_j = 1$. Therefore,

$$(9.43) \qquad \sum_{j \in \mathscr{J}(k)} \beta_j q(j,z) \le \max_{j \in \mathscr{J}(k) \cup \{0\}} q(j,z) \quad \text{for each } k \in \mathscr{K}.$$

Finally, because the arrival processes are assumed to be Poisson under Assumption 2.1, it follows from Proposition B.8 that, with probability one,

$$(9.44) \qquad \lim_{n \to \infty} \frac{1}{n} \max_{1 \le \ell \le n} u_i(\ell) = 0 \text{ for each } i \in \mathscr{I},$$

where $\{u_i(\ell), \ell \ge 1\}$ is the sequence of interarrival times for class $i \in \mathscr{I}$.

Proof of Theorem 9.13. Let $(\hat{D}, \hat{F}, \hat{T}, \hat{Z}, \hat{Y})$ be a fluid limit path with a corresponding sequence $\{x_n\} \subset \mathscr{X}$ and a corresponding $\omega \in \Omega$ such that (2.14), (2.15), (6.38), (9.44), (6.39), and (9.32) all hold. As noted earlier, we only need to show that the fluid equation (9.31) is satisfied. The proof is similar to that of Theorem 9.8. Let ω be as before, and fix a time $t > 0$. Suppose that $\beta \in \mathcal{N}$ is such that

$$(9.45) \qquad p(\beta, \hat{Z}(t)) < \max_{\alpha \in \mathcal{A}} p(\alpha, \hat{Z}(t)).$$

We would like to prove that

$$(9.46) \qquad \dot{\hat{Y}}^{\beta}(t) = 0.$$

Set $r_n = |x_n|$. We first show that there exists an allocation $\beta^* \in \mathcal{N}$ and an $\epsilon > 0$ such that, for any large number n and each time $\tau \in [r_n(t - \epsilon), r_n(t + \epsilon)]$, one has

$$(9.47) \qquad p(\beta, Z(\tau)) < p(\beta^*, Z(\tau))$$

and

$$(9.48) \qquad \beta^* \text{ is a feasible allocation at } \tau \text{ whenever } \beta \text{ is.}$$

To prove (9.47) and (9.48), we first have from (9.42) and (9.43) that

$$\max_{\alpha \in \mathcal{N}} p(\alpha, \hat{Z}(t)) = \sum_{k \in \mathcal{K}} \max_{j \in \mathcal{J}(k) \cup \{0\}} q(j, \hat{Z}(t)).$$

To construct a $\beta^* \in \mathcal{N}$ that satisfies (9.47) and (9.48), one observes that the inequality (9.45) implies existence of a server $k \in \mathcal{K}$ such that

$$(9.49) \qquad \sum_{j \in \mathcal{J}(k)} \beta_j q(j, \hat{Z}(t)) < \max_{j \in \mathcal{J}(k) \cup \{0\}} q(j, \hat{Z}(t)).$$

If

$$(9.50) \qquad \max_{j \in \mathcal{J}(k)} q(j, \hat{Z}(t)) > 0,$$

let $j^* \in \mathcal{J}(k)$ satisfy

$$q(j^*, \hat{Z}(t)) = \max_{j \in \mathcal{J}(k)} q(j, \hat{Z}(t))$$

and define β^* as follows:

$$\beta_j^* := \begin{cases} \beta_j & \text{if } j \notin \mathcal{J}(k), \\ 1 & \text{if } j = j^*, \\ 0 & \text{if } j \in \mathcal{J}(k) \setminus \{j^*\}. \end{cases}$$

Otherwise, let

$$
(9.51) \qquad \beta_j^* := \begin{cases} \beta_j & \text{if } j \notin \mathcal{J}(k), \\ 0 & \text{if } j \in \mathcal{J}(k). \end{cases}
$$

It follows that $\beta^* \in \mathcal{N}$ and

$$
(9.52) \qquad p(\beta, \hat{Z}(t)) < p(\beta^*, \hat{Z}(t)).
$$

We now prove that (9.47) holds for the allocation vector $\beta \in \mathcal{N}$ that satisfies (9.45) and the $\beta^* \in \mathcal{N}$ constructed earlier. The continuity of $\hat{Z}(\cdot)$ and inequality (9.52) imply the following: there exist an $\epsilon > 0$ and a $\delta > 0$ such that, for each $\tau \in [t - \epsilon, t + \epsilon]$,

$$
p(\beta, \hat{Z}(\tau)) + \delta \leq p(\beta^*, \hat{Z}(\tau)).
$$

Thus, when n is sufficiently large,

$$
p(\beta, Z(r_n \tau)) + r_n \delta/2 \leq p(\beta^*, Z(r_n \tau))
$$

for each $\tau \in [t - \epsilon, t + \epsilon]$, which implies (9.47).

Our next task is to prove (9.48) for $\beta \in \mathcal{N}$ and $\beta^* \in \mathcal{N}$ as before. When (9.50) does not hold, the definition (9.51) ensures that β^* satisfies (9.48), because server k will not be employed under allocation β^*. Let us therefore assume that (9.50) *does* hold. We first claim that

$$
(9.53) \qquad \hat{Z}_{i^*}(t) > 0,
$$

where $i^* \in \mathcal{I}$ is the unique buffer i satisfying $R_{ij^*} > 0$; such an i^* exists by Assumption 9.1. Suppose that (9.53) does not hold. Then

$$
q(j^*, \hat{Z}(t)) - R_{i^* j^*} \hat{Z}_{i^*}(t) + \sum_{i \neq i^*} R_{ij} \hat{Z}_i(t) \leq \sum_{i \neq i^*} R_{ij} \hat{Z}_i(t) \leq 0,
$$

contradicting (9.50). Again, by the continuity of $\hat{Z}(\cdot)$ and the inequality (9.53), by choosing $\epsilon > 0$ and $\delta > 0$ in the preceding paragraph small enough,

$$
\hat{Z}_{i^*}(\tau) \geq \delta \text{ for each } \tau \in [t - \epsilon, t + \epsilon].
$$

Thus, when n is sufficiently large,

$$
Z_{i^*}(r_n \tau) \geq r_n \delta/2
$$

for each $\tau \in [t-\epsilon, t+\epsilon]$. Choosing $r_n > 2\delta$, we then have the following for each $\tau \in [r_n(t-\epsilon), r_n(t+\epsilon)]$:

$$Z_{i^*}(\tau) \geq 1,$$

which implies (9.48). Thus, we have proved (9.47) and (9.48).

Given the definition of the basic back-pressure policy, conditions (9.47) and (9.48) imply the following: if allocation β is not employed at time $r_n(t-\epsilon)$, it will never be employed at any time during the interval $[r_n(t-\epsilon), r_n(t+\epsilon)]$. Suppose that allocation β *is* employed at time $r_n(t-\epsilon)$. We now argue in two separate cases that the allocation β will not be employed after a "short time," because β^* is preferred over β. Case (i) is that where server k is employed at time $r_n(t-\epsilon)$ and k satisfies (9.45). In this case, the server must be working on an activity $j \in \mathcal{J}(k) \setminus \{j^*\}$ at $r_n(t-\epsilon)$; otherwise, $\beta^* = \beta$, contradicting (9.47). Let $\hat{v}_j(t)$ be the residual service time of activity j at time t. (If there is no activity j in service at time t, define $\hat{v}_j(t)$ to be the service time of the next activity j.) At $r_n(t-\epsilon) + \hat{v}_j(r_n(t-\epsilon))$, server k completes a type j service, and an allocation decision needs to be made at this time. Conditions (9.47) and (9.48) show that β^* is preferred over β at this time. Thus β will not be employed during the time interval $[r_n(t-\epsilon) + \hat{v}_j(r_n(t-\epsilon)), r_n(t+\epsilon)]$.

Case (ii) is that where server k is idle at time $r_n(t-\epsilon)$. In this case, allocation β^* is preferred over allocation β at the first decision time after $r_n(t-\epsilon)$. Thus β will not be employed during the time interval $[r_n(t-\epsilon) + \eta(r_n(t-\epsilon)), r_n(t+\epsilon)]$, where $t + \eta(t)$ is the first decision time after t. Clearly, $\eta(t)$ is upper bounded by $\min_{i \in \mathcal{I}} \hat{u}_i(t)$, where $\hat{u}_i(t)$ is the residual interarrival time for class i. Therefore, in all cases we have

$$
(9.54) \qquad
\begin{aligned}
Y^\beta(r_n(t+\epsilon)) - Y^\beta(r_n(t-\epsilon)) &\leq \max_{j \in \mathcal{J}} \hat{v}_j(r_n(t-\epsilon)) \\
&\quad + \min_{i \in \mathcal{I}} \hat{u}_i(r_n(t-\epsilon)).
\end{aligned}
$$

It follows from (9.54) and Lemma 9.14 that

$$\lim_{n \to \infty} r_n^{-1}(Y^\beta(r_n(t+\epsilon)) - Y^\beta(r_n(t-\epsilon))) = 0.$$

Thus $\hat{Y}^\beta(t+\epsilon) - \hat{Y}^\beta(t-\epsilon) = 0$, and hence $\dot{\hat{Y}}^\beta(t) = 0$. $\qquad \square$

Lemma 9.14. *Let $\hat{v}_j(t, \omega)$ to be the residual service time for the service of type $j \in \mathcal{J}$ that is under way at time t. (If there is no such service under* *way at t, let $\hat{v}_j(t,\omega)$ to be the service time for the next type j service to start after t.) Define $u_i(t,\omega)$ as the residual interarrival time for class $i \in \mathscr{I}$ at time t. Then, for any sample path ω satisfying (2.14), (2.15), (6.38), and (9.44),*

$$(9.55) \qquad \lim_{t\to\infty} \hat{v}_j(t,\omega)/t = 0,$$

$$(9.56) \qquad \lim_{t\to\infty} u_i(s,\omega)/t = 0.$$

Proof. Fix a sample path ω that satisfies (2.14), (2.15), (6.38), and (9.44). In the following, we drop the dependence on ω to simplify notation. To prove (9.55), one can verify that

$$\hat{v}_j(t) \leq \max_{0\leq\ell\leq S_j(t)+1} v_j(\ell),$$

where $S_j(t) := \max\{n \geq 0 : \sum_{0\leq\ell\leq n} v_j(\ell) \leq t\}$ is the number of type j service completions in t units of time devoted to type j services. Equation (2.15) implies that

$$(9.57) \qquad \lim_{t\to\infty} \frac{1}{t} S_j(t) = \frac{1}{m_j}.$$

The equality (9.55) follows from this and (6.38). Similarly, (9.56) follows from (9.44). $\qquad\square$

9.8 Sources and Literature

Proposition 9.4 is essentially equivalent to what is called the *substitution theorem*, or Samuelson's substitution theorem, for the general Leontief model; it follows easily from theorems 9.1 and 9.6 of Gale (1960). See Berman and Plemmons (1994) and Koehler et al. (1975) for more on Leontief matrices and Leontief substitution systems.

The back-pressure algorithm was developed by Tassiulas and Ephremides (1992), specifically for a slotted-time model of a multihop packet radio network. (Slotted-time communication network models are treated in Chapter 12 of this book.) They did not call their algorithm "back-pressure." Rather, that name was proposed later by Tassiulas (1995) for multihop systems. It was further popularized by Neely et al. (2005) and by others, although many authors, perhaps even most, use the name "max-weight" in both single-hop and multihop contexts. A large literature on max-weight and back-pressure scheduling has developed since the pioneering work of Tassiulas and Ephremides (1992), including an influential paper by McKeown et al. (1999) on max-weight scheduling of an input-queued switch.

This chapter is largely based on the work of Dai and Lin (2005), although those authors focused on rate stability, which is a weaker notion of stability than positive recurrence. A primary goal of Dai and Lin (2005) was to extend the analysis of max-weight and back-pressure control policies to a larger class of models than those considered in the pioneering work of Tassiulas and Ephremides (1992), Tassiulas (1995), and Tassiulas and Bhattacharya (2000), all of which focused on models of the kind described in Section 9.7. See the second column on page 198 of Dai and Lin (2005) for a detailed discussion of the differing model classes.

Dai and Lin (2005) introduced the term *strict Leontief networks* to mean networks that satisfy Assumption 9.1, but not necessarily Assumption 9.2. Our Lemma 9.10 states that Assumption 9.1 implies what Dai and Lin (2005) called the extreme-allocation-available (EAA) property; its proof follows that of theorem 6 in Dai and Lin (2005). The proof of our Theorem 9.8 follows that of lemma 4 in Dai and Lin (2005). The statement of our Theorem 9.12 is analogous to that of theorem 2 in Dai and Lin (2005), but the latter also covers critically loaded networks, reflecting the fact that positive recurrence implies rate stability but not vice versa.

10

Proportionally Fair Resource Allocation

In this chapter, we introduce what is called proportionally fair resource allocation, or just *proportional fairness* (PF) for brevity. To begin, Section 10.1 contains a preparatory discussion of a related optimization problem. Thereafter, mimicking the historical development of the subject, Sections 10.2 and 10.3 describe proportional fairness first in a static model setting. Sections 10.4 through 10.6 then define and analyze a dynamic control policy based on the same idea, called the *PF control policy*, specifically in the context of a relaxed unitary network (that is, a unitary network subject to relaxed control).

The PF control policy is shown to be maximally stable for a relaxed unitary network, which yields the following two corollaries (see Section 10.6): proportionally fair capacity allocation is a maximally stable policy for a bandwidth sharing network, and HLPPS control is maximally stable for a queueing network. As usual, the chapter concludes with brief comments on sources and literature.

As we shall explain in Section 10.3, the PF control policy has an attractive property that we call *resource-relevant aggregation*. The analysis undertaken in this chapter is framed so as to also cover a PF control policy for multihop packet networks (see Section 12.6), and in that setting the aggregation property is crucial.

10.1 A Concave Optimization Problem

Let $I > 0$ be a fixed integer and define $\mathscr{I} := \{1,\ldots,I\}$. Let $\mathscr{A}$ be a bounded, closed, and convex subset of $\mathbb{R}^I_+$ that is *monotone* in the following sense:

(10.1)   If $x \in \mathscr{A}$, $\tilde{x} \in \mathbb{R}^I_+$ and $\tilde{x} \leq x$, then $\tilde{x} \in \mathscr{A}$.

We also assume there exists an $x \in \mathscr{A}$ such that $x_i > 0$ for each $i \in \mathscr{I}$. It follows from this and (10.1) that $\mathscr{A}$ contains the origin.

The next section will provide a general interpretation of the problem structure developed here, but for the moment, readers can think of I as the number of item classes or job classes in a processing network, and for each $x \in \mathscr{A}$ and $i \in \mathscr{I}$, think of x_i as the service rate allocated to class i. For each $z \in \mathbb{R}_+^I$ and $x \in \mathscr{A}$, we define

$$(10.2) \qquad f(z,x) := \sum_{i \in \mathscr{I}} z_i \log(x_i),$$

where, by convention, $\log(0) = -\infty$ and $0\log(0) = 0$. Thus the function f maps $\mathbb{R}_+^I \times \mathscr{A}$ into $[-\infty, \infty)$. For each $z \in \mathbb{R}_+^I$, consider the following optimization problem:

$$(10.3) \qquad \max_{x \in \mathscr{A}} f(z,x).$$

Because $f(z,x)$ is a concave function of x for each fixed $z \in \mathbb{R}_+^I$, we refer to (10.3) as a concave optimization problem.

It will be argued in Lemma 10.1 that there exists an $x^* \in \mathscr{A}$ that attains the maximum in (10.3). The maximizer x^* is a function of z in general, and it is not necessarily unique. Despite the potential nonuniqueness, this paragraph will give meaning to the following definition:

$$(10.4) \qquad \psi(z) = \mathrm{argmax}\left\{ \sum_{i \in \mathscr{I}} z_i \log(x_i) : x \in \mathscr{A} \right\}, \quad z \in \mathbb{R}_+^I.$$

Lemma 10.1 shows that $\psi_i(z)$ is uniquely defined by (10.4) for each i such that $z_i > 0$. On the other hand, one sees from the definition (10.2) that if $z_i = 0$, then all nonnegative values for x_i give the same value for $f(z,x)$, holding constant the other components of x. Given the convention $0\log(0) = 0$, and the assumed monotonicity of $\mathscr{A}$ we can therefore define

$$(10.5) \qquad \psi_i(z) := 0 \quad \text{if } z_i = 0.$$

For the proof of the following lemma, it will be useful to define the constant

$$(10.6) \qquad a_{\max} := \max_{i \in \mathscr{I}} \max_{x \in \mathscr{A}} x_i.$$

Because $\mathscr{A}$ is bounded and nontrivial, one has $0 < a_{\max} < \infty$.

Lemma 10.1. *For each $z \in \mathbb{R}_+^I$ let*

$$\mathscr{I}_+(z) := \{i \in \mathscr{I} : z_i > 0\}.$$

(a) For each $z \in \mathbb{R}_+^I$, there exists a $\psi(z) \in \mathscr{A}$ that achieves the maximum in (10.4). Furthermore, for each $i \in \mathscr{I}_+(z)$, $\psi_i(z)$ is uniquely determined and is strictly positive. (b) If $\mathscr{I}_+(z)$ is nonempty, then $\psi(z)$ is extreme in the following sense: for each $x \in \mathscr{A}$ there exists an $i \in \mathscr{I}_+(z)$ such that

$$(10.7) \qquad\qquad x_i \le \psi_i(z).$$

(c) For each $i \in \mathscr{I}_+(z)$ and each $r > 0$, one has

$$(10.8) \qquad\qquad \psi_i(rz) = \psi_i(z).$$

(d) For each $z \in \mathbb{R}_+^I$ and $i \in \mathscr{I}_+(z)$, the function $\psi_i(\cdot)$ is continuous at z. (e) Assume $z \ne 0$. For each x in the interior of $\mathscr{A}$, the following holds: $f(z,x) < \max_{a \in \mathscr{A}} f(z,a)$. (f) The function $\max_{x \in \mathscr{A}} f(z,x)$ is continuous in z on $\mathbb{R}_+^I$.

Proof. Fix a $z \in \mathbb{R}_+^I$. It will be convenient to establish the following notation for the objective function in (10.3):

$$g(x) := f(z,x) \quad \text{for } x \in \mathscr{A}.$$

Part (a). We first prove the existence of a maximizer $\psi(z)$. By assumption, there exists an $\hat{x} \in \mathscr{A}$ such that $\hat{x}_i > 0$ for all $i \in \mathscr{I}$. For each $i \in \mathscr{I}_+(z)$, because $\lim_{x_i \downarrow 0} z_i \log(x_i) = -\infty$, there exists a $\delta_i \in (0, \hat{x}_i)$ such that

$$(10.9) \qquad z_i \log(x_i) + \sum_{k \in \mathscr{I} \setminus \{i\}} z_k \log(a_{\max}) < g(\hat{x})$$

for $x_i \in (0, \delta_i)$. Define $\mathscr{A}(\delta) := \{x \in \mathscr{A} : x_i \ge \delta_i \text{ for } i \in \mathscr{I}_+(z)\}$. Then $\mathscr{A}(\delta)$ is a bounded, closed set that contains $\hat{x}$. Because $g(\cdot)$ is continuous on $\mathscr{A}(\delta)$, there exists a $\psi(z) \in \mathscr{A}(\delta)$ such that

$$g(\psi(z)) = \max_{x \in \mathscr{A}(\delta)} g(x) \ge g(\hat{x}).$$

We will prove shortly that

$$(10.10) \qquad g(x) < g(\hat{x}) \quad \text{for each } x \in \mathscr{A} \setminus \mathscr{A}(\delta),$$

which imples that

$$g(\psi(z)) = \max_{x \in \mathscr{A}(\delta)} g(x) = \max_{x \in \mathscr{A}} g(x).$$

This proves the existence of $\psi(z)$. To prove (10.10), we first note the following: for each $x \in \mathscr{A} \setminus \mathscr{A}(\delta)$, there exists an $i \in \mathscr{I}_+(z)$ such that $x_i < \delta_i$. Therefore,

$$g(x) = z_i \log(x_i) + \sum_{k \in \mathscr{I} \setminus \{i\}} z_k \log(x_k) \leq z_i \log(x_i)$$

$$+ \sum_{k \in \mathscr{I} \setminus \{i\}} z_k \log(a_{\max}) < g(\hat{x}),$$

where the first inequality holds because $x_k \leq a_{\max}$ for each $k \in \mathscr{I}$, and the second inequality follows from (10.9).

To prove that $\psi_i(z) > 0$ for $i \in \mathscr{I}_+(z)$, first observe that $z_i \log(\psi_i(z)) > -\infty$ for each $i \in \mathscr{I}$, because

$$\sum_{i \in \mathscr{I}} z_i \log(\psi_i(z)) = g(\psi(z)) \geq g(\hat{x}) > -\infty.$$

It follows that $\psi_i(z) > 0$ for each $i \in \mathscr{I}_+(z)$.

Finally, we prove uniqueness. To see this, suppose that $x^k \in \mathscr{A}$ for $k = 1, 2$, and that $g(x^1) = g(x^2) = \max_{x \in \mathscr{A}} g(x)$. Assume that $x_{i_0}^1 \neq x_{i_0}^2$ for some $i_0 \in \mathscr{I}_+(z)$. Then for $\theta = 1/2$,

$$\log(\theta x_i^1 + (1-\theta)x_i^2) \geq \theta \log(x_i^1) + (1-\theta) \log(x_i^2)$$

for each $i \in \mathscr{I}$, and the inequality is strict when $i = i_0$. It follows that

$$g\left(\theta x^1 + (1-\theta)x^2\right) = \sum_{i \in \mathscr{I}} z_i \log\left(\theta x_i^1 + (1-\theta)x_i^2\right) > g(x^1) = g(x^2),$$

contradicting our assumption that $g(x^1)$ has the maximum value. This concludes the proof for part (a).

Part (b).   Suppose that $\psi_i(z) < x_i$ for each $i \in \mathscr{I}_+(z)$. Because $\log(\cdot)$ is strictly increasing,

$$\log\left(\psi_i(z)\right) < \log(x_i)   \text{ for each } i \in \mathscr{I}_+(z).$$

It follows that

$$g(\psi(z)) < g(x),$$

contradicting our assumption that $g(\psi(z))$ has the maximum value. This proves part (b).

Part (c). This follows immediately from part (a).

Part (d). Fix $z \in \mathbb{R}_+^I$ and $i \in \mathscr{I}_+(z)$, and let $\{z^k, k \geq 1\} \subset \mathbb{R}_+^I$ be any sequence converging to z. We need to prove that

$$\lim_{k \to \infty} \psi_i(z^k) = \psi_i(z).$$

Because $\{\psi(z^k), k \geq 1\}$ is a bounded sequence, it suffices to prove that each limit point $\psi^* \in \mathscr{A}$ of $\{\psi(z^k), k \geq 1\}$ satisfies

$$(10.11) \qquad\qquad\qquad \psi_i^* = \psi_i(z).$$

For notational simplicity, we assume the sequence $\{\psi(z^k), k \geq 1\}$ itself converges to ψ^*. By part (a), $\psi_i(z^k) > 0$ for k large enough and each $i \in \mathscr{I}_+(z)$. Fix an $x \in \mathscr{A}$. Then

$$\sum_{i \in \mathscr{I}} z_i \log(\psi_i^*/a_{\max}) = \sum_{i \in \mathscr{I}_+(z)} z_i \log(\psi_i^*/a_{\max})$$

$$= \lim_{k \to \infty} \sum_{i \in \mathscr{I}_+(z)} z_i^k \log\left(\psi_i(z^k)/a_{\max}\right)$$

$$\geq \lim_{k \to \infty} \sum_{i \in \mathscr{I}} z_i^k \log\left(\psi_i(z^k)/a_{\max}\right)$$

$$\geq \lim_{k \to \infty} \sum_{i \in \mathscr{I}} z_i^k \log(x_i/a_{\max}) = \sum_{i \in \mathscr{I}} z_i \log(x_i/a_{\max}).$$

Here the second equality follows from the convergence of $\{z^k\}$ and $\psi(z^k)$, and the fact that $\lim_{k \to \infty} z_i^k > 0$ for each $i \in \mathscr{I}_+(z)$. The first inequality follows from the fact that $\log(\psi_i(z^k)/a_{\max}) \leq 0$ for each $i \in \mathscr{I}$, and the second inequality follows from the definition of $\psi(z^k)$. The last equality follows again from the convergence of $\{z^k\}$. Thus we have proved that

$$g(\psi^*) \geq g(x)$$

for each $x \in \mathscr{A}$. Therefore, $g(\psi^*)$ achieves the maximum on the right side of (10.4). This fact and part (a) imply (10.11), proving part (d).

Part (e). Because x is in the interior of $\mathscr{A}$, one has that $x + \delta e \in \mathscr{A}$ for small enough $\delta > 0$, where e is the vector of ones. Thus,

$$f(z,x) = \sum_{i \in \mathscr{I}_+(z)} z_i \log(x_i)$$

$$< \sum_{i \in \mathscr{I}_+(z)} z_i \log(x_i + \delta) = f(z, x + \delta e) \le \max_{a \in \mathscr{A}} f(z,a).$$

Part (f). For each fixed $x \in \mathscr{A}$, $f(z,x)$ is linear in $z \in \mathbb{R}_+^I$ and is therefore convex in z. One can easily check that $z \in \mathbb{R}_+^I \to \max_{x \in \mathscr{A}} f(z,x) \in \mathbb{R}$ is a convex function. Define

$$h(z) = \begin{cases} \max_{x \in \mathscr{A}} f(z,x) & \text{if } z \in \mathbb{R}_+^I, \\ \infty & \text{otherwise.} \end{cases}$$

Clearly, $h : z \in \mathbb{R}^I \to h(z) \in (-\infty, \infty]$ is also a convex function. Now we prove that h is lower semicontinuous. For that assume that $\{z_n\} \subset \mathbb{R}^I$ is a sequence that converges to $z \in \mathbb{R}^I$. We need to prove that

$$(10.12) \qquad \liminf_{n \to \infty} h(z_n) \ge h(z).$$

If $z \notin \mathbb{R}_+^I$, then $z^n \notin \mathbb{R}_+^I$ for large enough n. Thus, (10.12) holds. Now assume $z \in \mathbb{R}_+^I$. By the proof of part (a), there exists an $x^* \in \mathscr{A}$ with $x_i^* > 0$ for each $i \in \mathscr{I}$ such that $f(z, x^*) = \max_{x \in \mathscr{A}} f(z,x)$. One can verify that $h(z_n) \ge f(z_n, x^*)$ for each n. Thus, we have

$$\liminf_{n \to \infty} h(z_n) \ge \lim_{n \to \infty} f(z_n, x^*) = f(z, x^*) = h(z),$$

proving (10.12).

Having proved that h is convex and lower semicontinuous on $\mathbb{R}^I$, we now invoke theorem 2.35 on page 59 in Rockafellar (1998) to conclude that h is continuous on $\mathbb{R}_+^I$, proving part (f). $\qquad\square$

10.2 Proportional Fairness in a Static Setting

As usual, we take as given a number I of job classes, or just *classes* for brevity, and define $\mathscr{I} := \{1, \ldots, I\}$. Processing resources are not directly represented in the model under discussion here. Rather, their associated capacity constraints are modeled by a bounded, closed, convex set $\mathscr{A} \subset \mathbb{R}_+^I$, which is taken as primitive. We assume that $\mathscr{A}$ contains the origin and is nontrivial in the following sense: there exists an $x \in \mathscr{A}$

such that $x_i > 0$ for each $i \in \mathscr{I}$. This rather abstract formulation will facilitate the application of ideas developed here in multiple contexts.

Static allocation framework.   The scenario we envision is one where a decision maker confronts a demand vector $z \in \mathbb{R}^I_+$ and must select a corresponding *allocation* vector $x \in \mathscr{A}$. Hereafter, when we say that an allocation vector x is *feasible*, this simply means that $x \in \mathscr{A}$. It is easiest to envision components of z as integer valued counts of "users" in the various classes, and one may vaguely describe x_i as the total amount of "service effort" allocated to users in class i, with the understanding that "service effort" can have different meanings in different contexts. Components of z are *not* actually assumed to be integers in the development that follows, so the ideas developed here can be applied later in both fluid model contexts and SPN contexts.

A notion of fairness.   Consider two allocation vectors x and x^* that are both feasible. Let us imagine that a change is made from x^* to x. It is natural to describe the ratio

(10.13)
$$\frac{x_i - x_i^*}{x_i^*}$$

as the corresponding *proportional change* in the allocation granted to class i. A feasible allocation x^* is said to be *proportionally fair* for the demand vector z if, given a change to any other feasible allocation x, one has

(10.14)
$$\sum_{i \in \mathscr{I}} z_i \left(\frac{x_i - x_i^*}{x_i^*} \right) \leq 0.$$

An implicit assumption underlying this definition is the following: whatever total allocation may be granted to class i, it will be divided equally among the z_i users in that class. Thus (10.13) is the proportional change for *each* class i user when the allocation vector changes from x^* to x, and the left side of (10.14) is the sum of those proportional changes over all users of all classes. By definition, x^* is proportionally fair for z if that sum of proportional changes is non-positive for every alternative feasible allocation vector x.

Our definition (10.14) of proportional fairness may appear to be different from the standard textbook definition, as in Srikant and Ying (2014), page 17, specifically because the factor of z_i inside the sum on the left side of (10.14) does not appear in the standard definition. That discrepancy occurs because we define x_i as the *total* allocation granted to a class containing z_i individual users, whereas the very similar notation $\{x_r, r \in \mathscr{R}\}$ is used in the standard treatment to denote the allocations granted to individual users. Our definition is in fact equivalent to the standard one.

Relationship to equal sharing. For many readers it will not be obvious that (10.14) expresses "fairness" in any sense, but in that regard it is useful to consider the following special case. Assuming $0 < K < I$, let $\{\mathscr{I}(k), k \in \mathscr{K}\}$ be an arbitrary partition of $\mathscr{I}$, and let $\mathscr{A} := \{u \in \mathbb{R}_+^K : u \le e\}$, where e is the K-vector of ones. One can interpret this scenario as follows.

The system's processing resources are K servers, each with capacity 1. Server 1 is responsible for processing jobs whose class designations fall in the set $\mathscr{I}(1), \ldots$, and server K is responsible for processing jobs whose class designations fall in the set $\mathscr{I}(K)$. Also, the total service effort allocated by a server to the jobs for which it is responsible cannot exceed the server capacity of 1. The following is a special case of the aggregation property explicated in Section 10.3: under the stated conditions, the proportionally fair resource allocations are $x_i^* = z_i/y_k$ for each $k \in \mathscr{K}$ and $i \in \mathscr{I}(k)$, where

$$(10.15) \qquad y_k := \sum_{\ell \in \mathscr{I}(k)} z_\ell \quad \text{for all } k \in \mathscr{K}.$$

That is, each server k should divide its capacity among the classes for which it is responsible in proportion to their job counts, and that can be accomplished by having server k share its capacity equally among the y_k jobs for which it is responsible. (The basic and relaxed SPN models formulated in Chapter 2 require that all service effort allocated to a given class be directed to one job, namely, the oldest job of that class currently present. However, that requirement is not present, and indeed, is not even meaningful in the static allocation framework currently under discussion.) Because such equal sharing represents a plausible notion of "fairness," this helps to justify the standard terminology used in this chapter.

The PF allocation function. Hereafter, when we speak of allocations, that will mean the vector

$$(10.16) \qquad x^* = \psi(z),$$

where $\psi(z)$ is the solution of the concave optimization problem (10.4) that was studied in the previous section. That is,

$$(10.17) \quad \psi(z) = \operatorname{argmax}\left\{\sum_{i \in \mathscr{I}} z_i \log(x_i) : x \in \mathscr{A}\right\} \quad \text{for each } z \in \mathbb{R}_+^I,$$

with uniqueness of the solution ensured by the convention (10.5). It will be shown in the next subsection that these allocations solve a certain utility optimization problem, and hence have property (10.14), which explains the name. Hereafter $\psi(\cdot)$ will be called the *PF allocation function*.

PF allocations as utility maximizers. Referring again to the case where components of z are integers, one may interpret or motivate the optimization problem (10.17) as follows. First, suppose that any class i user, if granted an allocation of ξ, will derive utility $u_i(\xi)$ from that allocation, where $u_i(\cdot)$ is a given function, strictly concave and increasing on $\mathbb{R}_+$. Let us assume that the decision maker wishes to make allocations to individual users so as to maximize the sum of all users' utilities. It follows from the concavity of $u_i(\cdot)$ that the total allocation x_i granted to class i should be divided equally among the z_i users in that class, so the total utility derived by all users from the allocation vector x will be

$$\sum_{i \in \mathscr{I}} z_i u_i\left(\frac{x_i}{z_i}\right).$$

The PF allocations (10.16) result from assuming identical logarithmic utility functions for all classes. That is, they are the eventual optimal choice if one ascribes to each class $i \in \mathscr{I}$ the utility function $u_i(\xi) = \log(\xi)$. With those identical utilities, the decision maker's objective is to

$$(10.18) \quad \text{maximize} \sum_{i \in \mathscr{I}} z_i \log\left(\frac{x_i}{z_i}\right) = \sum_{i \in \mathscr{I}} z_i \log(x_i) - \sum_{i \in \mathscr{I}} z_i \log(z_i).$$

Of course, the second sum on the right side of (10.18) can be ignored, because it does not depend on x, so (10.18) is equivalent to the optimization (10.17) that defines $\psi(\cdot)$.

We now show that (10.14) does indeed hold for the allocation vector x^* defined via (10.16). First, if $f(x)$ is a concave function over a domain $\mathscr{D}$, it is well known that the following optimality condition (the Karush–Kuhn–Tucker condition, or KKT condition) holds: $\nabla f(x^*)(x - x^*) \leq 0$ for all $x \in \mathcal{D}$, where x^* is the maximizer of $f(\cdot)$. Our vector x^* of PF allocations is the optimizer in the case

$$f(x) = \sum_{i \in \mathcal{I}} z_i \log\left(\frac{x_i}{z_i}\right),$$

and in that case, the KKT condition specializes to give (10.14).

Bandwidth sharing example. To illustrate the application of proportional fairness in a dynamic setting, let us consider again the bandwidth sharing (BWS) example pictured in Figure 1.4, assuming for simplicity that its two links have equal capacity, which we normalize to 1 (that is, $b_1 = b_2 = 1$). Fix a time t and let $z = (z_1, z_2, z_3)$ be the vector of job counts in the three classes at that time. (Recall that "jobs" in this example represent files awaiting transfer over one of three routes.) As stated in Section 1.4, the system manager must allocate flow rates x_1, x_2, x_3 to the three classes, and one possible approach is to choose PF allocations based on current system status, recomputing the allocation vector x each time the job count vector z changes due to either an arrival or a service completion.

Making connection with the static allocation framework described earlier in this section, we have $I = 3$ job classes, and the capacity constraint set $\mathcal{A}$ for this example is defined via (1.3), meaning that

$$(10.19) \qquad \mathcal{A} = \left\{x \in \mathbb{R}^3_+ : x_1 + x_3 \leq 1 \text{ and } x_2 + x_3 \leq 1\right\}.$$

The PF optimization problem (10.17) thus comes down to the following: choose $x \in \mathbb{R}^3_+$ to maximize $z_1 \log(x_1) + z_2 \log(x_2) + z_3 \log(x_3)$, subject to the two capacity constraints in (10.19). To avoid trivialities, let us assume that all three components of z are positive. Both constraints must then be binding at the optimal solution x^*, because otherwise either x_1 or x_2 could be increased. The substitutions $x_1 = 1 - x_3$ and $x_2 = 1 - x_3$ reduce the optimization problem to a single variable x_3, and then elementary calculus yields the following final solution (the PF allocations):

$$(10.20) \qquad x_1^* = x_2^* = \frac{z_1 + z_2}{z_1 + z_2 + z_3} \quad \text{and} \quad x_3^* = \frac{z_3}{z_1 + z_2 + z_3}.$$

Dynamic versus static resource allocation. At least on the surface of things, one has little reason to expect a priori that the PF allocations (10.20) will perform well in our dynamic bandwidth sharing problem, where jobs come and go over time. Indeed, the justification for PF allocations, in terms of either utility maximization or the fairness property (10.14), attaches primary significance to service delivery *rates*, and those rates have no intrinsic significance in our conception of SPN performance. That apparent mismatch is highlighted in the following quotation from pages 189 through 190 of Massoulié and Roberts (2000): "User perceived quality of service may be measured by the response time of a given document transfer ... The fact that this [transfer was achieved] 'fairly' is largely irrelevant and, moreover, totally unverifiable by the user." Nonetheless, it will be seen that dynamic control using PF allocations is a maximally stable policy for a unitary network with relaxed control.

10.3 Aggregation Property of the PF Allocation Function

In this section, we consider again the PF allocation function $\psi(\cdot)$ defined via (10.17), assuming that the convex set $\mathscr{A}$ has a certain special structure. To describe that structure, we take as given a positive integer $L \leq I$ and define $\mathscr{L} := \{1, \ldots, L\}$. Also given is a partition $\{\mathscr{I}(\ell), \ell \in \mathscr{L}\}$ of $\mathscr{I}$, by which we mean that (a) each $\mathscr{I}(\ell)$ is a nonempty subset of $\mathscr{I}$, (b) $\mathscr{I}(\ell) \cap \mathscr{I}(\ell') = \emptyset$ for $\ell \neq \ell'$, and (c) $\bigcup_{\ell \in \mathscr{L}} \mathscr{I}(\ell) = \mathscr{I}$. Cells of the partition are called *demand groups*, the defining feature of which is that the resources required to process or serve any two classes in the same group are identical (we elaborate on this later in this section). Thus, a more fully descriptive name for cells of the partition $\{\mathscr{I}(\ell), \ell \in \mathscr{L}\}$ is *resource-relevant demand groups*.

To repeat, the partition $\{\mathscr{I}(\ell), \ell \in \mathscr{L}\}$ is viewed initially as a primitive model element, but later in this section we shall consider the application of proportional fairness in an SPN setting, and there the partition will be derived from more basic model data; specifically, it will be derived from the capacity consumption matrix A of the SPN. To simplify future notation, we define an $L \times I$ matrix G that encodes the given partition, setting $G_{\ell i} := 1$ if $i \in \mathscr{I}(\ell)$ and $G_{\ell i} := 0$ otherwise.

We also take as given a set $\tilde{\mathscr{A}} \subset \mathbb{R}_+^L$ that has all the properties assumed earlier for $\mathscr{A}$, namely, it is bounded, closed, and convex, it contains the origin, and there exists an $a \in \tilde{\mathscr{A}}$ such that $a_\ell > 0$ for each $\ell \in \mathscr{L}$. The assumed special structure of $\mathscr{A}$ is that

$$(10.21) \qquad \mathscr{A} = \{x \in \mathbb{R}_+^I : Gx \in \tilde{\mathscr{A}}\}.$$

To make clear the meaning of (10.21), let us agree to write $a(x) = Gx$ for $x \in \mathbb{R}_+^I$, so that

$$(10.22) \qquad a_\ell(x) = \sum_{i \in \mathscr{I}(\ell)} x_i \quad \text{for each } \ell \in \mathscr{L}.$$

Viewing x as a vector of service rate allocations to the various job classes, $a_\ell(x)$ is the *aggregate service rate* devoted to demand group ℓ, and (10.21) says that the feasibility of x depends only on those aggregate quantities. We interpret this to mean that classes belonging to the same demand group have identical resource requirements for their processing. A concrete example of this will be given shortly.

To state the aggregation property referred to in the title of this section, we define $\tilde{\psi} : \mathbb{R}_+^L \to \mathbb{R}_+^L$ just as ψ was defined via (10.17), but with $\tilde{\mathscr{A}}$ in place of $\mathscr{A}$, as follows:

$$(10.23) \quad \tilde{\psi}(y) := \operatorname{argmax}\left\{ \sum_{\ell \in \mathscr{L}} y_\ell \log(a_\ell) : a \in \tilde{\mathscr{A}} \right\} \quad \text{for each } y \in \mathbb{R}_+^L.$$

The aggregation property is expressed by the following proposition.

Proposition 10.2. *Given a vector $z \in \mathbb{R}_+^L$, let $y := Gz$, or equivalently, $y_\ell := \sum_{i \in \mathscr{I}(\ell)} z_i$ for each $\ell \in \mathscr{L}$. Then*

$$(10.24) \qquad \psi_i(z) = \tilde{\psi}_\ell(y)\frac{z_i}{y_\ell} \quad \text{for each } i \in \mathscr{I}(\ell) \text{ and } \ell \in \mathscr{L},$$

with the convention that $0/0 = 0$ in (10.24).

Proof. Define $x^* = (x_i^*) \in \mathbb{R}_+^I$ via

$$x_i^* := \tilde{\psi}_\ell(y)\frac{z_i}{y_\ell} \quad \text{for each } i \in \mathscr{I}(\ell) \text{ and } \ell \in \mathscr{L},$$

with the convention that $0/0 = 0$ in (10.24). From the definitions of G, y, and $\tilde{\psi}(\cdot)$, one has that $Gx^* = \tilde{\psi}(y) \in \tilde{\mathscr{A}}$. Therefore, we have from (10.21) that $x^* \in \mathscr{A}$. To prove that $x^* = \psi(z)$, and therefore to prove the lemma, it suffices to show that

$$(10.25) \qquad f(z,x) \leq f(z,x^*) \quad \text{for each } x \in \mathscr{A}$$

where f is defined via (10.2). To prove (10.25), let $x \in \mathscr{A}$ be arbitrary and let $a := Gx$. Then $a \in \tilde{\mathscr{A}}$ by (10.21). One can verify that

$$x_i = a_\ell \frac{z_i}{y_\ell} \quad \text{for each } i \in \mathscr{I}(\ell) \text{ and } \ell \in \mathscr{L}.$$

We then have

$$
f(z,x) = \sum_{i \in \mathscr{I}} z_i \log(x_i) = \sum_{\ell \in \mathscr{L}} \log(a_\ell) y_\ell + \sum_{\ell \in \mathscr{L}} \sum_{i \in \mathscr{I}(\ell)} z_i \log \left(\frac{z_i}{y_\ell} \right)
$$

$$
\leq \sum_{\ell \in \mathscr{L}} \log(\tilde{\psi}_\ell(y)) y_\ell + \sum_{\ell \in \mathscr{L}} \sum_{i \in \mathscr{I}(\ell)} z_i \log \left(\frac{z_i}{y_\ell} \right)
$$

$$
= \sum_{i \in \mathscr{I}} z_i \log(x_i^*) = f(z, x^*),
$$

where the inequality is justified by two facts: first, $a \in \tilde{\mathscr{A}}$, as noted earlier, and second, $\tilde{\psi}(y)$ solves the optimization problem (10.23) by definition.   □

Components of the vector y in Proposition 10.2 are interpreted as group-level *aggregate demands*, and components of $\tilde{\psi}(y)$ as group-level *aggregate allocations*. The proposition tells us that aggregate allocations depend on z only through the vector $y := Gz$ of aggregate demands, and moreover, once the aggregate allocation $\tilde{\psi}_\ell(y)$ has been determined for any group ℓ, it is divided equally among the y_ℓ individual users belonging to that group. Thus, in the static allocation framework currently under discussion, implementation of the PF allocations does not actually require that users be sorted into their respective classes, only into their resource-relevant demand groups.

10.4 Unitary Network with PF Control

As the bandwidth sharing example in Section 10.2 illustrates, one can apply the notion of proportional fairness to *dynamically* allocate resources in an SPN. Let us consider specifically a *unitary* network, as defined in Section 2.6, where we have a one-to-one correspondence between job classes and service types, so both are indexed by $i \in \mathscr{I}$. Recall that the data of a unitary network include an I-vector λ of external arrival rates, an I-vector m of mean service times, and an

$I \times I$ routing matrix P that is substochastic and transient. From those data, we define via (2.38) an I-vector α of total arrival rates into the various classes, and then define the network's load vector ρ via (2.40). Proposition 5.1 states that a unitary network is subcritical if and only if it satisfies the standard load condition $\rho < b$.

PF allocation function for a relaxed unitary SPN.   Also, recall that we denote by A the $K \times I$ capacity consumption matrix for a unitary network, and denote by b the K-vector of server capacities. To apply the notion of proportional fairness developed in this chapter, we adopt the following obvious definition of the convex set $\mathscr{A}$

$$(10.26) \qquad \mathscr{A} := \{x \in \mathbb{R}_+^I : Ax \le b\}.$$

We assume that the unitary network operates under a relaxed control policy having the special structure studied in Section 7.4, so control actions are determined by a policy function $h : \mathbb{R}_+^I \to \mathbb{R}_+^I$. More specifically, we take as our policy function h the PF allocation function ψ defined via (10.17) and (10.26).

The aggregation property discussed in Section 10.3 then takes the following form. Let L be the number of distinct columns in A, and define $\mathscr{L} := \{1, \ldots, L\}$ as usual. The partition of interest here is such that two classes $i, j \in \mathscr{I}$ belong to the same cell $\mathscr{I}(\ell)$ if and only if columns i and j of A are identical. That is, two classes belong to the same cell of the partition (or equivalently, belong to the same demand group) if and only if they have identical resource requirements for their next service. We encode this partition in an $L \times I$ matrix G as before, setting $G_{\ell i} = 1$ if $i \in \mathscr{I}(\ell)$ and $G_{\ell i} = 0$ otherwise. Defining the process $Y(t) = GZ(t), t \ge 0$, one can then describe its components $Y_\ell(t)$ as *group-level aggregate job counts*.

Let $\tilde{A}$ be the nonnegative $K \times L$ matrix whose ℓth column is the common column of A for classes $i \in \mathscr{I}(\ell)$. With $\tilde{A}$ defined in this way, a vector $x \in \mathbb{R}_+^I$ satisfies $Ax \le b$ if and only if $\tilde{A}y \le b$, where $y = Gx$. It follows that (10.21) holds with

$$(10.27) \qquad \tilde{\mathscr{A}} := \{y \in \mathbb{R}_+^L : \tilde{A}y \le b\},$$

and thus Proposition 10.2 holds with $\tilde{\psi}(\cdot)$ defined via (10.23).

Modified version of BWS example.   To illustrate these ideas, consider the system pictured in Figure 10.1, which is a modified version of

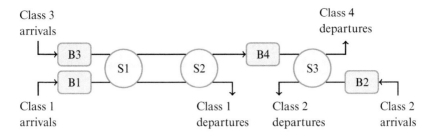

Figure 10.1 A unitary network with two demand groups.

the bandwidth sharing example (hereafter, **BWS** example) that was introduced in Section 1.4, pictured in Figure 1.4, and revisited earlier in this chapter (see Section 10.2). In this modified system, we have three servers labeled S1, S2, and S3 that are most naturally interpreted as communication links. External arrivals are of three different classes, two of which must be transmitted from their point of origin to their eventual destination without intermediate buffering: class 1 requires simultaneous possession of S1 and S2 for its service, and class 2 requires service from S3 only.

In contrast, arriving jobs of class 3 are transmitted from origin to destination in two hops, that is, by means of two sequential services; the first service requires simultaneous possession of S1 and S2, and the second one requires only S3; jobs that have completed the first service and await the second one are designated as class 4, and their associated buffer is labeled B4. There are four processing activities or service types in this example, one for each of the three input classes and one that takes its inputs from the intermediate storage buffer B4. Numbering those service types $1, \ldots, 4$ in the obvious way, one sees that they fall into two groups as we defined that term: services of types 1 and 3 each require simultaneous possession of S1 and S2, while services of types 2 and 4 are each executed by S3 alone. Thus the matrices A and $\tilde{A}$ are as follows for this example:

$$A = \begin{pmatrix} 1 & 0 & 1 & 0 \\ 1 & 0 & 1 & 0 \\ 0 & 1 & 0 & 1 \end{pmatrix} \quad \text{and} \quad \tilde{A} = \begin{pmatrix} 1 & 0 \\ 1 & 0 \\ 0 & 1 \end{pmatrix}.$$

Readers can easily verify that this example meets all the requirements defining a unitary network in Section 2.6. It is *not* a queueing network as we defined that term in Section 2.6, because some of its services require simultaneous resource possession. Also, our definition of a bandwidth sharing network (see Section 4.5) assumes or requires single-hop processing of all input classes, so this example does not fall into that subcategory either.

PF control policy. Applying Proposition 10.2, we have the following vector β of class-level service rates going forward from any given decision time, where $\hat{z}$ is the updated job count vector as usual, $\hat{y} = G\hat{z}$, and $\tilde{\psi}$ is the aggregate allocation function defined via (10.23):

$$(10.28) \quad \beta_i = \psi_i(\hat{z}) = \tilde{\psi}_\ell(\hat{y})\frac{\hat{z}_i}{\hat{y}_\ell} \quad \text{for each } \ell \in \mathcal{L} \text{ and each } i \in \mathcal{I}(\ell),$$

with the convention that $0/0 = 0$ in (10.28). Hereafter this will be called *the PF control policy*. It represents a dynamic application of proportionally fair capacity allocation, generalizing the BWS example pictured in Figure 1.4 and discussed in Section 10.2.

PF fluid model. As readers can easily verify, the policy function $h = \psi$ satisfies Assumption 7.6, so Theorem 7.8 can be applied to obtain a fluid equation characterizing the PF control policy. Specifically, by combining Theorems 6.5 and 7.8, but dropping the hat notation for fluid model processes throughout this chapter, we arrive at the following fluid model for a unitary network operating under the PF control policy (here $M = \text{diag}(m)$ as usual):

$$(10.29) \quad Z(t) = Z(0) + A(t) - D(t), \quad t \geq 0,$$

$$(10.30) \quad Z(t) \geq 0, \quad t \geq 0,$$

$$(10.31) \quad A(t) = \lambda t + P'D(t), \quad t \geq 0,$$

$$(10.32) \quad D(t) = M^{-1}T(t), \quad t \geq 0,$$

$$(10.33) \quad T(\cdot) \text{ is nondecreasing and globally Lipschitz with } T(0) = 0,$$

and for each $\ell \in \mathcal{L}$, each $i \in \mathcal{I}(\ell)$, and each $t > 0$,

$$(10.34) \quad Z_i(t) > 0 \text{ implies } \frac{d}{dt}T_i(t) = \tilde{\psi}_\ell(Y(t))\frac{Z_i(t)}{Y_\ell(t)},$$

where

$$(10.35) \quad Y(t) := GZ(t), \quad t \geq 0.$$

In this recapitulation of earlier results, we have replaced the original fluid equations (6.5) and (6.6) with (10.33), using the obvious fact that (6.6) implies global Lipschitz continuity of T. Also, by Proposition 10.2, (10.34) is equivalent to (7.21) in the current context.

In a unitary network context, the fluid processes A, D, T, and Z are all I-dimensional, and their interpretations are as explained in Chapter 6. The fluid process Y in (10.35) is L-dimensional, and we interpret $Y_\ell(t)$ as the aggregate amount of fluid in demand group ℓ at time t.

Definition 10.3. The *PF fluid model* consists of (10.29) through (10.35).

10.5 Proof of Fluid Model Stability Using Entropy Lyapunov Function

To review, the data of our PF fluid model are I-vectors λ and m, an $I \times I$ matrix P, a partition $\{\mathscr{I}(\ell), \ell \in \mathscr{L}\}$ of the index set $\mathscr{I}$, an $L \times I$ matrix G that encodes the partition, and a nonnegative $K \times L$ matrix $\tilde{A}$ that enters the fluid model through the definition (10.23) of the aggregate allocation function $\tilde{\psi}(\cdot)$.

In this section, we prove that the PF fluid model is stable under the standard load condition $\rho < b$, which is equivalent to subcriticality by Proposition 5.1. To state an alternative form of the standard load condition, let

$$(10.36) \qquad \gamma_\ell := \sum_{i \in \mathscr{I}(\ell)} \alpha_i m_i, \quad \ell \in \mathscr{L},$$

and $\gamma = (\gamma_1, \ldots, \gamma_L)'$. One interprets γ_ℓ as the aggregate service effort required per time unit to process job classes in demand group ℓ. The proof of the following is left as an exercise.

Proposition 10.4. *The standard load condition* (5.1) *holds for a unitary network (or equivalently, the unitary network is subcritical) if and only if*

$$(10.37) \qquad \gamma < a \quad \text{for some } a \in \tilde{\mathscr{A}}.$$

Theorem 10.5. *If* (10.37) *holds, then the PF fluid model is stable.*

The remainder of this section is devoted to the proof of Theorem 10.5, which is long and intricate, as noted earlier in Section 1.7. What distinguishes this proof is its use of a Lyapunov function that is *not* absolutely continuous (and therefore not locally Lipschitz); most stability proofs in the literature involve Lyapunov functions having that property, and greater care must be taken in its absence.

Let (D, T, Z) be a solution to the fluid model equations (10.29) through (10.35). We have that T is globally Lipschitz by (10.33). It then follows from (10.31) through (10.33) that A and D are nondecreasing and Lipschitz continuous, implying that Z is Lipschitz continuous by (10.29). Recall that t is a regular point for the fluid model solution (D, T, Z) if each component is differentiable at t. We denote by $\dot{f}(t)$ the derivative of a function f at t when it exists.

To prove Theorem 10.5, we define the Lyapunov function

$$(10.38) \qquad \phi(t) := \sum_{i \in \mathscr{I}} Z_i(t) \log \left(\dot{D}_i(t) / \alpha_i \right) \quad \text{for each } t \geq 0.$$

By (10.34) and part (a) of Lemma 10.1, $Z_i(t) > 0$ implies that $\dot{D}_i(t) = \mu_i \dot{T}_i(t)$ exists and is positive, so the term

$$(10.39) \qquad Z_i(t) \log \left(\dot{D}_i(t) / \alpha_i \right)$$

in (10.38) is well defined when $Z_i(t) > 0$. By convention, we define (10.39) to be zero when $Z_i(t) = 0$. Therefore, (10.38) is a meaningful definition for each $t \geq 0$. The function ϕ in (10.38) is known as the *entropy Lyapunov function* for a PF fluid solution (D, T, Z).

Unlike most Lyapunov functions encountered in this book, we are unable to prove that $\phi : \mathbb{R}_+ \to \mathbb{R}$ is Lipschitz continuous. Nevertheless, it follows from Definition 6.1 and Lemma 8.11 that the following five lemmas will suffice to prove Theorem 10.5. Their proofs rely on preliminary results developed in Section B.4.

Lemma 10.6. *For each $t \geq 0$ one has $\phi(t) \geq 0$. Furthermore, $Z(t) \neq 0$ implies that $\phi(t) > 0$.*

Lemma 10.7. *The function ϕ is continuous on $(0, \infty)$.*

Lemma 10.8. *There exists a constant $M > 0$ such that*

$$(10.40) \qquad D^+ \phi(t) \leq M \quad \text{for each } t \geq 0,$$

where $D^+ \phi(t)$ is the Dini derivative defined via (A.9).

Lemma 10.9. *For each regular point $t > 0$ of the fluid model solution (D, T, Z),*

$$(10.41) \qquad D^+ \phi(t) \leq \sum_{i \in \mathscr{I}} \dot{Z}_i(t) \log \left(\dot{D}_i(t) / \alpha_i \right).$$

Remark 10.10.   The right side of (10.41) is well defined at each regular point t, because $\dot{D}_i(t) > 0$ for all classes i by (10.34) and Lemma 10.15.

Lemma 10.11.   *Given that* (10.37) *holds, there exists an $\epsilon > 0$ such that, for each regular point $t > 0$,*

$$(10.42) \qquad Z(t) \neq 0 \quad \textit{implies} \quad \sum_{i \in \mathscr{I}} \dot{Z}_i(t) \log \left(\dot{D}_i(t)/\alpha_i \right) \leq -\epsilon.$$

The rest of this section is devoted to the proofs of these five lemmas. For that purpose, it is useful to have an alternative expression for ϕ. Recall that we are given, as part of the data specifying the PF fluid model, a partition $\{ \mathscr{I}(\ell), \ell \in \mathscr{L} \}$ of the job classes $\mathscr{I}$. For each $t \geq 0$, one can write

$$(10.43) \qquad\qquad \phi(t) = \sum_{\ell \in \mathscr{L}} \phi_\ell(t),$$

where

$$(10.44) \quad \phi_\ell(t) := \sum_{i \in \mathscr{I}(\ell)} Z_i(t) \log \left(\dot{D}_i(t)/\alpha_i \right) \quad \text{for each group } \ell \in \mathscr{L}.$$

For each $t \geq 0$, one has

$$(10.45) \qquad \phi_\ell(t) = \sum_{i \in \mathscr{I}(\ell): Z_i(t) > 0} Z_i(t) \log \left(\dot{D}_i(t)/\alpha_i \right)$$

$$= \sum_{i \in \mathscr{I}(\ell): Z_i(t) > 0} Z_i(t) \log \left(\frac{Z_i(t) \tilde{\psi}_\ell(Y(t))}{Y_\ell(t) \alpha_i m_i} \right)$$

$$= \sum_{i \in \mathscr{I}(\ell)} Z_i(t) \log \left(\frac{Z_i(t) \tilde{\psi}_\ell(Y(t))}{Y_\ell(t) \alpha_i m_i} \right),$$

where the first equality follows from our convention that (10.39) equals zero when $Z_i(t) = 0$, the second equality follows from (10.32) and (10.34), and the last equality derives from the following conventions:

$$(10.46) \qquad\qquad \frac{0}{0} = 0 \quad \text{and} \quad 0 \log(0) = 0.$$

Proof of Lemma 10.6.   Fix $t \geq 0$ and $\ell \in \mathcal{L}$. We first assume that $Y_\ell(t) > 0$. Then (10.45) can be rewritten as

(10.47)

$$\phi_\ell(t) = Y_\ell(t) \sum_{i \in \mathscr{I}(\ell)} \frac{Z_i(t)}{Y_\ell(t)} \log\left(\frac{Z_i(t)}{Y_\ell(t)} \frac{\gamma_\ell}{\alpha_i m_i} \frac{\tilde{\psi}_\ell(Y(t))}{\gamma_\ell}\right)$$

$$= Y_\ell(t) \sum_{i \in \mathscr{I}(\ell)} \frac{Z_i(t)}{Y_\ell(t)} \log\left(\frac{Z_i(t)}{Y_\ell(t)} \frac{\gamma_\ell}{\alpha_i m_i}\right) + Y_\ell(t) \log\left(\frac{\tilde{\psi}_\ell(Y(t))}{\gamma_\ell}\right),$$

where γ_ℓ is defined via (10.36). Let us define two probability distributions $p = (p_i)$ and $q = (q_i)$ on the finite set $\mathscr{I}(\ell)$ by setting $p_i := Z_i(t)/Y_\ell(t)$ and $q_i := \alpha_i m_i / \gamma_\ell$, $i \in \mathscr{I}(\ell)$. From the definition (B.23), one sees that the final sum in (10.47) is the relative entropy $D(p\|q)$, which is nonnegative by Lemma B.11 (Pinsker's inequality). Thus we have from (10.47) that

(10.48)
$$\phi_\ell(t) \geq Y_\ell(t) \log\left(\frac{\tilde{\psi}_\ell(Y(t))}{\gamma_\ell}\right).$$

If $Y_\ell(t) = 0$, on the other hand, then $\phi_\ell(t) = 0$ by convention, and hence (10.48) continues to hold. Therefore,

(10.49)
$$\phi(t) = \sum_{\ell \in \mathcal{L}} \phi_\ell(t) \geq \sum_{\ell \in \mathcal{L}} Y_\ell(t) \log\left(\frac{\tilde{\psi}_\ell(Y(t))}{\gamma_\ell}\right)$$

$$= \sum_{\ell \in \mathcal{L}} Y_\ell(t) \log\left(\tilde{\psi}_\ell(Y(t))\right) - \sum_{\ell \in \mathcal{L}} Y_\ell(t) \log(\gamma_\ell)$$

$$\geq 0,$$

where the last inequality follows from two facts: first, γ satisfies (10.37) and is therefore an element of $\tilde{\mathscr{A}}$; and second, $\tilde{\psi}(\cdot)$ is by definition an optimal solution of problem (10.23). Clearly, $Z(t) = 0$ implies $\phi(t) = 0$. Now assume that $Z(t) \neq 0$. Then $Y_\ell(t) > 0$ for at least one $\ell \in \mathcal{L}$. Then by part (e) of Lemma 10.1, the inequality (10.49) must be strict, proving that $\phi(t) > 0$.   □

To prepare for the proofs of Lemmas 10.7 and 10.8, we define

(10.50)
$$f(t) := \sum_{\ell \in \mathcal{L}} \sum_{i \in \mathscr{I}(\ell)} Z_i(t) \log \frac{Z_i(t)}{Y_\ell(t)}, \quad t \geq 0,$$

with the conventions adopted in (10.46). Also, let

$$h(t) := \sum_{\ell \in \mathcal{L}} \sum_{i \in \mathcal{I}(\ell)} Z_i(t) \log\left(\tilde{\psi}_\ell\left(Y(t)\right)\right)$$

(10.51)

$$= \sum_{\ell \in \mathcal{L}} Y_\ell(t) \log\left(\tilde{\psi}_\ell\left(Y(t)\right)\right), \quad t \geq 0.$$

It follows from (10.43) and (10.45) that

(10.52) $$\phi(t) = f(t) + h(t) - \sum_{i \in \mathcal{I}} Z_i(t) \log(\alpha_i m_i).$$

In the proof of Lemma 10.7, we use the following preparatory result.

Lemma 10.12. *The function $f : \mathbb{R}_+ \to \mathbb{R}$ in (10.50) is continuous.*

Proof. It will suffice to show that, for each $\ell \in \mathcal{L}$ and each class $i \in \mathcal{I}(\ell)$, the function

$$f_i(t) = Z_i(t) \log \frac{Z_i(t)}{Y_\ell(t)}$$

is continuous in $t > 0$. That is, fixing $\ell \in \mathcal{L}$, $i \in \mathcal{I}(\ell)$ and $t_0 > 0$, it will suffice to show that

(10.53) $$\lim_{\delta \to 0} f_i(t_0 + \delta) = f_i(t_0).$$

The following three cases will be considered: (a) $Z_i(t_0) > 0$, (b) $Y_\ell(t_0) = 0$, and (c) $Z_i(t_0) = 0$ and $Y_\ell(t_0) > 0$. In case (a), (10.53) clearly holds. In case (b), $Y_\ell(t_0) = 0$ and $f_i(t_0) = 0$, so for each $\delta \in \mathbb{R}$ with $|\delta| < t_0$ one has

$$|f_i(t_0 + \delta)| = \left| Z_i(t_0 + \delta) \log\left(\frac{Z_i(t_0 + \delta)}{Y_\ell(t_0 + \delta)}\right) \right|$$

$$= \left| Y_\ell(t_0 + \delta) \frac{Z_i(t_0 + \delta)}{Y_\ell(t_0 + \delta)} \log\left(\frac{Z_i(t_0 + \delta)}{Y_\ell(t_0 + \delta)}\right) \right|$$

$$\leq Y_\ell(t_0 + \delta) e^{-1},$$

where the last inequality follows from Lemma B.15. Because $\lim_{\delta \to 0} Y_\ell(t_0 + \delta) = Y_\ell(t_0) = 0$, we have $\lim_{\delta \to 0} f_i(t_0 + \delta) = 0$, proving (10.53). Finally, in case (c) one has $Y_\ell(t_0) > 0$, $Z_i(t_0) = 0$, and $f_i(t_0) = 0$. Thus, for each $\delta \in \mathbb{R}$ with $|\delta| < t_0$,

$$f_i(t_0 + \delta) = Z_i(t_0 + \delta) \log(Z_i(t_0 + \delta)) - Z_i(t_0 + \delta) \log(Y_\ell(t_0 + \delta)).$$

Because $Y_\ell(t_0) > 0$,

$$(10.54) \qquad \lim_{\delta \to 0} Z_i(t_0 + \delta) \log(Y_\ell(t_0 + \delta)) = Z_i(t_0) \log(Y_\ell(t_0)) = 0.$$

Observe that $Z_i(t_0 + \delta) \geq 0$, $Z_i(t_0 + \delta) \to Z_i(t_0) = 0$, and $x \log(x)$ is right continuous at $x = 0$ by Lemma B.15. It follows that

$$(10.55) \qquad \lim_{\delta \to 0} Z_i(t_0 + \delta) \log(Z_i(t_0 + \delta)) = 0.$$

Concluding the proof, one sees that (10.53) follows from (10.54) and (10.55) in this case. $\qquad \square$

Proof of Lemma 10.7. Because $Z(t)$ is continuous in $t > 0$, the third term in (10.52) is continuous. By Lemma 10.12, the function f is continuous. It remains to prove that $h(t)$ is continuous in $t > 0$. This follows from the definition (10.51) of $h(\cdot)$, part (f) of Lemma 10.1, and the continuity of $Y(t)$. $\qquad \square$

To prove Lemma 10.8, we first establish the following result. Recall the definitions (A.9) and (A.10) of the upper and lower Dini derivatives, denoted D^+ and D^-, respectively. Lemma A.12 is used repeatedly in the proof.

Lemma 10.13. *Recall the definition (10.50) of f. For each $t > 0$,*

$$(10.56) \quad D^+ f(t) \leq \sum_{\ell \in \mathscr{L}} \sum_{i \in \mathscr{I}(\ell): Z_i(t) > 0} \left(D^- Z_i(t) \log \frac{Z_i(t)}{Y_\ell(t)} + D^+ Z_i(t) \right.$$
$$\left. - D^- Y_\ell(t) \frac{Z_i(t)}{Y_\ell(t)} \right).$$

Proof. Fix a $t \geq 0$. For any $\delta > 0$, one has

$$f(t + \delta) - f(t)$$
$$= \sum_{\ell \in \mathscr{L}} \sum_{i \in \mathscr{I}(\ell)} \left(Z_i(t + \delta) \log \frac{Z_i(t + \delta)}{Y_\ell(t + \delta)} - Z_i(t) \log \frac{Z_i(t)}{Y_\ell(t)} \right)$$
$$= \sum_{\ell \in \mathscr{L}} \sum_{i \in \mathscr{I}(\ell): Z_i(t) > 0} \left(Z_i(t + \delta) \log \frac{Z_i(t + \delta)}{Y_\ell(t + \delta)} - Z_i(t) \log \frac{Z_i(t)}{Y_\ell(t)} \right)$$

$$+ \sum_{\ell \in \mathscr{L}} \sum_{i \in \mathscr{I}(\ell) : Z_i(t) = 0} Z_i(t+\delta) \log \frac{Z_i(t+\delta)}{Y_\ell(t+\delta)}$$

$$\leq \sum_{\ell \in \mathscr{L}} \sum_{i \in \mathscr{I}(\ell) : Z_i(t) > 0} \left(Z_i(t+\delta) \log \frac{Z_i(t+\delta)}{Y_\ell(t+\delta)} - Z_i(t) \log \frac{Z_i(t)}{Y_\ell(t)} \right)$$

$$= \sum_{\ell \in \mathscr{L}} \sum_{i \in \mathscr{I}(\ell) : Z_i(t) > 0} \left[(Z_i(t+\delta) - Z_i(t)) \log \frac{Z_i(t+\delta)}{Y_\ell(t+\delta)} \right.$$

$$\left. + Z_i(t) \left(\log \frac{Z_i(t+\delta)}{Y_\ell(t+\delta)} - \log \frac{Z_i(t)}{Y_\ell(t)} \right) \right].$$

It follows from Lemma A.12 that

$$(10.57) \qquad D^+ f(t) \leq \sum_{\ell \in \mathscr{L}} \sum_{i \in \mathscr{I}(\ell) : Z_i(t) > 0} \left[D^- Z_i(t) \log \frac{Z_i(t)}{Y_\ell(t)} \right.$$

$$\left. + Z_i(t) D^+ \left(\log \frac{Z_i(t)}{Y_\ell(t)} \right) \right].$$

For $i \in \mathscr{I}(\ell)$ with $Z_i(t) > 0$, again by Lemma A.12,

$$D^+ \left(\log \frac{Z_i(t)}{Y_\ell(t)} \right) = D^+ \left(\log(Z_i(t)) - \log(Y_\ell(t)) \right)$$

$$\leq D^+ \left(\log(Z_i(t)) \right) - D^- \left(\log(Y_\ell(t)) \right)$$

$$= \frac{D^+ \left(Z_i(t) \right)}{Z_i(t)} - \frac{D^- \left(Y_\ell(t) \right)}{Y_\ell(t)},$$

which implies that

$$(10.58) \qquad Z_i(t) D^+ \left(\log \frac{Z_i(t)}{Y_\ell(t)} \right) \leq D^+ \left(Z_i(t) \right) - D^- \left(Y_\ell(t) \right) \frac{Z_i(t)}{Y_\ell(t)}.$$

Substituting (10.58) in (10.57) gives (10.56). $\qquad \square$

It will now be convenient to define

$$(10.59) \qquad \hat{a}_{\max} := 1 \vee a_{\max},$$

where $a_{\max}$ is defined via (10.6). Lemma A.12 is used repeatedly in the following proof.

Proof of Lemma 10.8.   Fix a $t > 0$. From (10.52), we have

$$(10.60) \quad D^+\phi(t) \le D^+f(t) + D^+h(t) + D^+\left(-\sum_{i\in\mathscr{I}} Z_i(t)\log(\alpha_i m_i)\right).$$

Recall the bound (10.56) for $D^+f(t)$. It will be shown that

$$(10.61) \quad D^+h(t) \le \sum_{\ell\in\mathscr{L}} \sum_{i\in\mathscr{I}(\ell):Z_i(t)>0} D^-Z_i(t)\log\left(\tilde{\psi}_\ell(Y(t))\right)$$

$$+ \log(\hat{a}_{\max})\left(\sum_{i\in\mathscr{I}} D^+Z_i(t) - \sum_{i\in\mathscr{I}:Z_i(t)>0} D^-Z_i(t)\right),$$

where $\hat{a}_{\max}$ is defined via (10.59). Assuming (10.61), we now complete the proof of Lemma 10.8. Combining (10.60) with (10.56) and (10.61), one has

(10.62)

$$D^+\phi(t) \le \sum_{\ell\in\mathscr{L}} \sum_{i\in\mathscr{I}(\ell):Z_i(t)>0} \left(D^-Z_i(t)\log\left(\frac{Z_i(t)}{Y_\ell(t)}\tilde{\psi}_\ell(Y(t))\right)\right.$$

$$\left. + D^+Z_i(t) - D^-Y_\ell(t)\frac{Z_i(t)}{Y_\ell(t)}\right)$$

$$+ \log(\hat{a}_{\max})\left(\sum_{\ell\in\mathscr{L}} D^+Y_\ell(t) - \sum_{i\in\mathscr{I}:Z_i(t)>0} D^-Z_i(t)\right)$$

$$+ D^+\left(-\sum_{i\in\mathscr{I}} Z_i(t)\log(\alpha_i m_i)\right)$$

$$= \sum_{\ell\in\mathscr{L}} \sum_{i\in\mathscr{I}(\ell):Z_i(t)>0} \left(D^-Z_i(t)\log\left(\dot{D}_i(t)/\alpha_i\right)\right.$$

$$\left. + D^+Z_i(t) - D^-Y_\ell(t)\frac{Z_i(t)}{Y_\ell(t)}\right)$$

$$+ \log(\hat{a}_{\max})\left(\sum_{\ell\in\mathscr{L}} D^+Y_\ell(t) - \sum_{i\in\mathscr{I}:Z_i(t)>0} D^-Z_i(t)\right)$$

$$+ D^+\left(-\sum_{i\in\mathscr{I}} Z_i(t)\log(\alpha_i m_i)\right) + \sum_{i\in\mathscr{I}:Z_i(t)>0} D^-Z_i(t)\log(\alpha_i m_i).$$

Because Z and Y are globally Lipschitz continuous with some Lipschitz constants $C_Z > 0$ and $C_Y > 0$, the following holds for all $t > 0$:

$$|D^+Z_i(t)| \le C_Z, \quad |D^-Z_i(t)| \le C_Z, \quad \text{and} \quad |D^+Y_\ell(t)| \le C_Y.$$

Thus, to prove the lemma, it suffices to show that

$$(10.63) \qquad \sum_{i \in \mathcal{I}: Z_i(t) > 0} D^- Z_i(t) \log\left(\dot{D}_i(t)\right) \leq M_1$$

for some constant M_1 independent of t. To prove (10.63), fix a class i with $Z_i(t) > 0$. Equation (10.34) implies that $\dot{D}_i(t)$ exists. It follows from the fluid model equation (10.29) that

$$D^- Z_i(t) = D^- A_i(t) - \dot{D}_i(t).$$

Clearly, $0 \leq D^- A_i(t) \leq C_A$, where C_A is the Lipschitz constant of A, and $0 < \dot{D}_i(t) \leq \mu_i a_{max}$ when $Z_i(t) > 0$. Therefore, by Lemma B.16 we have

$$(10.64) \qquad \left(D^- Z_i(t)\right) \log\left(\dot{D}_i(t)\right) \leq C_A \log\left(1 \vee (\mu_i a_{max})\right) + \frac{1}{e},$$

which proves (10.63). Therefore, the lemma is proved assuming that (10.61) holds.

To prove (10.61), let us again fix a $t \geq 0$. Then for $\delta > 0$, we have

$$(10.65)$$

$$
\begin{aligned}
h(t+\delta) - h(t) &= \sum_{\ell \in \mathcal{L}} Y_\ell(t+\delta) \log(\tilde{\psi}_\ell(Y(t+\delta))) \\
&\quad - \sum_{\ell \in \mathcal{L}} Y_\ell(t) \log(\tilde{\psi}_\ell(Y(t))) \\
&\leq \sum_{\ell \in \mathcal{L}} Y_\ell(t+\delta) \log(\tilde{\psi}_\ell(Y(t+\delta))) \\
&\quad - \sum_{\ell \in \mathcal{L}} Y_\ell(t) \log(\tilde{\psi}_\ell(Y(t+\delta))) \\
&= \sum_{\ell \in \mathcal{L}} \left(Y_\ell(t+\delta) - Y_\ell(t)\right) \log\left(\frac{\tilde{\psi}_\ell(Y(t+\delta))}{\hat{a}_{max}}\right) \\
&\quad + \sum_{\ell \in \mathcal{L}} \left(Y_\ell(t+\delta) - Y_\ell(t)\right) \log(\hat{a}_{max}) \\
&\leq \sum_{\ell \in \mathcal{L}} \sum_{i \in \mathcal{I}(\ell): Z_i(t) > 0} \left(Z_i(t+\delta) - Z_i(t)\right) \log\left(\frac{\tilde{\psi}_\ell(Y(t+\delta))}{\hat{a}_{max}}\right) \\
&\quad + \sum_{\ell \in \mathcal{L}} \left(Y_\ell(t+\delta) - Y_\ell(t)\right) \log(\hat{a}_{max}).
\end{aligned}
$$

Here the first inequality follows from the fact that $\tilde{\psi}(Y(t))$ is an optimal solution to the optimization problem (10.4) and that $\tilde{\psi}(Y(t+\delta)) \in \mathcal{A}$

is a feasible solution for that problem; the last inequality follows from the fact that $\tilde{\psi}_\ell(Y(t+\delta)) \leq \hat{a}_{max}$, which implies

$$\log\left(\frac{\tilde{\psi}_\ell(Y(t+\delta))}{\hat{a}_{max}}\right) \leq 0.$$

Because $\hat{a}_{max}$ is defined to be at least 1, we have $\log(\hat{a}_{max}) \geq 0$. Also, $Z_i(t) > 0$ implies $Y_\ell(t) > 0$, and therefore, by part (c) of Lemma 10.1, $\tilde{\psi}_\ell(Y(t+\delta)) \to \tilde{\psi}_\ell(Y(t))$ as $\delta \to 0$. Dividing both sides of (10.65) by $\delta > 0$ and taking lim sup as $\delta \to 0$ on both sides, we have

$$D^+ h(t) \leq \sum_{\ell \in \mathscr{L}} \sum_{i \in \mathscr{I}(\ell):Z_i(t)>0} \left(D^- Z_i(t)\right) \log\left(\frac{\tilde{\psi}_\ell(Y(t))}{\hat{a}_{max}}\right)$$
$$+ \log(\hat{a}_{max}) \sum_{\ell \in \mathscr{L}} D^+ Y_\ell(t),$$

proving (10.61).   □

To prove Lemma 10.9, we use the following.

Lemma 10.14. *At each regular point $t > 0$,*

(10.66) $$D^+ f(t) \leq \sum_{\ell \in \mathscr{L}} \sum_{i \in \mathscr{I}(\ell):Z_i(t)>0} \dot{Z}_i(t) \log \frac{Z_i(t)}{Y_\ell(t)}.$$

Proof. Assume $t > 0$ is a regular point of (D, T, Z). Then

(10.67) $$D^+ Z_i(t) = D^- Z_i(t) = \dot{Z}_i(t), \quad D^- Y_\ell(t) = \dot{Y}_\ell(t).$$

From Lemma 8.9, we have that $Z_i(t) = 0$ implies $D^+ Z_i(t) = \dot{Z}_i(t) = 0$. Therefore, the last two terms on the right side of (10.56) are equal to

(10.68) $$\sum_{\ell \in \mathscr{L}} \sum_{i \in \mathscr{I}(\ell):Z_i(t)>0} \left(D^+ Z_i(t) - D^- Y_\ell(t)\frac{Z_i(t)}{Y_\ell(t)}\right)$$
$$= \sum_{\ell \in \mathscr{L}} \sum_{i \in \mathscr{I}(\ell):Z_i(t)>0} \left(\dot{Z}_i(t) - \dot{Y}_\ell(t)\frac{Z_i(t)}{Y_\ell(t)}\right)$$
$$= \sum_{\ell \in \mathscr{L}} \sum_{i \in \mathscr{I}(\ell)} \left(\dot{Z}_i(t) - \dot{Y}_\ell(t)\frac{Z_i(t)}{Y_\ell(t)}\right) = \sum_{\ell \in \mathscr{L}} \left(\dot{Y}_\ell(t) - \dot{Y}_\ell(t)\right) = 0.$$

Finally, (10.66) follows from (10.56), (10.67), and (10.68).   □

Proof of Lemma 10.9. Our proof of (10.41) relies on the inequality (10.60). Assume $t > 0$ is a regular point. Because $D^+ Z_i(t) = D^- Z_i(t) = \dot{Z}_i(t)$, one has $D^+ Y_\ell(t) = \sum_{i \in \mathscr{I}(\ell)} \dot{Z}_i(t)$. Also, $Z_i(t) = 0$ implies $\dot{Z}_i(t) = 0$, so from (10.61) we have

$$(10.69) \qquad D^+ h(t) \leq \sum_{i \in \mathscr{I}} \dot{Z}_i(t) \log \left(\tilde{\psi}_\ell(Y(t)) \right).$$

Thus, by (10.60), (10.66), and (10.69),

$$D^+ \phi(t) \leq \sum_{i \in \mathscr{I}} \dot{Z}_i(t) \log \frac{Z_i(t)}{Y_\ell(t)}$$
$$+ \sum_{i \in \mathscr{I}} \dot{Z}_i(t) \log \left(\tilde{\psi}_\ell(Y(t)) \right) - \sum_{i \in \mathscr{I}} \dot{Z}_i(t) \log(\alpha_i m_i)$$
$$= \sum_{i \in \mathscr{I}} \dot{Z}_i(t) \log \left(\frac{Z_i(t)}{Y_\ell(t)} \frac{\tilde{\psi}_\ell(Y(t))}{\alpha_i m_i} \right) = \sum_{i \in \mathscr{I}} \dot{Z}_i(t) \log(\dot{D}_i(t)/\alpha_i),$$

proving the lemma. □

Our proof of Lemma 10.11 relies on the following lemma, whose proof does not use fluid model equation (10.34).

Lemma 10.15. *Let $t > 0$ be a regular point of (D, T, Z), and further assume that $\dot{D}_i(t) > 0$ for each class $i \in \mathscr{I}$ such that $Z_i(t) > 0$. Then*

$$(10.70) \qquad \dot{D}_i(t) > 0 \quad \text{for each class } i \in \mathscr{I}.$$

Proof. Under the assumptions of our PF fluid model (see Section 10.4), the I-vector λ and $I \times I$ matrix P jointly satisfy (B.13) and (B.17), where α is defined via (B.16). Given that (B.17) holds, Lemma B.9 asserts the existence of an integer $p \geq 1$ and classes $i_k \in \mathscr{I}$, $k = 0, \ldots, p$, that satisfy (B.19). We first argue that

$$(10.71) \qquad \dot{D}_{i_0}(t) > 0.$$

If $Z_{i_0}(t) > 0$, then (10.71) follows from the assumptions of the current lemma. If $Z_{i_0}(t) = 0$, Lemma 8.9 implies that $\dot{Z}_{i_0}(t) = 0$. Thus, by (10.29) and (10.31),

$$(10.72) \qquad \dot{D}_{i_0}(t) = \dot{A}_{i_0}(t) \geq \lambda_{i_0} > 0,$$

where the last inequality holds because $i_0 \in \mathscr{I}_0$. If $Z_{i_1}(t) > 0$, then $\dot{D}_{i_1}(t) > 0$ holds by the assumptions of the current lemma. Otherwise, similar to (10.72), we have

$$\dot{D}_{i_1}(t) = \dot{A}_{i_1}(t) = \lambda_{i_1} + \sum_{j \in \mathscr{I}} P_{j i_1} \dot{D}_j(t)$$

$$\geq P_{i_0 i_1} \dot{D}_{i_0}(t) > 0.$$

Continuing in this manner, the proof is completed by induction. $\qquad\square$

Proof of Lemma 10.11. Fix a regular point $t > 0$. From (10.29) and (10.31) we have $\dot{Z}(t) = \dot{A}(t) - \dot{D}(t)$, where $\dot{A}(t) = \lambda + P'\dot{D}(t)$. Also, (10.34) and Lemma 10.15 ensure that $\dot{D}_i(t) > 0$ for each $i \in \mathscr{I}$. We now apply Lemma B.13 with $d = \dot{D}(t)$, which implies $d - a = -\dot{Z}(t)$, so Lemma B.13 gives the following:

(10.73)

$$\sum_{i \in \mathscr{I}} \dot{Z}_i(t) \log\left(\dot{D}_i(t)/\alpha_i\right) \leq -\frac{1}{2\sum_{i \in \mathscr{I}}(\lambda_i + \dot{D}_i(t))}\left(\sum_{i \in \mathscr{I}} \lambda_i |1 - \dot{D}_i(t)/\alpha_i| + \right.$$

$$\left. + \sum_{i \in \mathscr{I}} \alpha_i |\dot{D}_i(t)/\alpha_i - \sum_{j \in \mathscr{I}_+} P_{ij} \dot{D}_j(t)/\alpha_j|\right)^2,$$

where $\mathscr{I}_+ = \mathscr{I} \cup \{0\}$, $P_{i0} = 1 - \sum_{j \in \mathscr{I}} P_{ij}$, and $\alpha_0 = \dot{D}_0(t) = 1$.

Recall that (10.37) is assumed to hold, where γ and $\tilde{\mathscr{A}}$ are defined by (10.36) and (10.27), respectively. Let $a \in \tilde{\mathscr{A}}$ be chosen to satisfy (10.37). Now define

(10.74) $$\delta := \min_{\ell \in \mathscr{L}}(a_\ell - \gamma_\ell) > 0.$$

Assume $Z(t) \neq 0$. From part (b) of Lemma 10.1 we have the following: there exists an $\ell \in \mathscr{L}$ such that $Y_\ell(t) > 0$ and $a_\ell \leq \tilde{\psi}_\ell(Y(t))$. Therefore,

(10.75) $$\gamma_\ell + \delta \leq a_\ell \leq \tilde{\psi}_\ell(Y(t)),$$

where the first inequality follows from the definition (10.74) of δ. From (10.75), one has that

$$\sum_{i \in \mathscr{I}(\ell)} \left(\alpha_i m_i + \frac{\delta}{L}\right) \leq \sum_{i \in \mathscr{I}(\ell)} \frac{Z_i(t)}{Y_\ell(t)} \tilde{\psi}_\ell(Y(t)),$$

which implies the existence of at least one class $i \in \mathscr{I}(\ell)$ such that

$$\alpha_i m_i + \frac{\delta}{L} \le \frac{Z_i(t)}{Y_\ell(t)} \tilde{\psi}_\ell(Y(t)).$$

Dividing both sides of that inequality by $\alpha_i m_i$ gives

$$1 + \frac{\delta}{\alpha_i m_i L} \le \frac{Z_i(t)}{\alpha_i m_i Y_\ell(t)} \tilde{\psi}_\ell(Y(t)) = \dot{D}_i(t)/\alpha_i,$$

where the equality follows from (10.32) and (10.34). Therefore,

(10.76) $$1 + \delta_1 \le \dot{D}_i(t)/\alpha_i,$$

where $\delta_1 := \delta \left(L \max_{i \in \mathscr{I}} \alpha_i m_i\right)^{-1}$. Because $\dot{D}_i(t) \le \mu_i a_{\max}$ for each $i \in \mathscr{I}$, the right side of (10.73) is less than or equal to the first expression immediately below, which can itself be upper bounded as follows:

$$-\frac{1}{2\sum_{i \in \mathscr{I}}(\lambda_i + \mu_i a_{\max})}\left(\sum_{i \in \mathscr{I}} \lambda_i |1 - \dot{D}_i(t)/\alpha_i| + \sum_{i \in \mathscr{I}} \alpha_i |\dot{D}_i(t)/\alpha_i \right.$$
$$\left. - \sum_{j \in \mathscr{I}_+} P_{ij}\dot{D}_j(t)/\alpha_j| \right)^2$$

$$\le -\frac{1}{2\sum_{i \in \mathscr{I}}(\lambda_i + \mu_i a_{\max})}\left(\sum_{i \in \mathscr{I}} \alpha_i |\dot{D}_i(t)/\alpha_i - \sum_{j \in \mathscr{I}_+} P_{ij}\dot{D}_j(t)/\alpha_j| \right)^2$$

$$\le -\frac{1}{2\sum_{i \in \mathscr{I}}(\lambda_i + \mu_i a_{\max})}\kappa^2(\alpha, \delta_1);$$

the first of these inequalities is obvious, and the second one follows from (10.76) and the definition of $\kappa(\alpha, \delta_1)$ via (B.32). From Lemma B.14, one has that $\kappa(\alpha, \delta_1) > 0$. Lemma 10.11 is thus proved with

$$\epsilon = \frac{1}{2\sum_{i \in \mathscr{I}}(\lambda_i + \mu_i a_{\max})}\kappa^2(\alpha, \delta_1). \qquad \square$$

10.6 Maximal Stability of the PF Control Policy

Combining Theorem 10.5 (the PF fluid model is stable in the subcritical case) with Theorem 6.2 (fluid model stability implies SPN stability), we conclude that a subcritical unitary network is stable under the PF control policy. Also, the PF control policy does not require knowledge of the arrival rate vector λ for its implementation, so we have the following from Corollary 5.6.

Corollary 10.16.   *The PF control policy is maximally stable for a unitary network.*

In the remainder of this section, we consider the implications of Corollary 10.16 for two families of specially structured unitary networks that have been introduced in earlier chapters.

Proportionally fair bandwidth sharing.   Let us consider a BWS network and its equivalent head-of-line model (equivalent HL model, or EHL model), both of which were defined in Section 4.5. The latter is an example of a unitary network.

Recall that a BWS network is a special case of the generic processor sharing model described in Section 4.3. Thus a control policy for the BWS network is specified by a policy function $h : \mathbb{R}_+^I \to \mathbb{R}_+^I$. Here we focus specifically on the *PF policy function*

(10.77)
$$h(\hat{z}) = \psi(\hat{z}) = \operatorname{argmax}\left\{ \sum_{i \in \mathscr{I}} \hat{z}_i \log(x_i) : x = (x_i) \in \mathbb{R}_+^I,\ Ax \leq b \right\}.$$

To be absolutely clear about the meaning of (10.77), let us denote by $\hat{z}$ the updated job count vector at a decision time t for the original BWS network, and denote by β the vector of service rate allocations to the various job classes going forward from t. Equation (4.19) specifies that $\beta = h(\hat{z})$, and here we are focusing on the policy function $h(\cdot)$ defined via (10.77). Hereafter, when we say that a BWS network is "operating under PF control," or simply is "under PF control," this means that attention is restricted to the PF policy function (10.77).

Starting with a BWS network operating under PF control, and following the model translation recipe laid out in Section 4.4, readers can verify via Proposition 10.2 the following fact: the policy function $\tilde{h}(\cdot)$ for the corresponding EHL model, which is defined in general by the relationship (4.27), is again the PF allocation function defined by (10.17), but now in the setting of Section 4.4, where service rate allocations to refined job classes depend on the job counts in those refined classes. That is, the EHL model for the original BWS network is a unitary network operating under the PF control policy, so Corollary 10.16 applies.

Corollary 10.17. *If the standard load condition $\rho < b$ is satisfied, then a BWS network is stable under PF control, and otherwise it is not stable under any control policy. In this sense, proportional fairness is maximally stable for a BWS network.*

Proof. Because the EHL model is a unitary network, Proposition 5.1 tells us that it is subcritical if and only if $\rho < b$. Thus, applying first Corollary 10.16 (a subcritical unitary network is stable under PF control) and then Theorem 5.2 (if an SPN is not subcritical, no stable control policy exists), we conclude the following: if $\rho < b$, the EHL model is stable under PF control; otherwise, it is not stable under any control policy. Finally, applying Proposition 4.4 (a processor sharing model is stable if and only if the corresponding EHL model is stable), we have the same conclusion for the original BWS network.   □

Queueing network with HLPPS control. A queueing network, as defined in Section 2.6, is another example of a unitary SPN. Let us consider now the form taken by the PF control policy in a queueing network setting. First, the partition specified in Section 10.4 has $L = K$ cells or demand groups (that is, one demand group for each server), with group k consisting of those classes $i \in \mathscr{I}(k)$ that are processed by server k. Second, the definition (10.26) of $\mathscr{A}$ specializes to give

$$(10.78) \qquad\qquad \tilde{\mathscr{A}} = \prod_{k \in \mathscr{K}} [0, b_k].$$

That is, the total service rate delivered by server k can be chosen anywhere in the interval $[0, b_k]$, without regard to the total service rates chosen for other servers. With $\tilde{\mathscr{A}}$ having this product structure, the aggregate allocation function $\tilde{\psi}(\cdot)$, defined in general via (10.23), is simply $\tilde{\psi}(\cdot) \equiv b$. That is, the total service rate allocated to group k is the full capacity of server k, regardless of system status. Next, as we have seen in Section 10.2 under the heading "relationship to equal sharing," the PF allocation rule calls for the capacity of each server k to be divided among the classes $i \in \mathscr{I}(k)$ in proportion to their job counts z_i. Finally, of course, the entire service effort provided to any given class i must be directed to the oldest current job in that class.

This is precisely the HLPPS control policy defined earlier in Section 4.6. Having concluded that the PF control policy coincides with HLPPS for a queueing network, we have the following conclusion from Corollary 10.16.

Corollary 10.18. *The HLPPS control policy is stable for any subcritical queueing network. Thus HLPPS is maximally stable for a queueing network.*

10.7 Sources and Literature

The notion of proportional fairness was originally propounded in a static model context (as in Section 10.2) by Kelly (1997), and it was further developed in that setting by Kelly et al. (1998). It is now a mainstream topic in network resource allocation, as indicated by the textbook treatment in Srikant and Ying (2014), section 2.2.

As mentioned earlier, it was Massoulié and Roberts (2000) who introduced the dynamic bandwidth sharing network described in Section 4.5, specifically as a model of Internet flows. Those authors studied a number of bandwidth allocation policies, including the dynamic version of proportional fairness described in Section 10.4. A more general concept of α-fair allocation was advanced and analyzed by Mo and Walrand (2000).

For a dynamic bandwidth sharing network with exponential file size distributions, De Veciana et al. (2001) proved the maximal stability of several bandwidth sharing policies, including proportional fairness (that is, proportionally fair capacity allocation). For the case where file size distributions are of phase type, Massoulié (2007) proved the maximal stability of proportional fairness. The proofs in Massoulié (2007) rely on fluid models and entropy Lyapunov functions, much as in our treatment (see Sections 10.4 and 10.5).

The inclusion of resource-relevant demand groups in the static allocation model of Section 10.2, and in its fluid model counterpart in Section 10.4, makes those models more general than what one typically finds in the literature. As noted in the text, the added generality allows us to treat simultaneously three families of fluid models, namely, those in Bramson (1996b), in Massoulié (2007), and in Walton (2015). Corollary 10.17, establishing the maximal stability of HLPPS control for queueing networks, was first stated and proved by Bramson (1996b).

In recent years, a lively area of research has been heavy traffic performance analysis of bandwidth sharing networks, especially under proportionally fair capacity allocation. That study was initiated by Kang et al. (2009), assuming exponential file size distributions and a "local traffic" condition, and was continued by Ye and Yao (2012) without the local traffic condition. More recent work by Vlasiou et al. (2014) and Wang et al. (2017) makes the weaker assumption of phase-type file size distributions, and among other things, is moving toward resolution of a long-standing open problem: proving that the steady-state performance in heavy traffic of a bandwidth sharing network with proportionally fair capacity allocation is insensitive to the file size distributions.

11

Task Allocation in Server Farms

In this chapter, we formulate and analyze a family of models abstracted from an important and still unsettled body of practice. The associated discussion serves to illustrate the process by which researchers formulate tractable models of complex systems, ignoring or suppressing some aspects of system structure, emphasizing other aspects that are judged to be essential.

The problem on which we focus is task allocation in a server farm. In that labeling of the problem, the word "server" should be interpreted narrowly to mean a general-purpose computer of standard design. The term "server farm" refers to a collection of many such machines, maintained by an organization to meet the computing needs of some user population. Server farms employing tens of thousands of individual machines are now quite common, those machines being typically arranged on *racks* that each house 10 to 50 servers, and located in a *data center* that further includes supporting equipment such as power sources, communication links, and environmental controls. Large data centers are industrial scale operations using as much electricity as a small town.

11.1 Data Locality

A problem of particular concern in current practice is how to organize server farms for *data-intensive applications*, by which we mean applications that draw upon large data sets; examples of "large" data sets are those associated with search engines, scientific research, online social networks, and the health care industry. There have emerged in recent years a number of distributed computing paradigms, such as the *MapReduce* framework originally developed by Google, to address data-intensive applications in a server-farm environment. In this approach, large data sets are broken into many "chunks," each of a size suitable for storage on a single server, but each chunk is replicated several times, and the replicas are stored on different servers; the MapReduce framework specifically dictates that there be three replicas of each data chunk.

When a job arrives that requires manipulation of the large data set, a "map" routine is run to break the original job into many "map tasks," each of which draws upon just one data chunk. Each map task is then allocated to some server for execution, and eventually a collection of "reduce tasks" may be required to assemble results of individual map tasks into a resolution of the original job.

In many settings, neither the "mapping" process (that is, breaking the original job into chunk-specific tasks) nor the execution of "reduce tasks" is computationally significant, so modeling attention focuses on the efficient execution of map tasks. In particular, for speedy execution of a map task it is highly desirable to place computation near data, that is, to assign that task to a server on which its input data are stored, or at least assign it to a server that is physically close to one that stores its input data. This is commonly referred to as the *data locality* issue. We shall consider alternative *routing algorithms*, that is, methods for allocating map tasks to servers, that take account of data locality.

11.2 A Map-Only Model with Three Levels of Proximity

The task allocation model to be considered here is one in which map tasks in L different *categories* (we explain that term in the next few paragraphs) are generated by an exogenous arrival process, and each such task must be allocated to one of K different servers for execution; by assumption, a map task simply departs from the system when its execution is complete. Thus, the decomposition of an original job into its constituent map tasks is not explicitly represented in our model, nor are the "reduce tasks" referred to earlier; the latter feature is sometimes expressed by saying that we consider a *map-only model*.

The category of a map task specifies the data chunk that serves as its input, and for each category $\ell = 1, \ldots, L$ we identify the following three sets of servers: $\mathcal{K}_1(\ell)$ consists of the servers (three of them in the standard MapReduce arrangement) on which data chunk ℓ is stored; $\mathcal{K}_2(\ell)$ consists of servers that do not belong to $\mathcal{K}_1(\ell)$ but are located in the same rack as a server belonging to $\mathcal{K}_1(\ell)$; and $\mathcal{K}_3(\ell)$ consists of servers that belong to neither $\mathcal{K}_1(\ell)$ nor $\mathcal{K}_2(\ell)$.

Servers in $\mathscr{K}_1(\ell)$ are said to be *local* for tasks in category ℓ, those in $\mathscr{K}_2(\ell)$ are said to be *rack local* for tasks in category ℓ; and those in $\mathscr{K}_3(\ell)$ are said to be *remote* for tasks in category ℓ. As explained in the next section, we take as given three *nominal service time* distributions, the first of which applies when a map task is allocated to a local server, the second of which applies when it is allocated to a rack local server, and the third of which applies when it is allocated to a remote server; those three distributions have means θ_1, θ_2, and θ_3, respectively, where $0 < \theta_1 < \theta_2 < \theta_3$. What distinguishes a nominal service time from an actual service time is an adjustment for the relative speed of the server performing the task; see the next section for details.

In this book, the word "category" is used only in discussion of task allocation problems, and always with the narrow meaning indicated in the preceding. In contrast, the word "class" is part of our general terminological system (see Section 2.1), and in Section 11.3, where we formulate a general model of task allocation, the notion of "class" is given a problem-specific meaning consistent with its more general role throughout the book.

An example of a task allocation model is pictured in Figure 11.1, where we have six servers, denoted S1, ... , S6, arranged in three racks containing two servers each. In this example, there are four distinct data chunks, denoted C1, ... , C4, that are each stored on two servers. To be specific, C1 is stored on S1 and S4, C2 is stored on S2 and S3, C3 is stored on S4 and S5, and C4 is stored on S1 and S6. Thus, for example, S1 is a local server for tasks in categories 1 and 4, is a rack local server for those in category 2, and is a remote server for category 3. More completely, the sets $\mathscr{K}_1(\ell)$, $\mathscr{K}_2(\ell)$, and $\mathscr{K}_3(\ell)$ are as follows for this example:

$$\mathscr{K}_1(1) = \{S1, S4\}, \quad \mathscr{K}_1(2) = \{S2, S3\}, \quad \mathscr{K}_1(3) = \{S4, S5\}$$
$$\text{and} \quad \mathscr{K}_1(4) = \{S1, S6\},$$
$$\mathscr{K}_2(1) = \{S2, S3\}, \quad \mathscr{K}_2(2) = \{S1, S4\}, \quad \mathscr{K}_2(3) = \{S3, S6\}$$
$$\text{and} \quad \mathscr{K}_2(4) = \{S2, S5\},$$
$$\mathscr{K}_3(1) = \{S5, S6\}, \quad \mathscr{K}_3(2) = \{S5, S6\}, \quad \mathscr{K}_3(3) = \{S1, S2\}$$
$$\text{and} \quad \mathscr{K}_3(4) = \{S3, S4\}.$$

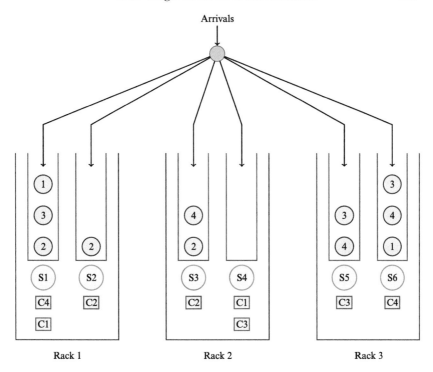

Figure 11.1 Illustrative server farm model.

The open-ended rectangle placed above each server in Figure 11.1 represents a buffer in which tasks that have been allocated to the server await completion of their processing; the tasks themselves are represented by small circles placed within the buffers, with numbering to indicate the category of the task (that is, the number on a task specifies the data chunk that serves as its input).

11.3 Augmented SPN Formulation

Hereafter, we shall restrict attention to task allocation models with *immediate commitment*, as that term was defined in Section 4.2. That is, we assume that each map task must be irrevocably allocated to some server at the moment of its arrival or creation, based on the category of the new task and the backlog of tasks already allocated to the various servers. A general model of task allocation with immediate commitment will be developed in the remainder of this chapter, where we first spell out the complete model formulation and then analyze one particular routing algorithm.

As an alternative to immediate commitment, one might imagine that map tasks are sorted into one of L different storage buffers as they arrive, depending on which data chunk they require as input, and that tasks are released from those buffers only when they are allocated to servers that are otherwise unoccupied. We call this a *delayed commitment* model of server farm operation, because in it no task is committed to a server until that server is actually ready to execute the task. In server farms where a high volume of map tasks must be processed, and where large data sets are stored in relatively small chunks, creating many distinct "categories" of map tasks, delayed commitment is impractical because of its extensive administrative overhead; there will be no further consideration of such models here.

For a precise mathematical specification of the task allocation problem, we adopt the augmented SPN formulation in Section 4.2, which allows for exogenous arrivals from L different "sources," and requires that each new arrival be immediately committed to one of several possible classes for subsequent processing. In our current context, what plays the role of a "source" is a map task "category," as defined earlier, and the commitment required at the time of a task's arrival is the allocation of that task to one of K servers. After a task from category $\ell \in \mathscr{L}$ has been allocated to a specific server $k \in \mathscr{K}$ we designate its class as the pair (ℓ, k). We denote by $i = (\ell, k)$ a generic class, and as usual denote by $I = LK$ the number of such classes.

More specifically, our general task allocation model is an augmented version of the *basic* SPN model (Section 2.3), so server sharing is disallowed. Each of the K servers is treated as a pool unto itself, meaning that

$$(11.1) \qquad\qquad b_1 = \cdots = b_K = 1.$$

There is exactly one activity or service type for each class (ℓ, k); that activity is undertaken by server k, and we associate with it one of the three nominal service time distributions mentioned in the previous section, depending on whether server k is local, rack local or remote for tasks in category ℓ (details are provided later). This general task assignment model has the structure of a parallel-server system, as defined in Section 4.7, except for its immediate commitment requirement.

Nominal service times and relative processing speeds. We take as primitive three mutually independent i.i.d. sequences of *nominal* service times, each having a phase-type distribution. These are denoted $\{\xi_1(n), n \in \mathbb{Z}_+\}$, $\{\xi_2(n), n \in \mathbb{Z}_+\}$, and $\{\xi_3(n), n \in \mathbb{Z}_+\}$, corresponding to local tasks, rack local tasks, and remote tasks, respectively. A "nominal" service time is the length of time that some *benchmark* server would require for execution of the task.

Also given as model data are constants $r_1, \ldots, r_K > 0$, which we call *relative processing speeds* of the various servers. By assumption, server k works at r_k times the speed of the nominal server, regardless of what task it may be performing. (The letter r can be viewed as mnemonic for "rate.") Let us denote by $v_{\ell k}(n)$ the actual service time for the nth task in category ℓ that is assigned to server k. Reflecting the model's distinctive special structure, we model those actual service times as follows:

$$(11.2) \qquad v_{\ell k}(n) = \begin{cases} \xi_1(n)/r_k & \text{if } k \in \mathcal{K}_1(\ell), \\ \xi_2(n)/r_k & \text{if } k \in \mathcal{K}_2(\ell), \\ \xi_3(n)/r_k & \text{if } k \in \mathcal{K}_2(\ell). \end{cases}$$

For purposes of the routing algorithm introduced in Section 11.6, we denote by $m_{\ell k}$ the expected (actual) service time for a class (ℓ, k) task. That is, $m_{\ell k} := \mathbb{E}(v_{\ell k}(n))$, so (11.2) gives the following formula, where $\theta_p = \mathbb{E}(\xi_p)$ for $p = 1, 2, 3$ as in Section 11.2:

$$(11.3) \qquad m_{\ell k} = \begin{cases} \theta_1/r_k & \text{if } k \in \mathcal{K}_1(\ell), \\ \theta_2/r_k & \text{if } k \in \mathcal{K}_2(\ell), \\ \theta_3/r_k & \text{if } k \in \mathcal{K}_2(\ell). \end{cases}$$

The scenario we have in mind is that where the many servers in a data center are of just a few distinct generations, with each successive generation being uniformly faster than its predecessor by some factor. Combined with the three-tier proximity model discussed in Section 11.2, this allows a vast number of class-specific service time distributions to be built up from just three nominal service time distributions and a handful of relative processing speeds. In a typical data center, where there are many thousands of servers and many more task categories, it would be impossible to directly estimate from operating data a service time distribution for each (ℓ, k) pair. Thus the multiplicative structure hypothesized here is important for practical use of the task assignment model.

Remark 11.1. Although the special structure (11.2) is important for applications, it is not used in the analysis that follows. To be specific, we use only the following hereafter: for each class (ℓ, k), the sequence $\{v_{\ell k}(n), n \in \mathbb{Z}_+\}$ is i.i.d. and has a phase-type distribution, with $m_{\ell k} := \mathbb{E}(v_{\ell k}(n))$; moreover, the sequences for different (ℓ, k) pairs are mutually independent. The latter condition can be relaxed to require only that the sequence of K-tuples $\{(v_{\ell 1}(n), \ldots, v_{\ell K}(n)), n \in \mathbb{Z}_+\}$ be i.i.d., allowing correlations among the service times associated with different servers.

FCFS service. For concreteness, we assume that tasks allocated to each server are processed by that server on a first-come-first-served (FCFS) basis, without inserted idleness. However, it is more or less obvious in this setting that decisions about order of service do not affect system stability, assuming that each server follows a nonidling policy. Finally, we restrict attention to simply structured routing policies, as that term was defined in Section 4.2.

11.4 Markov Representation

To model task allocation in server farms realistically, it is *not* acceptable to assume independent Poisson arrival streams for tasks in different categories. A more realistic model would be one in which *batches of tasks* arrive according to a Poisson process, and the composition of each batch (that is, the numbers of constituent tasks in different categories) is drawn from an arbitrary discrete distribution: in such a model, one can interpret an arrival as an externally generated job that draws upon a large, distributed data set, and the corresponding batch as the collection of map tasks, each drawing upon just one data chunk, that are generated from the job.

Thus, in formulating a general task allocation model, we shall modify our baseline stochastic assumptions (see Section 2.2) by allowing the L-dimensional uncommitted arrival process $U = \{U(t), t \geq 0\}$ to be a general Markovian arrival process (MArP), as discussed in Sections 4.1 and 4.2. This includes as a special case the batch-Poisson model of arrivals referred to earlier. As before, we denote by $Y = \{Y(t), t \geq 0\}$ the arrivals-regulating Markov chain (ARMC) that underlies U. Proposition E.7 provides a SLLN for U, but as in Section 4.2, the vector of long-run average arrival rates in the various categories, which is denoted by λ in the statement of Proposition E.7, will here be denoted by $v = (v_1, \ldots, v_L)$. Also, readers are reminded that, as part of our baseline stochastic assumptions, each of the service time distributions in our task allocation model is assumed to be of phase type.

With these assumptions, one can construct a Markov state description x for the total system at any given time by combining ideas developed in Sections 4.1 and 4.2 with those developed via two examples in Chapter 3, namely, the single-pool example with FCFS control (Section 3.3) and the single-pool example with phase-type distributions (Section 3.4). To be specific, under any simply structured routing policy, an adequate Markov state description is obtained by including in x the current state of the ARMC Y, plus the following information for each server $k \in \mathcal{K}$: the number of tasks z_k currently allocated to the server; the category $\ell \in \mathcal{L}$ of the task in each queue position $1, \ldots, z_k$; and the service phase for the task currently in service.

11.5 Simplified Criterion for Subcriticality

The task allocation model formulated in this chapter is a single-hop SPN with alternate routing and immediate commitment, where external arrivals in categories $1, \ldots, L$ follow a MArP with arrival rate vector $v = (v_1, \ldots, v_L)$. In the last paragraph of Section 5.2, the subcritical parameter regime is defined for such networks, and Corollary 5.5 states that stability is only achievable in the subcritical case. In this section, we show that, given the special structure of the task allocation model, the general definition of subcriticality can be restated in much simpler terms.

Recall from Section 11.3 that after a task from category $\ell \in \mathcal{L}$ has been allocated to a specific server $k \in \mathcal{K}$, we designate its class as the pair (ℓ, k), so the number of classes in the task allocation model is $I = LK$. Also, the service time distribution for each class (ℓ, k) is of phase type with mean $m_{\ell k}$. In Lemma 11.2 $\lambda_{\ell k}$ should be interpreted as the arrival rate into class (ℓ, k), or equivalently, as the long-run average rate at which tasks in category ℓ are assigned to server k under the routing policy adopted by the system manager.

Lemma 11.2. *The task allocation model is subcritical, as that term was defined in the last paragraph of Section 5.2, if and only if there exists an I-vector $\lambda = (\lambda_{\ell k}) \geq 0$ such that*

$$(11.4) \qquad \sum_{k \in \mathcal{K}} \lambda_{\ell k} = v_\ell \quad \text{for each category } \ell \in \mathcal{L}, \text{ and}$$

$$(11.5) \qquad \sum_{\ell \in \mathcal{L}} m_{\ell k} \lambda_{\ell k} < 1 \quad \text{for each server } k \in \mathcal{K}.$$

Proof.   As noted in the proof of Corollary 5.5, an SPN with alternate routing and immediate commitment is subcritical if and only if the following holds: there exist $\lambda \in \mathbb{R}_+^I$ and $x \in \mathbb{R}_+^J$ such that

$$(11.6) \qquad G\lambda = \nu, \quad Rx = \lambda, \quad \text{and} \quad Ax < b.$$

For our task allocation model, the $L \times I$ source-buffer matrix G has $G_{\ell i} = 1$ if $i = (\ell, k)$ for some $k \in \mathcal{K}$, and $G_{\ell i} = 0$ otherwise. Thus one sees that the first equation in (11.6) is equivalent to (11.4). Also, $J = I$ in the current setting, and the general definition (5.3) of the input–output matrix R reduces to $R = M^{-1}$, because $\Gamma = 0$ (all tasks exit after completing their one service) and B is the I-dimensional identity matrix; here M is the $I \times I$ diagonal matrix whose diagonal elements $m_{\ell k}$ are given by formula (11.3). Thus the second equation in (11.6) simply dictates that $x = M\lambda$. Next, our task allocation model has $A_{ki} = 1$ if and only if $i = (\ell, k)$ for some $\ell \in \mathcal{L}$, and b is the K-vector of ones by (11.1). Combining these factors, one sees that the final inequality in (11.6) is equivalent to (11.5). $\qquad\qquad\qquad\qquad\qquad\qquad\square$

11.6 Workload-Weighted Task Allocation (WWTA)

Consistent with the general notation established in Section 2.1, we denote by $Z_{\ell k}(t)$ the number of class (ℓ, k) tasks in the system at time t, that is, the number of tasks in category ℓ that have been allocated to server k and are waiting for service or currently in service at time t. Recall that $m_{\ell k}$ denotes the mean service time for such tasks. We define the *workload* for server k at time t as

$$(11.7) \qquad W_k(t) := \sum_{\ell \in \mathcal{L}} m_{\ell k} Z_{\ell k}(t).$$

Definition 11.3.   *Workload-weighted task allocation (WWTA)* is the simply structured routing policy that allocates an arriving task in category ℓ to any server k in the set

$$(11.8) \qquad \mathrm{argmin}_{k \in \mathcal{K}} \, m_{\ell k} W_k(t-)$$

(ties are broken arbitrarily), where $W(t-)$ is the left limit of W at time t.

This definition implicitly assumes that arrivals are completely ordered: allocating a newly arrived task to any given server will change the workload of that server, which may affect how the next arrival is allocated, so if two tasks arrive simultaneously, different results may be obtained depending on which of them is designated as the "earlier" arrival. To assure that the definition remains meaningful in models with batch arrivals, we simply assume that members of any such batch are unambiguously ordered by some exogenous means, such as random ordering.

11.7 Fluid Model

In this section, we develop fluid model equations, numbered (11.11) through (11.15), that hold for our task allocation model under any simply structured routing policy, plus an additional equation (11.16) that characterizes the WWTA policy. As usual, these fluid model equations are justified through a fluid limit procedure, which was first developed in Chapter 6 under our standard assumptions. As noted earlier in the preamble to Chapter 6, our task allocation model differs in two substantial ways from the standard assumptions of Chapter 2: it features alternate routing with immediate commitment, and it models the external arrival process U as a MArP rather than a Poisson process. The fluid limit procedure developed in Chapter 6 must be modified to accommodate these model differences, but the required changes are minor. The fluid limit justification is stated and proved in Theorem 11.4.

Before taking fluid limits, we record two system equations that describe alternate routing with immediate commitment. Let $E_{\ell k}(t)$ to be the cumulative number of tasks in category ℓ that are routed to server k in $[0, t]$, and define $D_{\ell k}(t)$ as the cumulative number of tasks in category ℓ that have been completed by server k in $[0, t]$. In our current context, (4.6) dictates that E and U jointly satisfy

$$(11.9) \qquad \sum_{k \in \mathcal{K}} E_{\ell k}(t) = U_\ell(t). \quad t \in \mathbb{R}_+,$$

and then, using $E_{\ell k}(\cdot)$ to replace the exogenous arrival process into class (ℓ, k) in (2.10), one has that

$$(11.10) \qquad Z_{\ell k}(t) = Z_{\ell k}(0) + E_{\ell k}(t) - D_{\ell k}(t) \geq 0, \quad t \in \mathbb{R}_+.$$

(In this model, the process F is identical to the process D, and thus is omitted in the rest of this section.) To introduce fluid limits, we define fluid scaled processes

$$\left(\hat{D}^x(\cdot, \omega), \hat{E}^x(\cdot, \omega), \hat{T}^x(\cdot, \omega), \hat{W}^x(\cdot, \omega), \hat{Z}^x(\cdot, \omega) \right)$$

for each $\omega \in \Omega$ and each initial state $x \in \mathcal{X}$ with $|x| > 0$, as in (6.37).

Fluid equations. The fluid model under the WWTA routing policy is defined by the following equations:

(11.11) $\quad \hat{Z}_{\ell k}(t) = \hat{Z}_{\ell k}(0) + \hat{E}_{\ell k}(t) - \hat{D}_{\ell k}(t) \geq 0;$

(11.12) $\quad \hat{E}_{\ell k}(\cdot)$ and $\hat{D}_{\ell k}(\cdot)$

$\qquad$ are nondecreasing and Lipschitz continuous;

(11.13) $\quad$ for each category $\ell \in \mathcal{L}$, $\displaystyle\sum_{k \in \mathcal{K}} \hat{E}_{\ell k}(t) = v_\ell t;$

(11.14) $\quad$ for each server $k \in \mathcal{K}$, $\displaystyle\hat{W}_k(t) = \sum_{\ell \in \mathcal{L}} m_{\ell k} \hat{Z}_{\ell k}(t);$

(11.15) $\quad$ for each server $k \in \mathcal{K}$, $\hat{W}_k(t) > 0$ implies that

$$\sum_{\ell \in \mathcal{L}} m_{\ell k} \dot{\hat{D}}_{\ell k}(t) = 1; \text{ and}$$

(11.16) $\quad$ for each category $\ell \in \mathcal{L}$, $\displaystyle\sum_{k \in \mathcal{K}} m_{\ell k} \hat{W}_k(t) \dot{\hat{E}}_{\ell k}(t)$

$$= v_\ell \min_{k \in \mathcal{K}} m_{\ell k} \hat{W}_k(t).$$

Theorem 11.4. *Consider the task allocation model with a simply structured routing policy. (a) For almost all $\omega \in \Omega$, there exist fluid limit paths as in Definition 6.6. (b) Each fluid limit path $(\hat{D}, \hat{E}, \hat{T}, \hat{W}, \hat{Z})$ satisfies the fluid model equations (11.11) through (11.15). (c) Under the WWTA routing policy, each fluid limit path further satisfies (11.16).*

Proof. For (a), we wish to prove a version of Theorem 6.5. For that purpose, fix $\omega \in \Omega$ such that (2.14), (2.15), and (6.38) hold for the core stochastic elements, with U_ℓ replacing E_i in (2.14). For any unbounded set $C \subset \mathcal{X}$ of initial states, there exists a sequence $\{x_n\} \subset C$ with $|x_n| \to \infty$ such that

(11.17)
$$\left(\hat{D}^{x_n}(\cdot, \omega), \hat{E}^{x_n}(\cdot, \omega), \hat{T}^{x_n}(\cdot, \omega), \hat{W}^{x_n}(\cdot, \omega), \hat{Z}^{x_n}(\cdot, \omega) \right)$$
$$\to (\hat{D}, \hat{E}, \hat{T}, \hat{W}, \hat{Z})$$

as $n \to \infty$. The argument supporting (11.17) is as follows. First, for each category $\ell \in \mathcal{L}$, we combine Lemma A.10 with Proposition E.7 (the SLLN for a MArP) to conclude, just as in the proof of (6.45), that

(11.18) $\qquad\qquad \hat{U}_\ell^{x_n}(\cdot,\omega) \to \hat{U}_\ell(\cdot) \quad$ u.o.c.

for any sequence $\{x_n\} \subset \mathscr{X}$ with $|x_n| \to \infty$, where $\hat{U}_\ell(t) = v_\ell t$ for $t \in \mathbb{R}_+$. Fix an ω that satisfies (2.15), (6.38), and (11.18) for all $\ell \in \mathscr{L}$. It follows from (11.9) that for each server $k \in \mathscr{K}$,

(11.19) $\qquad \hat{E}_{\ell k}^x(t,\omega) - \hat{E}_{\ell k}^x(s,\omega) \le \hat{U}_\ell^x(t,\omega) - \hat{U}_\ell^x(s,\omega), \quad 0 \le s \le t$

for any $x \in \mathscr{X}$ with $|x| \ne 0$. It follows from (11.18), (11.19), and Lemma A.11 that there exists a sequence $\{x_n\} \subset C$ with $|x_n| \to \infty$ such that for each $k \in \mathscr{K}$

(11.20) $\qquad\qquad \hat{E}_{\ell k}^{x_n}(\cdot,\omega) \to \hat{E}_{\ell k}(\cdot) \quad$ u.o.c.,

where $\hat{E}_{\ell k}$ is some continuous, nondecreasing function. Given that (11.20) holds, it follows exactly as in the proof of Theorem 6.5 that there exists a subsequence of $\{x_n\}$, still denoted by $\{x_n\}$ for notational simplicity, such that

$$\left(\hat{D}^{x_n}(\cdot,\omega), \hat{E}^{x_n}(\cdot,\omega), \hat{T}^{x_n}(\cdot,\omega), \hat{Z}^{x_n}(\cdot,\omega) \right)$$

converges. Because

(11.21) $\qquad\qquad \hat{W}_k^{x_n}(t) = \sum_{\ell \in \mathscr{L}} m_{\ell k} \hat{Z}_{\ell k}^{x_n}(t),$

the convergence in (11.17) readily follows.

To prove (b), (11.11) and (11.13) follow from (11.10) and (11.9), respectively. Property (11.12) follows from the corresponding property of the prelimit. Equation (11.14) follows from the definition (11.7) of workload. To prove (11.15), it follows from (2.9) that $\hat{D}^{x_n}(t,\omega) = \hat{F}^{x_n}(t,\omega)$ for each n, t, and ω. It then follows from the proof of Lemma 6.8 that

$$\hat{D}_{\ell k}(t) = \hat{F}_{\ell k}(t) = \frac{1}{m_{\ell k}} \hat{T}_{\ell k}(t)$$

for each $t \ge 0$ and each class (ℓ,k). For a $k \in \mathscr{K}$, assume $\hat{W}_k(t) > 0$. Now (11.15) follows from

$$\sum_{\ell \in \mathscr{L}} \dot{\hat{T}}_{\ell,k}(t) = 1,$$

which is proved exactly as was (7.1).

For part (c), it suffices by (11.13) to prove the following: for each $k \in \mathcal{K}$, $\ell \in \mathcal{L}$, and $t > 0$,

$$(11.22) \qquad m_{\ell k} \hat{W}_k(t) > \min_{k' \in \mathcal{K}} m_{\ell k'} \hat{W}_{k'}(t) \quad \text{implies} \quad \dot{\hat{E}}_{\ell k}(t) = 0.$$

Assume that $m_{\ell k} \hat{W}_k(t) > \min_{k' \in \mathcal{K}} m_{\ell k'} \hat{W}_{k'}(t)$. By the continuity of $\hat{W}$, there exists an $\epsilon > 0$ and a $\delta > 0$ such that for each $u \in [t - \epsilon, t + \epsilon]$,

$$m_{\ell k} \hat{W}_k(u) - \delta > \min_{k' \in \mathcal{K}} m_{\ell k'} \hat{W}_{k'}(u).$$

Thus, when n is sufficiently large,

$$m_{\ell k} W_k^{x_n}(u) - |x_n| \delta > \min_{k' \in \mathcal{K}} m_{\ell k'} W_{k'}^{x_n}(u)$$
$$\text{for all } u \in \big[|x_n|(t - \epsilon), |x_n|(t + \epsilon)\big].$$

By the workload-weighted routing algorithm (11.8), for sufficiently large n, server k will not be allocated any category ℓ tasks during the time interval $\big[|x_n|(t - \epsilon), |x_n|(t + \epsilon)\big]$, and therefore

$$E_{\ell k}^{x_n}(u_2) - E_{\ell k}^{x_n}(u_1) = 0$$
$$\text{for all } u_1, u_2 \in \big[|x_n|(t - \epsilon), |x_n|(t + \epsilon)\big] \text{ with } u_1 \le u_2,$$

which implies that

$$\hat{E}_{\ell k}^{x_n}(u_2) - \hat{E}_{\ell k}^{x_n}(u_1) = 0 \quad \text{for all } u_1, u_2 \in \big[(t - \epsilon), (t + \epsilon)\big] \text{ with } u_1 \le u_2.$$

Taking $n \to \infty$, one has

$$\hat{E}_{\ell k}(u_2) - \hat{E}_{\ell k}(u_1) = 0 \quad \text{for all } u_1, u_2 \in \big[(t - \epsilon), (t + \epsilon)\big] \text{ with } u_1 \le u_2,$$

which implies $\dot{\hat{E}}_{\ell k}(t) = 0$, proving (11.22). $\qquad \square$

11.8 Maximal Stability of WWTA (Quadratic Lyapunov Function)

The main result of this section, Theorem 11.6, concerns stability of the task allocation model under the routing policy described earlier in Section 11.6 and there called WWTA. As with all other stability results in this book, we rely on a fluid-based argument to prove that theorem, but in the current context we cannot directly cite Theorem 6.2, because it was developed under the assumption that (a) external arrivals occur according to independent Poisson processes, and (b) there are no routing decisions that require immediate commitment. Neither of those assumptions hold for our task allocation model, so the following extension of Theorem 6.2 is needed.

Theorem 11.5. *Consider a task allocation model satisfying all the assumptions enunciated earlier, and operating under a simply structured routing policy. If its associated fluid limit is stable, then the task allocation model is also stable (that is, the ambient Markov chain X described in Section 11.4 is positive recurrent).*

Proof. We need only modify in minor ways the proof of Theorem 6.2. Assuming the WWTA routing policy is used, Theorem 11.4 establishes that the desired fluid limit is well defined, and furthermore, that each fluid limit path $(\hat{D}, \hat{E}, \hat{T}, \hat{W}, \hat{Z})$ satisfies the fluid model equations (11.11) through (11.16). Using exactly the same proof, one sees that the fluid limit is well defined under *any* simply structured routing policy, and furthermore, each fluid limit path satisfies equations (11.11) through (11.15), plus a policy-dependent equation analogous to (11.16).

It remains to check that Lemma 6.10 continues to hold in the current setting. Clearly, for each time $t \geq 0$, the total number of tasks in the system at time t is upper bounded by the initial total number and the cumulative number of arrivals by time t. Thus, for each $x \in \mathcal{X}$ with $|x| \neq 0$, one has

$$(11.23) \qquad \frac{1}{|x|}|Z^x(|x|h)| \leq 1 + \sum_{\ell \in \mathcal{L}} \frac{1}{|x|} U_\ell(|x|h).$$

The lemma now follows from (11.23) and part (b) of Proposition E.7 (uniform integrability of a MArP). $\qquad \square$

Theorem 11.6. *Suppose there exists an I-vector $\lambda = (\lambda_{\ell k}) \geq 0$ satisfying (11.4) and (11.5). Then the WWTA fluid model, defined by (11.11) through (11.16), is stable. Thus, by Theorem 11.5, the task allocation model itself is stable under the WWTA routing policy.*

Proof. Let (D, E, W, Z) be a fluid model solution to (11.11) through (11.16), and define the quadratic Lyapunov function

$$f(t) := \frac{1}{2} \sum_{k \in \mathcal{K}} \left(W_k(t) \right)^2.$$

Then

$$
\begin{aligned}
\dot{f}(t) &= \sum_{k \in \mathscr{K}} W_k(t) \dot{W}_k(t) = \sum_{k \in \mathscr{K}} W_k(t) \sum_{\ell \in \mathscr{L}} m_{\ell k} \dot{Z}_{\ell k}(t) \\
&= \sum_{k \in \mathscr{K}} W_k(t) \sum_{\ell \in \mathscr{L}} m_{\ell k} \big(\dot{E}_{\ell k}(t) - \dot{D}_{\ell k}(t) \big) \\
&= \sum_{k \in \mathscr{K}} W_k(t) \sum_{\ell \in \mathscr{L}} m_{\ell k} \dot{E}_{\ell k}(t) - \sum_{k \in \mathscr{K}} W_k(t) \sum_{\ell \in \mathscr{L}} m_{\ell k} \dot{D}_{\ell k}(t) \\
&= \sum_{k \in \mathscr{K}} W_k(t) \sum_{\ell \in \mathscr{L}} m_{\ell k} \dot{E}_{\ell k}(t) - \sum_{k \in \mathscr{K}} W_k(t) \\
&= \sum_{\ell \in \mathscr{L}} \sum_{k \in \mathscr{K}} W_k(t) m_{\ell k} \dot{E}_{\ell k}(t) - \sum_{k \in \mathscr{K}} W_k(t) \\
&= \sum_{\ell \in \mathscr{L}} v_\ell \min_{k \in \mathscr{K}} W_k(t) m_{\ell k} - \sum_{k \in \mathscr{K}} W_k(t),
\end{aligned}
$$

where the second equality follows from (11.14), the third equality follows from (11.11), the fifth equality follows from (11.15), and the last equality follows from (11.16). Let $(\lambda_{\ell k})$ satisfy (11.4) and (11.5). Now,

$$
\begin{aligned}
&\sum_{\ell \in \mathscr{L}} v_\ell \min_{k \in \mathscr{K}} W_k(t) m_{\ell k} - \sum_{k \in \mathscr{K}} W_k(t) \\
&= \sum_{\ell \in \mathscr{L}} \Big(\sum_{k \in \mathscr{K}} \lambda_{\ell k} \Big) \Big(\min_{k' \in \mathscr{K}} W_{k'}(t) m_{\ell k'} \Big) - \sum_{k \in \mathscr{K}} W_k(t) \\
&= \sum_{\ell \in \mathscr{L}} \sum_{k \in \mathscr{K}} \Big(\lambda_{\ell k} \big(\min_{k' \in \mathscr{K}} W_{k'}(t) m_{\ell k'} \big) \Big) - \sum_{k \in \mathscr{K}} W_k(t) \\
&\le \sum_{\ell \in \mathscr{L}} \sum_{k \in \mathscr{K}} \Big(\lambda_{\ell k} \big(W_k(t) m_{\ell k} \big) \Big) - \sum_{k \in \mathscr{K}} W_k(t) \\
&= \sum_{k \in \mathscr{K}} W_k(t) \sum_{\ell \in \mathscr{L}} \lambda_{\ell k} m_{\ell k} - \sum_{k \in \mathscr{K}} W_k(t) \\
&\le -\epsilon \sum_{k \in \mathscr{K}} W_k(t) \le -\epsilon \sqrt{f(t)},
\end{aligned}
$$

where

$$
\epsilon = \min_k \Big(1 - \sum_{\ell \in \mathscr{L}} \lambda_{\ell k} m_{\ell k} \Big) > 0.
$$

It then follows from Lemma 8.6 that $f(t) = 0$ for $t \ge \sqrt{f(0)}/\epsilon$, proving that the fluid model is stable. $\square$

Corollary 11.7. *The WWTA routing policy is maximally stable in the following sense: if there exists any simply structured routing policy under which the task allocation model is stable, then it is stable under WWTA.*

Proof. This is an immediate consequence of Corollary 5.5 (stability is only achievable in the subcritical case), plus Lemma 11.2 and Theorem 11.6, which together show that WWTA is stable in the subcritical case. □

11.9 Sources and Literature

Dean and Ghemawat (2008) is the standard reference on the Map-Reduce framework for data-intensive applications (Section 11.1); it is written for a computer engineering audience.

The map-only model of task allocation described in Section 11.2, with its attendant notion of three-level data locality, is due to Xie et al. (2016), as is the WWTA routing policy (Section 11.6). Their paper extends earlier work by Xie and Lu (2015), who propounded a two-level model of data locality.

Xie et al. (2016) proved what we call maximal stability of the WWTA routing policy, using the equivalent term "throughput optimality," and further showed that WWTA, coupled with a priority policy for sequencing the tasks assigned to any given server, achieves what is called "heavy traffic delay optimality." Xie et al. (2016) go on to argue convincingly, through both theoretical analysis and numerical work, the superiority of WWTA-with-priority-sequencing over a policy called JSQ-MaxWeight, which was raised to prominence by the work of Wang et al. (2014).

The quadratic Lyapunov function used here to prove maximal stability of the WWTA routing policy (Section 11.8) is identical to one used by Xie et al. (2016), but in their case the quadratic Lyapunov function was used in a direct analysis of the discrete stochastic model, whereas we use it to prove stability of the corresponding fluid model.

12

Multihop Packet Networks

We consider in this chapter a stylized model of a digital communication network, in which inputs from the external world are treated as an uncontrollable source of uncertainty. That aspect of the model is inconsistent with standard network protocols like TCP, which may suppress or delay external inputs as a means of congestion control, but it is widely viewed as an acceptable approximation over moderate time spans. In other regards, the model we present is quite general.

An important theme in this chapter is decentralized control of large networks, by which we mean that decision making in any given part of the network is based solely on "local" information. Even a communication network that is interior to a single organization or a single data center may encompass many communication links and many layers of switching, and decentralization is essential for scalability of such systems.

To begin, let us consider the input-queued switch model discussed earlier in Section 1.4. A 2 × 2 switch was pictured in Figure 1.5, which for ease of reference is reproduced here as Figure 12.1. Like that example, our general model of a communication network operates in discrete *timeslots*, and its units of flow are data *packets* of uniform size. For lack of a better term, members of our slotted-time model family will be referred to generically as *packet networks*.

As in the continuous-time models of Chapter 2, the processing activities in our packet network models are "services" of various types. In the current context, all service types can be envisioned as data transfers over communication links that play the role of processing resources. Thus we use the verbs "to serve" and "to transfer" interchangeably throughout this chapter. As examples, a service may consist of transferring a packet from an input port of a switch to an output port, or of transferring a packet across a wired or wireless communication link. By assumption, all services require one timeslot in our model.

268

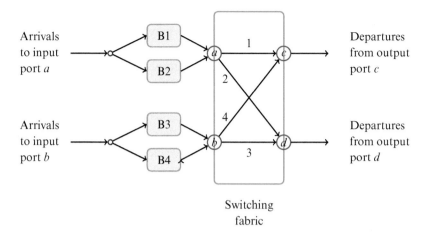

Figure 12.1 A 2×2 data switch (identical to Figure 1.5).

Elements of the general model are described, illustrated, and discussed in Sections 12.1 and 12.2. Later sections focus on two control policies that have attractive stability properties: *back-pressure control*, which is maximally stable in a very general setting but has extensive information requirements; and the *proportional scheduler*, which has modest information requirements and is maximally stable in networks where packet routes are fixed.

We begin in Section 12.1 with an abstract formulation that makes no explicit mention of processing resources. Rather, it takes as primitive a complete list of feasible actions by the system manager in any given timeslot; those actions are called *schedules*, and the list of feasible schedules is denoted $\mathscr{S}$. Compiling such a list would be impractical in a system of realistic size, but the abstract formulation minimizes notation and focuses attention on essential logic. In Section 12.2, we consider a more detailed model structure in which processing resources (that is, communication links) are explicitly identified, and explain how $\mathscr{S}$ is derived in that setting from more basic system data.

12.1 General Slotted-Time Model

Timeslots are indexed by $\tau = 1, 2, \ldots$ and packet classes by $i = 1, \ldots, I$. Broadly speaking, the class designation of a packet encodes all information about the packet that is relevant for future processing decisions, but that can mean different things depending on context, as later examples will show. Throughout this chapter, readers may imagine a packet's class designation as specifying (at least) its current location and eventual destination, both of which are nodes of a graph representing a physical network. However, no such graph will be formally introduced. Rather, we shall speak in terms of services that move a packet from one class to another, with the understanding that each such movement implies a physical transfer from one location to another.

As with the continuous-time models described in Chapter 2, our formulation in terms of packet classes and service types has the advantage of generality and the disadvantage of abstraction. It provides a unifying framework for problems with and without discretionary routing (this is illustrated by the multihop wired network example discussed in Section 12.2), and our class-centered notational system allows a compact expression of both packet network equations and the corresponding fluid models. As in previous chapters, we imagine each packet class as being stored in its own dedicated buffer; the words "buffer" and "class" are used essentially as synonyms. The notion of packet class, as well as other related modeling concepts, will be illustrated with examples in Sections 12.2 and 12.6.

By assumption, the numbers of packets of classes $1, \ldots, I$ that arrive from the outside world in successive timeslots (as in earlier chapters, we call these *external arrivals*) form mutually independent i.i.d. sequences with means $\lambda_1, \ldots, \lambda_I$ respectively. We denote by $E_i(\tau)$ the cumulative number of class i packets that arrive from such external sources through the end of timeslot τ, and impose the following mild assumption throughout:

$$(12.1) \qquad\qquad \mathbb{P}\{E(1) = 0\} > 0.$$

Service types and processing plans.   As in Chapter 2, we denote by $\mathscr{J} = \{1, \cdots, J\}$ the set of processing activities, or equivalently, service types, in our general model. As a matter of definition, a "service" of any given type $j \in \mathscr{J}$ transfers a single packet of a particular class, called the *input class* for activity j and denoted $u(j)$, either removing it from the network or else transferring it deterministically to a different class, called the *output class* for activity j and denoted $d(j)$; in the case where type j services remove packets from the network, we set $d(j) = 0$. In the obvious way, removal of a packet is interpreted as a transfer to its ultimate destination.

In our current notational system, the letters u and d are mnemonic for "upstream" and "downstream," respectively. That is, activity j is envisioned as transferring packets from an associated upstream buffer $u(j)$ to an associated downstream buffer $d(j)$, or transferring them out of the network if $d(j) = 0$.

Input–output relationships for the various activities are encoded by an $I \times J$ matrix R defined as follows: for $i \in \mathscr{I} = \{1, \ldots, I\}$ and $j \in \mathscr{J}$, set $R_{ij} = 1$ if $i = u(j)$, set $R_{ij} = -1$ if $i = d(j)$, and set $R_{ij} = 0$ otherwise. That is, the jth column of R contains a single 1 in the row corresponding to that activity's input class, and the remaining components of column j are either all zero if packets served by activity j leave the network after completing service, or else they include a single -1 in the row corresponding to that activity's output class, and the rest zeros. Earlier, in Chapter 5, we defined an input–output matrix R for our general SPN model via (5.3), and the current definition of R is consistent with that.

To avoid trivial complications, we shall assume that activities available to the system manager are restricted to ensure an acyclic flow of packets. To state this assumption in precise mathematical terms, we define a *cycle* as a sequence of activities $j_1, \ldots, j_n \in \mathscr{J}$ such that

$$d(j_1) = u(j_2), \ldots, d(j_{n-1}) = u(j_n) \quad \text{and} \quad d(j_n) = u(j_1).$$

Assumption 12.1. The data of the packet network are such that (a) each packet class $i \in \mathscr{I}$ is the input class for *some* activity, and (b) no cycles exist.

Definition 12.2. A *processing plan* for a packet class $i \in \mathscr{I}$ is a sequence of activities $j_1, \ldots, j_n \in \mathscr{J}$ such that

$$(12.2) \quad u(j_1) = i, \; d(j_1) = u(j_2), \; \ldots, \; d(j_{n-1}) = u(j_n) \text{ and } d(j_n) = 0.$$

The integer n is said to be the *length* of this processing plan. Under Assumption 12.1, there exists at least one processing plan for each packet class, and the length of every processing plan is at most I.

For future reference, let us denote by B the $I \times J$ matrix that has ones in the same components as R and all other components zero. This matrix B is a precise analog of the material requirements matrix B that was defined in Chapter 2.

Data switch example. To illustrate the notions of packet class and service type, let us first consider the 2×2 data switch pictured in

Figure 12.1. One can say that the switch consists of the four "links" that are labeled 1 through 4 in the figure. By assumption, each packet that arrives from the outside world requires a single transfer over one of those four links before exiting, so external arrivals need only be separated into $I = 4$ classes, depending on which of the system's four links they will traverse in their one processing operation. Also, we identify $J = 4$ service types: type j services are those that transfer packets of class j over link j ($j = 1, \ldots, 4$). The model's routing matrix R is thus the 4×4 identity matrix, and B is identical to R.

Single-hop and multihop routes.   A packet network is said to be *single-hop* if each column of R contains a single 1 and the rest zeros. In words, this means that every packet class leaves the system after receiving just one service. A packet network is described as *multihop* if it is not single-hop. We use the term "route" to mean the sequence of classes visited by a packet during its sojourn in the network, starting with the class into which it arrives. The "stages" of a packet's route are the successive classes it visits. If each row of R contains a single 1 (that is, if there is just one activity available for processing each packet class), then we say that the network model is one with *fixed routing*; otherwise, it has *discretionary routing*.

The term "routing" is consistently used, both in this book and in the broader technical literature, to mean decision making about *which* links will be used to transfer packets. In contrast, "scheduling" is used sometimes with a narrow meaning and sometimes with a broader one. The narrow usage refers to the *order* in which packets will be processed, for example, through a switch or over a wireless link. The broader use of "schedule" and "scheduling" encompasses *all* decision making about packet processing, including both routing decisions and order-of-service decisions. It is the second, broader meaning that predominates throughout this chapter.

Schedules.   In each timeslot, the system manager chooses a *schedule* $s = (s_1, \ldots, s_J) \in \mathbb{Z}_+^J$ from a given set $\mathscr{S}$ of feasible schedules. We interpret s_j as the number of type j services (or the number of type j transfers) executed in a timeslot. If schedule s is employed for a particular timeslot, then $(Bs)_i$ packets are transferred from buffer i during that timeslot ($i = 1, \ldots, I$), each of which either leaves the network or occupies another buffer at the end of the timeslot. Thus, denoting by $z = (z_1, \ldots, z_I)$ the vector of class-level packet counts at the beginning of a timeslot, the system manager's choice of a schedule s for that timeslot must satisfy the packet availability constraint

$$(12.3) \qquad\qquad Bs \leq z.$$

We assume that $\mathscr{S}$ is *monotone* in the following sense:

$$(12.4) \qquad \text{If } s \in \mathscr{S},\ \tilde{s} \in \mathbb{Z}_+^J \text{ and } \tilde{s} \leq s, \text{ then } \tilde{s} \in \mathscr{S}.$$

We denote by $\langle\mathscr{S}\rangle$ the convex hull of $\mathscr{S}$, and by $\hat{s}$ a generic element of $\langle\mathscr{S}\rangle$. An element of $\langle\mathscr{S}\rangle$ can be interpreted as a randomized choice of a processing schedule, or as a long-run average of the schedules employed under a given control policy. That is, components of a vector $\hat{s} \in \langle\mathscr{S}\rangle$ can be viewed as long-run average activity rates under some control policy, and we use the term "activity mix" to indicate that latter interpretation.

Main system equation. Let us denote by $s(\tau)$ the schedule chosen in timeslot τ, remembering that components of this J-vector represent *actual* numbers of packets transferred during the timeslot, as opposed to available or potential transfers, and let $Z(\tau) = (Z_1(\tau),\ldots,Z_I(\tau))$ be the vector of class-level packet counts at the conclusion of timeslot τ (or equivalently, at the beginning of timeslot $\tau + 1$). Defining

$$(12.5) \qquad D(\tau) = s(1) + \cdots + s(\tau) \quad \text{for} \quad \tau = 1, 2, \ldots,$$

we then have the following fundamental system equation:

$$(12.6) \qquad Z(\tau) = Z(0) + E(\tau) - RD(\tau) \quad \text{for} \quad \tau = 1, 2, \ldots,$$

Admissible policies and Markov chains. From (12.6), we have

$$(12.7) \qquad Z(\tau) = Z(\tau - 1) + [E(\tau) - E(\tau - 1)] - Rs(\tau)$$

for $\tau = 1, 2, \ldots$. A control policy is needed to dictate the schedule $s(\tau)$ in timeslot τ. In this chapter, we restrict attention to *Markovian* policies, which means that $s(\tau)$ is a function, either deterministic or randomized, of $Z(\tau - 1)$ only. To be precise, a Markovian policy is defined by the relationship

$$(12.8) \qquad\qquad s(\tau) = f(Z(\tau - 1), U(\tau)),$$

where $f : \mathbb{Z}_+^I \times [0, 1] \to \mathscr{S}$ is some function and $\{U(\tau) : \tau = 1, 2, \ldots\}$ is a sequence of i.i.d. random variables uniformly distributed on $(0, 1)$; by allowing the chosen schedule $s(\tau)$ to depend on $U(\tau)$, we are including randomized policies. Under such a Markovian control policy f, the class-level packet count process $Z = \{Z(\tau) : \tau \in \mathbb{Z}_+\}$ is a discrete-time Markov chain (DTMC).

Definition 12.3. A Markovian control policy f is said to be *admissible* if $Bf(z,u) \leq z$ for all $z \in \mathbb{Z}_+^I$ and all $u \in (0,1)$. An admissible Markovian policy f is said to be *stable* if the associated Markov chain Z is irreducible and positive recurrent.

12.2 Additional Structure of Links and Link Configurations

Rather than viewing the set $\mathscr{S}$ as a primitive model element, it is natural in most applications to take as given the following more basic information. First, there are *links* indexed by $k \in \mathscr{K} = \{1, \ldots, K\}$ and a $K \times J$ *link usage matrix* A that has a single one in each column and all the rest zeros. We interpret $A_{kj} = 1$ to mean that activity j transfers packets over link k. Thus A plays the same role as does the capacity consumption matrix defined in Chapter 2. (By assuming that each column of A contains a single one in the current model, we rule out activities that transfer packets over two or more links in a single timeslot. This contrasts with the bandwidth sharing model in Section 4.5, where an activity may use multiple links simultaneously.) Second, a *link configuration*, or just *configuration* for short, is a vector

$$c = (c_1, \ldots, c_K) \in \mathbb{Z}_+^K,$$

and we take as primitive a set $\mathscr{C}$ of *feasible configurations*. The system manager must choose a configuration $c \in \mathscr{C}$ at the beginning of each time slot, and one interprets component c_k as the maximum number of packets that can be transferred over link k during that timeslot; the examples that follow show that $\mathscr{C}$ is *not* necessarily a singleton. That is, we do not necessarily have separate and independent single-timeslot capacities for the various links. To avoid trivialities, we assume hereafter that each link $k \in \mathscr{K}$ has positive capacity under some feasible configuration. That is stated mathematically as follows.

Assumption 12.4. For each $k \in \mathscr{K}$, there exists $c \in \mathscr{C}$ such that $c_k > 0$.

Having chosen a configuration c, the system manager is then restricted to schedules s that satisfy the capacity constraint $As \leq c$, where A is the link usage matrix defined in the previous paragraph.

With this additional structure, the schedule set $\mathscr{S}$ (a collection of J-vectors, with one component for each activity or service type) is defined in terms of A and the configuration set $\mathscr{C}$ (a collection of K-vectors, with one component for each communication link) as follows. First, let

$$(12.9) \qquad \mathscr{S}_c := \left\{ s \in \mathbb{Z}_+^J : As \leq c \right\} \quad \text{for } c \in \mathscr{C}.$$

Then

$$(12.10) \qquad \mathscr{S} := \bigcup_{c \in \mathscr{C}} \mathscr{S}_c = \left\{ s \in \mathbb{Z}_+^J : As \leq c \text{ for some } c \in \mathscr{C} \right\}.$$

Remark 12.5. In some parts of this chapter, such as the analysis of back-pressure control in Section 12.5, we make no use of the special structure (12.10), but at other points it is assumed without comment. Mathematically, that assumption is not actually restrictive, because any set $\mathscr{S} \subset \mathbb{Z}_+^J$ that satisfies (12.4) can be written in the form (12.10), as follows: one can set $K = J$, let A be the $J \times J$ identity matrix, and let $\mathscr{C}$ the the set of undominated $s \in \mathscr{S}$ (meaning that there exists no $\tilde{s} \in \mathscr{S}$ such that $\tilde{s} \geq s$ and $\tilde{s} \neq s$). With these definitions, (12.10) always holds.

Link configurations for the data switch example. For the 2×2 data switch pictured in Figure 12.1, type j services transfer packets over link j $(j = 1,\ldots,4)$, so A is the 4×4 identify matrix, and we identify the following configuration set:

$$(12.11) \qquad \mathscr{C} = \{(1,0,1,0),(0,1,0,1)\}.$$

That is, we identify just two "feasible configurations" in formulating the switch model, each of which makes available two packet transfers. Those "feasible configurations" are identical to the two "feasible matchings" that were identified in Section 1.4. The set $\mathscr{S}$ of feasible schedules is then defined via (12.10).

Wireless network example with interference. Consider the network depicted in Figure 12.2, where blue circles represent physical locations, or endpoints, that are connected by wireless links. The links are numbered $1,\ldots,5$, as shown in the diagram, and for the moment we shall say nothing about external arrivals and the routes they follow, instead focusing exclusively on the feasible link configurations

In each timeslot, each link is capable of transmitting 1 packet, except that link 3 is capable of transmitting two packets, as indicated by the

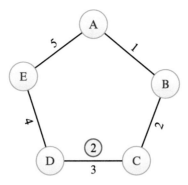

Figure 12.2 Five-link wireless network.

red circle in the figure. In addition, interference considerations dictate that no two neighboring links can transmit in the same time slot. Given the five-sided network topology shown in the figure, the interference constraint can be equivalently stated as follows: only two links can transfer packets in any given timeslot, and those two active links cannot be neighbors. Combining link capacities and interference constraints, we conclude that there are five feasible link configurations, as follows:

(12.12)
$$\mathscr{C} = \{(1,0,2,0,0),(1,0,0,1,0),(0,1,0,1,0),$$
$$(0,1,0,0,1),(0,0,2,0,1)\}.$$

The set $\mathscr{S}$ of feasible schedules is defined from $\mathscr{C}$ via (12.10), but that relationship involves the link usage matrix A, and A cannot be specified until packet classes and service types are spelled out. See Section 12.4 for a continuation of this example.

Multihop wired network example. Consider the wired network pictured in Figure 12.3. Its unidirectional links are numbered $1,\dots,5$, as shown in the diagram, and here again blue circles represent physical locations, or endpoints. Let us assume that each link can transfer at most one packet per timeslot. There are no interference considerations, so there is just one feasible link configuration, namely,

(12.13)
$$c = (1,1,1,1,1).$$

This describes the potential supply of packet transfers. With regard to demand, there are three distinct flows to be accommodated, which we designate as α, β, and γ. Packets following flow α, hereafter referred to

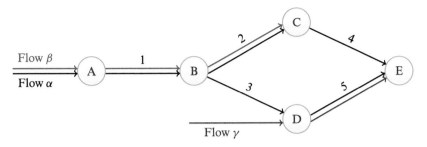

Figure 12.3 Five-link wired network.

as *α-packets*, arrive at location A and exit at location E; they can follow
either of two paths, namely, ABCE or ABDE. That is, α-packets can be
transferred from A to E using either links 1, 2, and 4 in that order, or
else using links 1, 3, and 5 in that order. On the other hand, β-packets
all follow path ABC (that is, they arrive at A and are transferred over
links 1 and 2 to exit at C), and γ-packets all follow path DE (that is,
they arrive at D and are transferred over link 5 to exit at E).

We assume that packets following any given path can be stored
temporarily at their point of arrival, and also at any intermediate
location on that path. To fit this example into our model framework,
we can define seven packet classes and eight processing activities as
in Table 12.1. This specification identifies two distinct ways to serve
a class 3 packet (that is, an α-packet stored at location B): it can be
transferred next over link 2 (this is activity 3, or service type 3), or it
can be transferred next over link 3 (this is activity 5).

Note that the last two packet classes identified in Table 12.1 have the
same location and the same future processing requirements. Specifically,
classes 6 and 7 are both stored at location D, and both require a next
transfer over link 5, after which they will exit. However, these two
packet classes come from different flows, and we might want to treat
them differently in scheduling transfers (for example, giving priority to
γ-packets over α-packets), in which case it is appropriate to recognize
their different origins with two distinct class designations. But if the
system manager is content to treat α-packets and γ-packets in undifferentiated fashion (for example, using FIFO processing at location D,
without regard to flow), then we can merge classes 6 and 7, and merge
activities 7 and 8 as well, giving a model formulation with just six packet
classes and seven service types.

Table 12.1 *Packet classes and processing activities.*

Class	Flow	Location	Activity	Input class	Output class	Link used
1	α	A	1	1	3	1
2	β	A	2	2	4	1
3	α	B	3	3	5	2
4	β	B	4	4	None	2
5	α	C	5	3	6	3
6	α	D	6	5	None	4
7	γ	D	7	6	None	5
			8	7	None	5

Finally, to illustrate the fixed-routing scenario analyzed in Section 12.6, suppose that each α-packet is randomly assigned to either the ABCE path or the ABDE path by means of a coin flip at the moment of its arrival from the outside world. To properly represent that situation, we need more packet classes and service types than shown in Table 12.1. Specifically, class 1 must be replaced by two classes, each consisting of α-packets stored at location A, but one assigned to path ABCE and the other assigned to path ABDE. Also, class 3 must be replaced by two classes, one to follow path BCE in the remainder of its processing and the other to follow path BDE. In similar fashion, activity 1 must be replaced by two distinct activities, one processing packets assigned to path ABCE and the other processing those assigned to ABDE. In the end, we arrive at a formulation with nine packet classes and 10 processing activities, or service types.

Characterization of the convex hull $\langle \mathscr{S} \rangle$. We denote by $\langle \mathscr{C} \rangle$ the convex hull of $\mathscr{C}$, and by $\hat{c}$ a generic element of $\langle \mathscr{C} \rangle$. As with schedules, an element of $\langle \mathscr{C} \rangle$ can be interpreted either as a randomized choice of a link configuration or as a long-run average of the configurations employed under a given policy, and we use the term "configuration mix" to indicate the latter interpretation. In the following proposition, we denote by $\langle \mathscr{S}_c \rangle$ the convex hull of $\mathscr{S}_c$.

Proposition 12.6. *For each $c \in \mathscr{C}$, one has $\langle \mathscr{S}_c \rangle = \left\{ x \in \mathbb{R}_+^J : Ax \leq c \right\}$.*

Proof. Fix a configuration $c \in \mathscr{C}$. From (12.9), it is obvious that every $\hat{s} \in \langle \mathscr{S}_c \rangle$ belongs to $\mathbb{R}_+^J$ and satisfies $A\hat{s} \leq c$. For the converse, fix a vector $x \in \mathbb{R}_+^J$ such that $Ax \leq c$. To prove that $x \in \langle \mathscr{S}_c \rangle$, we construct a probability distribution $\{\pi(s), s \in \mathscr{S}_c\}$ such that

(12.14)
$$x = \sum_{s \in \mathscr{S}_c} s\pi(s).$$

For each link $k \in \mathscr{K}$, we denote by $\mathscr{J}(k)$ the set of activities $j \in \mathscr{J}$ that use link k, that is, the set of $j \in \mathscr{J}$ such that $A_{kj} = 1$. A schedule $s \in \mathscr{S}_c$ can be partitioned into segments corresponding to activities that use different links, and we shall denote those segments in the following, somewhat abusive fashion: $s^k = \{s_j, j \in \mathscr{J}(k)\}$. The distribution $\pi(\cdot)$ to be constructed has the form

(12.15)
$$\pi(s) = \prod_{k \in \mathscr{K}} \pi_k(s^k).$$

Viewing $\pi(\cdot)$ as a mechanism for randomly selecting a schedule, (12.15) says that our mechanism is one that randomly *and independently* selects the segments corresponding to different links.

To specify the segment distributions $\pi_k(\cdot)$, let us first define, for each $k \in \mathscr{K}$,

(12.16) $p_j := x_j / c_k$ for $j \in \mathscr{J}(k)$ and $q_k := 1 - \displaystyle\sum_{j \in \mathscr{J}(k)} p_j.$

It follows from the inequality $Ax \leq c$ that $0 \leq q_k \leq 1$. Now let

(12.17) $\pi_k(s^k) := \begin{cases} p_j & \text{if } s^k = s^{kj} \text{ for some } j \in \mathscr{J}(k), \\ q_k & \text{if } s^k = 0, \\ 0 & \text{otherwise,} \end{cases}$

where, for each $j \in \mathscr{J}(k)$, s^{kj} has c_k as its jth component and all other components are equal to zero. That is, our random mechanism schedules services of at most one type using each link k: with probability p_j, it is type $j \in \mathscr{J}(k)$, and with probability q_k, no transfers over link k are scheduled; also, whatever service type j may be selected for link k, the number of packets transferred is c_k. Readers can easily verify that the distribution $\pi(\cdot)$ thus constructed does indeed satisfy (12.14), which completes the proof. □

Corollary 12.7. $\langle \mathscr{S} \rangle = \{x \in \mathbb{R}_+^J : Ax \leq \hat{c} \text{ for some } \hat{c} \in \langle \mathscr{C} \rangle\}.$

Proof. First suppose that $\hat{s} \in \langle \mathcal{S} \rangle$. Then we have $\hat{s} = \sum_{s \in \mathcal{S}} s\pi_s$ for some probability distribution $\{\pi_s, s \in \mathcal{S}\}$. By definition, each $s \in \mathcal{S}$ satisfies $As \leq c(s)$ for some $c(s) \in \mathcal{C}$. Multiplying each side of that inequality by π_s and summing over $s \in \mathcal{S}$ gives

$$(12.18) \qquad A\hat{s} \leq \sum_{s \in \mathcal{S}} c(s)\pi_s.$$

The right side of (12.18) is an element of $\langle \mathcal{C} \rangle$, so $\hat{s}$ belongs to the set specified after the equal sign in the statement of the corollary.

To prove the converse, fix a vector $x \in \mathbb{R}_+^J$ such that $Ax \leq \hat{c}$ for some $\hat{c} \in \langle \mathcal{C} \rangle$. By definition,

$$(12.19) \qquad \hat{c} = \sum_{c \in \mathcal{C}} cp_c$$

for some probability distribution $\{p_c, c \in \mathcal{C}\}$. In the next paragraph, we shall construct vectors $x(c) \in \mathbb{R}_+^J$, $c \in \mathcal{C}$, such that

$$(12.20) \qquad Ax(c) \leq c \text{ for each } c \in \mathcal{C}$$

and

$$(12.21) \qquad \sum_{c \in \mathcal{C}} x(c)p_c = x.$$

Combining Lemma 12.6 with (12.20), we conclude that $x(c) \in \langle \mathcal{S}_c \rangle$ for each $c \in \mathcal{C}$. Thus (12.21) says that x equals a convex combination of vectors lying in $\cup_{c \in \mathcal{C}} \langle \mathcal{S}_c \rangle$. It follows easily from the definitions of $\langle \mathcal{S}_c \rangle$ and $\langle \mathcal{S} \rangle$ that any such convex combination is an element of $\langle \mathcal{S} \rangle$, which completes the proof of the corollary.

To construct vectors $x(c) \in \mathbb{R}_+^J$ satisfying (12.20) and (12.21), let us assume initially that $\hat{c}_k > 0$ for all $k \in \mathcal{K}$. With that restriction, our proposed solution is

$$(12.22) \qquad x_j(c) = \left(\frac{x_j}{\hat{c}_k} \right) c_k \quad \text{for all } k \in \mathcal{K}, j \in \mathcal{J}(k), \text{ and } c \in \mathcal{C}.$$

The first factor on the right side of (12.22) does not depend on c. Thus, multiplying (12.22) by p_c, summing over $c \in \mathcal{C}$, and invoking (12.19), we conclude that (12.21) holds for the proposed solution. To verify (12.20), first recall that $A_{kj} = 1$ for $j \in \mathcal{J}(k)$ and $A_{kj} = 0$ otherwise. Also, let us define $y := Ax$ and $y(c) := Ax(c)$ for $c \in \mathcal{C}$. Using the special structure of A, we have from (12.22) that

$$(12.23) \qquad y_k(c) = \left(\frac{y_k}{\hat{c}_k}\right) c_k \quad \text{for all } k \in \mathcal{K} \text{ and } c \in \mathcal{C}.$$

From the assumed relationship $Ax \leq \hat{c}$, the first factor on the right side of (12.23) is ≤ 1 for all $k \in \mathcal{K}$. Thus $y_k(c) \leq c_k$ for all $k \in \mathcal{K}$ and $c \in \mathcal{C}$, which is equivalent to (12.20). The adjustment required for the case where $\hat{c}_k = 0$ for some $k \in \mathcal{K}$ is essentially trivial: for all such k, one simply sets $x_j(c) = 0$ for all $j \in \mathcal{J}(k)$. □

Subcritical region and maximal stability. In the context of our continuous-time SPN model, the subcritical region Λ was defined via (5.16), or equivalently via (5.17). In the current slotted-time model, a suitable analog is the following:

$$(12.24) \qquad \begin{aligned} \Lambda := \{\lambda \in \mathbb{R}_+^I : R\hat{s} = \lambda \text{ for some } \hat{s} \in \langle \mathcal{S} \rangle \\ \text{and } \hat{c} \in \langle \mathcal{C} \rangle \text{ satisfying } A\hat{s} < \hat{c}\}. \end{aligned}$$

That is, an arrival rate vector λ is said to be subcritical if there exist an activity mix $\hat{s}$ and a configuration mix $\hat{c}$ that satisfy two conditions. The first, $R\hat{s} = \lambda$, says that for each buffer $i \in \mathcal{I}$, the specified activity mix generates a net output rate equal to the external arrival rate into that buffer. The second condition, $A\hat{s} < \hat{c}$, says that the average link capacities required by the activity mix are strictly less than the average link capacities provided by the configuration mix.

As in our continuous-time SPN models, a packet network control policy is called *maximally stable* if it is stable for all $\lambda \in \Lambda$. The following analog of Theorem 5.2 provides justification for that terminology, showing that stability is only achievable in the current slotted-time model if $\lambda \in \Lambda$. Its proof is almost identical to, but simpler than, that of Theorem 5.2.

Theorem 12.8. *If there is a Markovian control policy under which $Z = \{Z_n : n \in \mathbb{Z}_+\}$ is an irreducible, positive recurrent Markov chain, then the arrival rate vector λ lies in the subcritical region Λ defined via (12.24).*

To conclude this section, we provide two alternative characterizations of the subcritical region Λ. First, it is immediate from Corollary 12.7 that (12.24) is equivalent to the following:

$$(12.25) \qquad \begin{aligned} \Lambda = \{\lambda \in \mathbb{R}_+^I : Rx = \lambda \text{ for some } x \in \mathbb{R}_+^J \\ \text{and } \hat{c} \in \langle \mathcal{C} \rangle \text{ satisfying } Ax < \hat{c}\}. \end{aligned}$$

Finally, one can state a *necessary* condition for subcriticality directly in terms of the convex hull $\langle \mathscr{S} \rangle$, without making reference to the link configurations that underlie $\mathscr{S}$, as follows.

Proposition 12.9. *If a vector* $\lambda \in \mathbb{R}_+^I$ *belongs to* Λ, *then*

$$(12.26) \qquad \lambda < R\hat{s} \quad \text{for some } \hat{s} \in \langle \mathscr{S} \rangle.$$

Proof. Let $x \in \mathbb{R}_+^J$ and $\hat{c} \in \langle \mathscr{C} \rangle$ be as in (12.25). For each $i \in \mathscr{I}$, we denote by e^i the I-vector whose ith component is 1 and other components are 0. Given a packet class i, let $j_1, \dots, j_n$ be a corresponding processing plan, as in Definition 12.2. Let s^i be the J-vector whose jth component is 1 if $j \in \{j_1, \dots, j_n\}$ and is 0 otherwise. One can easily verify that $Rs^i = e^i$. That is, the net effect of performing one service of each type $j_1, \dots, j_n$ sequentially is to remove one packet of class i from the system. Define

$$\tilde{s} = s^1 + \cdots + s^I \in \mathbb{Z}_+^J.$$

Then $R\tilde{s} = e$, where e is the I-vector of ones, so we have $R(x + \epsilon\tilde{s}) > \lambda$ and $A(x + \epsilon\tilde{s}) < \hat{c}$ for sufficiently small ϵ. From the latter inequality and Corollary 12.7, it follows that $(x + \epsilon\tilde{s}) \in \langle \mathscr{S} \rangle$, and thus (12.26) holds with $\hat{s} = x + \epsilon\tilde{s}$. $\qquad \square$

12.3 Fluid-Based Criterion for Positive Recurrence

Throughout this section, we fix an admissible Markovian control policy f as in (12.8). Paralleling the development in Chapter 6 for continuous-time SPNs, we shall define the fluid limit corresponding to policy f (see Definition 12.14) and define fluid limit stability (Definition 12.15). Our major result is then the following theorem, the proof of which mimics that of Theorem 6.2 and is thus omitted.

Theorem 12.10. *Assume that the DTMC Z under policy f is irreducible and the corresponding fluid limit is stable. Then the DTMC Z is positive recurrent.*

To prove fluid limit stability, one uses the following, now-familiar approach: a *fluid model* under policy f is defined, consisting of all solutions for a certain system of equations, and it is shown that all fluid limit paths are fluid model solutions; then the fluid model is shown to be stable, which implies fluid limit stability.

To define fluid limits and fluid models, first recall that $s(\tau) \in \mathscr{S}$ is the schedule used in timeslot τ, and $s(\tau)$ is determined using the policy function f via (12.8). For each $s \in \mathscr{S}$, let $T_s(\tau)$ be the cumulative number of times that schedule s has been used through the end of timeslot τ, that is,

$$T_s(\tau) = \sum_{\ell=1}^{\tau} 1_{\{s(\ell)=s\}} \text{ for each } \tau \in \mathbb{Z}_+.$$

One can verify that

(12.27) $$D(\tau) = \sum_{s \in \mathscr{S}} s T_s(\tau) \text{ for each } \tau \in \mathbb{Z}_+.$$

Clearly, $T_s(\tau)$ depends on the control policy employed. Observant readers will see that results developed in the rest of this section are actually valid under *any* control policy (even anticipating policies can be allowed), but for concreteness we continue to focus on the Markovian policy f that is fixed throughout. For each $\tau \in \mathbb{Z}_+$, let $T(\tau)$ denote the vector

$$(T_s(\tau), s \in \mathscr{S}).$$

Recall that the external arrival process $E = \{E(\tau) : \tau \in \mathbb{Z}_+\}$ is assumed to be an I-dimensional random walk defined on some probability space $(\Omega, \mathbb{P})$. Because $T(\tau)$ depends on the arrival process, it depends on the sample path $\omega \in \Omega$, and we write $T(\tau, \omega)$ rather than $T(\tau)$ when that fact requires emphasis. The following lemma is obvious.

Lemma 12.11. *For each $s \in \mathscr{S}$ and each $\omega \in \Omega$, $T_s(\cdot, \omega)$ is Lipschitz continuous in the following sense:*

(12.28) $$T_s(\tau_2, \omega) - T_s(\tau_1, \omega) \leq \tau_2 - \tau_1$$

for any $\tau_1, \tau_2 \in \mathbb{Z}_+$ with $\tau_1 \leq \tau_2$.

Convention. The stochastic processes D, E, T, and Z have a discrete time parameter. In particular, each of their sample paths is a function $g : \mathbb{Z}_+ \to \mathbb{R}^d$ for some integer $d > 0$. In the rest of this chapter, we adopt the following convention: each sample path is viewed as a right-continuous, piecewise constant function $g : \mathbb{R}_+ \to \mathbb{R}^d$ with $g(t) = g(\lfloor t \rfloor)$ for $t \geq 0$, where $\lfloor t \rfloor$ is the largest integer less than or equal to t.

Recall that the processes D, T, and Z all depend on the initial packet count vector $z \in \mathbb{Z}_+^I$, but E does not. In the following, we use D^z,

T^z, and Z^z to denote these processes in a network with initial packet count z. For each $z \in \mathbb{Z}_+^I$ with with $|z| > 0$, and for each sample path $\omega \in \Omega$, let

$$\hat{E}^z(t,\omega) := \frac{1}{|z|} E(|z|t,\omega), \quad \hat{D}^z(t,\omega) := \frac{1}{|z|} D^z(|z|t,\omega), \quad t \geq 0,$$

$$\hat{T}^z(t,\omega) := \frac{1}{|z|} T^z(|z|t,\omega), \quad \hat{Z}^z(t,\omega) := \frac{1}{|z|} Z^z(|z|t,\omega), \quad t \geq 0.$$

The following lemma is the familiar functional SLLN for the external arrival process E.

Lemma 12.12. *With probability one, for each $M > 0$,*

$$(12.29) \qquad \lim_{|z| \to \infty} \sup_{0 \leq t \leq M} |\hat{E}^z(t,\omega) - \lambda t| = 0.$$

For future reference, let Ω_1 denote the set of $\omega \in \Omega$ on which (12.29) holds. Also, a set $G \subset \mathbb{Z}_+^I$ is said to be unbounded if the set $\{|z| : z \in G\} \subset \mathbb{R}$ is unbounded.

Theorem 12.13. *Fix an $\omega \in \Omega_1$ and an unbounded set $G \subset \mathbb{Z}_+^I$. There exists a sequence $\{z^\ell\} \subset G$ and functions $\hat{D}$, $\hat{T}$, $\hat{Z}$ such that*

$$(12.30) \qquad \left(\hat{D}^{z^\ell}(\cdot,\omega), \hat{T}^{z^\ell}(\cdot,\omega), \hat{Z}^{z^\ell}(\cdot,\omega)\right) \to \left(\hat{D}(\cdot), \hat{T}(\cdot), \hat{Z}(\cdot)\right) \text{ u.o.c.}$$

as $\ell \to \infty$. Furthermore, each limit $(\hat{D}(\cdot), \hat{T}(\cdot), \hat{Z}(\cdot))$ satisfies the following:

$$(12.31) \qquad \hat{Z}(t) = \hat{Z}(0) + \lambda t - R\hat{D}(t), \quad t \geq 0,$$

$$(12.32) \qquad |Z(0)| = 1,$$

$$(12.33) \qquad \hat{Z}(t) \geq 0, \quad t \geq 0,$$

$$(12.34) \qquad \hat{D}(t) = \sum_{s \in \mathscr{S}} s\hat{T}_s(t), \quad t \geq 0,$$

$$(12.35) \qquad \sum_{s \in \mathscr{S}} \hat{T}_s(t) = t, \quad t \geq 0,$$

$$(12.36) \qquad \textit{each component of } \hat{T}(\cdot)$$
$$\textit{is nondecreasing and Lipschitz continuous.}$$

We leave the proof of Theorem 12.13 to the end of this section.

Definition 12.14. A *fluid limit path* is a triple $(\hat{D}, \hat{T}, \hat{Z})$ that occurs as a limit in (12.30) for some $\omega \in \Omega_1$, and the *fluid limit* under policy f consists of all fluid limit paths.

Definition 12.15. The fluid limit is said to be *stable* if there exists a $\delta > 0$ such that $\hat{Z}(t) = 0$ for each $t \geq \delta$ and each fluid limit path $(\hat{D}, \hat{T}, \hat{Z})$.

Fluid models. The *fluid model* under policy f, to which reference was made earlier, consists of (12.31) through (12.36), which do not actually depend on or reflect the policy under consideration, plus one or more additional equations that serve to characterize the policy. In the remainder of this chapter, two specific control policies will be considered, namely, back-pressure control (see Sections 12.4 and 12.5) and the random proportional scheduler (Sections 12.6 and 12.7); the additional fluid equations characterizing those policies are (12.44) and (12.65), respectively.

Proof of Theorem 12.13. Recall that for each $z \in \mathbb{Z}_+^I$ with $|z| > 0$, and for each sample path $\omega \in \Omega$, $\hat{T}^z(\cdot, \omega)$ is a piecewise constant function on $\mathbb{R}_+$. It has jumps at times $\ell/|z|$ for $\ell \in \mathbb{Z}_+$. We now define a continuous version of the function via linear interpolation between jump times. Namely, for each $\ell \in \mathbb{Z}_+$ and each $t \in [\ell/|z|, (\ell+1)/|z|)$, define

$$\tilde{T}^z(t,\omega) := \frac{1}{|z|}\Big(T^z(\ell,\omega)(\ell+1-|z|t) + T^z(\ell+1,\omega)(|z|t-\ell)\Big).$$

Then $\tilde{T}^z(\cdot,\omega) \in \mathbb{C}(\mathbb{R}_+, \mathbb{R}^d)$. Fix an $\omega \in \Omega$. It follows from (12.28) that the set of functions $\{\tilde{T}^z(\cdot,\omega), z \in \mathbb{Z}_+^I \setminus \{0\}\}$ is equi-Lipschitz in the sense of Corollary A.9. Because the set $G \subset \mathbb{Z}_+^I$ is unbounded, it follows from Corollary A.9 that there exists a sequence $\{z^\ell : \ell \in \mathbb{Z}_+\} \subset G$ with $\lim_{\ell\to\infty}|z^\ell| = \infty$ and a function $\hat{T}$ such that

$$(12.37) \qquad \tilde{T}^{z^\ell}(\cdot,\omega) \to \hat{T}(\cdot) \quad u.o.c. \text{ as } \ell \to \infty.$$

One can check that for each $z \in \mathbb{Z}_+^I$ with $|z| \neq 0$, and each $\omega \in \Omega$,

$$(12.38) \qquad \sup_{t\geq 0}|\tilde{T}^z(t,\omega) - \hat{T}^z(t,\omega)| \leq \frac{1}{|z|}.$$

Because $\lim_{\ell\to\infty}|z^\ell| = \infty$, (12.37) and (12.38) imply that

$$(12.39) \qquad \hat{T}^{z^\ell}(\cdot,\omega) \to \hat{T}(\cdot) \quad u.o.c. \text{ as } k \to \infty.$$

The convergence of $\hat{D}^{z^\ell}(\cdot,\omega)$ to $\hat{D}(\cdot)$, with $\hat{D}(\cdot)$ given by (12.34), follows from (12.27). The convergence of $\hat{Z}^{z^\ell}(\cdot,\omega)$ to $\hat{Z}(\cdot)$, with $\hat{Z}(\cdot)$ given by (12.31), follows from (12.6), (12.29), and the fact that $\omega \in \Omega_1$. One can easily verify that (12.32), (12.33), (12.35), and (12.36) are all satisfied by $(\hat{D},\hat{T},\hat{Z})$. □

12.4 Max-Weight and Back-Pressure Control

To formulate the system manager's dynamic control problem in mathematical terms, we have assumed that system status at the start of a timeslot is summarized by the I-vector z of class-level packet counts. (For reasons explained at the end of this section, that seemingly natural assumption is unrealistic in at least some contexts.) The manager must then choose a schedule $s \in \mathscr{S}$ as a function of z, and the chosen schedule s must satisfy the packet availability constraints (12.3).

Consider the following variant of the back-pressure control policy that was described previously (in a different model context) in Chapter 9: in each timeslot, given the observed vector z, choose $s \in \mathscr{S}$ to maximize $z \cdot Rs$, subject to the packet availability constraints (12.3). That is, defining

$$(12.40) \qquad \mathscr{S}(z) = \left\{ s \in \mathscr{S} : Bs \leq z \right\}, \quad z \in \mathbb{Z}_+^I,$$

the system manager solves the following optimization problem after observing the buffer contents vector z:

$$(12.41) \qquad \max_{s \in \mathscr{S}(z)} z \cdot Rs.$$

As in Chapter 9, the control policy that chooses schedules via (12.41) will be described as either the *max-weight* policy or the *back-pressure* policy, depending on whether the network under discussion is single-hop or multihop; the abbreviation MW/BP will often be used in statements that refer to both settings. The mechanics of MW/BP calculations will be illustrated shortly using the three examples that were introduced earlier in this chapter.

To rewrite the MW/BP optimization problem in a revealing alternative form, recall that $u(j)$ and $d(j)$ denote the input buffer and output buffer, respectively, for activity or service type $j \in \mathscr{J}$, with $d(j) = 0$ if the activity has no output buffer. Now define the "weight" associated with a type j service completion as

(12.42) $w_j(z) := z_{u(j)} - z_{d(j)}$ for all $j \in \mathcal{J}$,

where $z_0 = 0$ by convention. Given the special structure for R assumed in this chapter, we can then rewrite the optimization problem (12.41) as

(12.43) $$\max_{s \in \mathscr{S}(z)} \sum_{j \in \mathcal{J}} w_j(z)s_j.$$

Data switch example.   For the 2×2 data switch depicted in Figure 12.1, suppose that the vector z of buffer contents at the start of a timeslot is $z = (8,4,1,6)$. As noted earlier, the routing matrix R for this example is the 4×4 identity matrix, so the max-weight objective value (12.41) for a schedule s is $z_1 s_1 + \cdots + z_4 s_4$. Because all buffers are non-empty in the hypothesized scenario, the only schedules worth considering are those that fully exploit one of the two feasible configurations identified in (12.11). That is, we need only consider schedules $s = (1,0,1,0)$ and $s = (0,1,0,1)$, which have max-weight objective values $z_1 + z_3 = 8 + 1 = 9$ and $z_2 + z_4 = 4 + 6 = 10$, respectively. Because the latter value is larger, the max-weight algorithm chooses schedule $(0,1,0,1)$.

Another policy that has been studied in the literature is largest buffer first (LBF), which chooses a schedule to decrease the largest of the current buffer contents. In the scenario hypothesized in the preceding paragraph, the largest buffer content is $z_1 = 8$, so the LBF criterion chooses $s = (1,0,1,0)$ in preference to $s = (0,1,0,1)$. Thus we see that the max-weight and LBF criteria do not generally coincide. Indeed, the max-weight policy is maximally stable in our slotted-time model (see Section 12.5), but LBF is known to lack that virtue.

Five-link wireless network example.   For the wireless network pictured in Figure 12.2, suppose that arriving packets fall into just five classes, with class i requiring a single transfer over link i in a clockwise direction before exiting. Thus, for example, class 1 arrives at node A, is transferred over link 1, and departs from node B, whereas class 5 arrives at node E, is transferred over link 5, and departs from node A. The network is therefore single-hop, with five service types (or processing activities) defined in the obvious way. Its routing matrix R is the 5×5 identity matrix, its link usage matrix A is also the 5×5 identity matrix, and the max-weight objective value (12.41) for a schedule s is $z_1 s_1 + \cdots + z_5 s_5$.

Suppose the vector of current packet counts for the five classes is $z = (3,7,4,1,6)$. Thus, for example, there are seven packets that wait to be transmitted over link 2 at the beginning of this timeslot. In this scenario, each buffer contains at least as many packets as the single-timeslot capacity of the corresponding link, so the only schedules worth considering are those that fully exploit one of the five feasible link configurations identified in (12.11). The max-weight objective value for schedule $(0, 0, 2, 0, 1)$ is $4 \times 2 + 6 = 14$, and readers can verify that all other feasible schedules give a smaller objective value. Thus the max-weight policy dictates that links 3 and 5 be active during this timeslot, with each transmitting its maximum number of packets (two on the former link and one on the latter).

Multihop wired network example with discretionary routing. Consider again the five-link wired network pictured in Figure 12.3, with packet classes and processing activities (or service types) defined as in Table 12.1. Because this is a multihop network, we shall speak in terms of back-pressure control rather than max-weight control. Recall from (12.13) that a maximum of one packet can be transferred over each link in each timeslot, regardless of how the other links may be used.

Given the input–output relationships detailed in Table 12.1, links 1, 4, and 5 can each be analyzed in isolation (that is, without considering how other links are to be used) for purposes of back-pressure scheduling, regardless of what the buffer contents vector z may be. In contrast, one cannot schedule links 2 and 3 separately, because of their common reliance on buffer 3 as a source of input; it is specifically in the case $z_3 = 1$ that the scheduling of links 2 and 3 must be considered jointly. Also, only the contents of buffers 3, 4, 5, and 6 are relevant for scheduling transfers on links 2 and 3, and we shall assume the following data:

$$(z_3, z_4, z_5, z_6) = (6, 4, 3, 1).$$

The only activities or service types that use links 2 and 3 are activities 3, 4, and 5, and we calculate the following weights for those activities using formula (12.42) and the hypothesized buffer contents:

$$w_3(z) = 6 - 3 = 3, \quad w_4(z) = 4 - 0 = 4, \quad \text{and} \quad w_5(z) = 6 - 1 = 5.$$

Given these weights, we choose nonnegative integer decision variables s_3, s_4, and s_5 to maximize $w_3(z)s_3 + w_4(z)s_4 + w_5(z)s_5$, subject to link capacity constraints $s_3 + s_4 \le 1$ and $s_5 \le 1$, plus packet availability constraints $s_3 + s_5 \le 6$ and $s_4 \le 4$. As readers may easily verify, the optimal solution is $s_3 = 0, s_4 = 1$, and $s_5 = 1$. That is, the back-pressure policy dictates that one packet of class 4 be transferred over link 2 and one packet of class 3 be transferred over link 3 during the next timeslot.

Information requirements for back-pressure control. To implement back-pressure control, one must maintain packet counts by class. This means, for example, that packets stored at a given location within a network must be logically sorted into separate virtual buffers depending on their eventual destinations, and perhaps depending on other factors as well. The number of potential destinations grows explosively as a network's size increases, so the requirement to track system status at such a fine level of detail may be impractical.

Separately, to calculate the back-pressure "weights" for service types that use any given link, one needs real-time packet counts not only for classes that provide inputs to those services, but also for classes that receive outputs from them, and such "downstream" status information simply may not be available. One would like to base scheduling decisions solely on "local" information, and such a scheme will be our focus later in Section 12.6.

12.5 Maximal Stability of Back-Pressure Control

We return now to the general model described in Section 12.1. It follows from Theorem 12.8, Proposition 12.9, and Theorem 12.16 that the back-pressure control policy is maximally stable under our standing assumptions.

Theorem 12.16. *Consider a packet network satisfying (12.1) and Assumption 12.1, operating under the back-pressure control policy. The DTMC $Z = \{Z_\tau : \tau = 0, 1, \ldots\}$ is aperiodic and irreducible. Furthermore, if the stability condition (12.26) is satisfied, then Z is positive recurrent.*

To prove this theorem, we first establish a series of lemmas.

Lemma 12.17. *For any $z \in \mathbb{Z}_+^I$ with $z \neq 0$, the optimization problem (12.41) has a solution $s \neq 0$.*

Proof. It suffices to prove that there exists a schedule $s \in \mathscr{S}(z)$ such that $z \cdot Rs > 0$. For each buffer $i \in \mathscr{I}$, we define the *hop count* of the buffer to be the smallest integer $k \geq 1$ such that there exist activities j_1, $j_2, \ldots, j_k$ satisfying

$$i = u(j_1), d(j_1) = u(j_2), \ldots, d(j_{k-1}) = u(j_k), \text{ and } d(j_k) = 0.$$

Let us denote by z_r the largest component of the given vector z. That is, $z_r = \max_{i \in \mathscr{I}} z_i > 0$. When there is a tie for largest component, we choose r to be a class or buffer having the smallest hop count. By part (a) of Assumption 12.1, there exists an activity or service type j such that $r = u(j)$. First suppose that $d(j) = 0$, and consider the schedule s with $s_j = 1$ and all other components zero. Clearly, $s \in \mathscr{S}(z)$ and $z \cdot Rs = z_r > 0$, proving the lemma.

Now suppose on the contrary that $d(j) \neq 0$ for each activity j with $r = u(j)$. Among those activities, choose an activity j such that $d(j)$ has the smallest hop count. Let $i := d(j) \in \mathscr{I}$. It must be true that $z_r > z_i$, because otherwise $z_i = z_r$ and buffer i has a smaller hop count than buffer r, contradicting the choice of buffer r. Now consider the schedule s with $s_j = 1$ and all other components zero. Then $s \in \mathscr{S}(z)$ and $z \cdot Rs = z_r - z_i > 0$, proving the lemma. $\square$

Lemma 12.18. *For a packet network satisfying (12.1) and Assumption 12.1, operating under the back-pressure control policy, the DTMC Z is irreducible and aperiodic.*

Proof. Under the assumptions of the lemma, starting from any state $z \in \mathbb{Z}_+^I$, we first claim that Z can reach state 0. Indeed, Lemma 12.17 gives the following conclusion under the back-pressure policy: if the network is not already empty, then at least one service will be conducted (that is, at least one packet will be transferred) in each timeslot. Also, it follows from Remark 12.1 that no packet can remain in the network through more than I services, so the network must be empty after timeslot $\kappa = I|z|$ if there are no external arrivals in those κ timeslots. Thus (12.1) gives

$$\mathbb{P}\{Z(\kappa) = 0 | Z(0) = z\} \geq \mathbb{P}\{E(n) - E(n-1) = 0 \text{ for } n = 1, \ldots, \kappa\} > 0.$$

Define $\mathscr{X} \subset \mathbb{Z}_+^I$ to be set of states that are reachable from state 0. Then Z is irreducible with state space $\mathscr{X}$. Because

$$\mathbb{P}\{Z(1) = 0 | Z(0) = 0\} = \mathbb{P}\{E(1) = 0\} > 0,$$

the period of state 0 is 1, and thus the DTMC is aperiodic. $\square$

Remark 12.19. Without part (b) of Assumption 12.1 (absence of cycles), it is not necessarily true that state 0 can be reached from any initial state z under any version of the back-pressure policy. (The meaning of the word "version" will be explained shortly.) To see this, consider a network with two packet classes, three links, and three activities or service types: activity 1 transfers class 1 packets to class 2 using link 1; activity 2 removes class 2 packets from the network using link 2; and activity 3 transfers class 2 packets back to class 1 using link 3. Thus activities 1 and 3 constitute a cycle, violating part (b) of Assumption 12.1. We assume that each link can transfer one packet per timeslot.

Let the initial state be $z = (0, 1)$ and suppose for the moment that there are no external arrivals. Further suppose that, as a tie-breaking rule, activity 3 is preferred to activity 2 when those two activities or service types have equal weight. (Different tie-breaking rules define different versions of the max-weight policy.) With no external arrivals, the system state will then cycle endlessly from $(0, 1)$ to $(1, 0)$ and back again, never reaching state $(0, 0)$. Furthermore, if the initial state is $(0, 0)$ and all external arrivals are into class 2, readers can verify that, under the same version of the back-pressure policy, state $(0, 0)$ will never again be reached after the first arrival occurs.

We now develop and analyze the fluid model of a packet network operating under back-pressure control, with heavy reliance on our earlier analysis (see Chapter 9) of continuous-time models with back-pressure control. In the following lemma, a point $t \geq 0$ is said to be a regular point of a fluid limit path $(\hat{D}, \hat{T}, \hat{Z})$ if the path is differentiable at t.

Lemma 12.20. *For a packet network operating under the back-pressure control policy, each fluid limit path $(\hat{D}, \hat{T}, \hat{Z})$ satisfies the following fluid model equation in addition to (12.31) through (12.36):*

(12.44) $\quad \hat{Z}(t) \cdot R \dfrac{d}{dt} \hat{D}(t) = \max_{s \in \langle \mathscr{S} \rangle} \hat{Z}(t) \cdot Rs \quad$ *for each regular point $t \geq 0$.*

Proof. Let $(\hat{D}, \hat{T}, \hat{Z})$ be a fluid limit path. Fix a regular point $t \geq 0$. First note that

$$\max_{s \in \langle \mathscr{S} \rangle} \hat{Z}(t) \cdot Rs = \max_{s \in \mathscr{S}} \hat{Z}(t) \cdot Rs,$$

because a linear function achieves its maximum over a bounded polytope at an extreme point of the polytope. Thus it suffices to prove that

$$(12.45) \qquad \hat{Z}(t) \cdot R \frac{d}{dt} \hat{D}(t) = \max_{s \in \mathscr{S}} \hat{Z}(t) \cdot Rs.$$

To prove (12.45), first observe the following: because $\mathscr{S}$ is finite, there exists an $s^* \in \mathscr{S}$ such that

$$\hat{Z}(t) \cdot Rs^* = \max_{s \in \mathscr{S}} \hat{Z}(t) \cdot Rs.$$

Recall that we denote by $u(j)$ the upstream buffer served by activity $j \in \mathscr{J}$. By (12.42) and (12.43), one can then choose s^* such that the following analog of Lemma 9.10 also holds:

$$(12.46) \qquad \begin{aligned} &\text{if } \hat{Z}_i(t) = 0 \text{ for a buffer } i \in \mathscr{I}, \text{ then } s_j^* = 0 \\ &\text{for each } j \in \mathscr{J} \text{ with } i = u(j), \end{aligned}$$

As in the proof of (9.31) in Lemma 9.11, one can prove that for any $s \in \mathscr{S}$

$$\hat{Z}(t) \cdot Rs < \hat{Z}(t) \cdot Rs^* \text{ implies that } \frac{d}{dt} \hat{T}_s(t) = 0.$$

It follows that

$$(12.47) \qquad (\hat{Z}(t) \cdot Rs) \frac{d}{dt} \hat{T}_s(t) = (\hat{Z}(t) \cdot Rs^*) \frac{d}{dt} \hat{T}_s(t) \text{ for any } s \in \mathscr{S}.$$

Therefore,

$$\begin{aligned} \hat{Z}(t) \cdot R \frac{d}{dt} \hat{D}(t) &= \sum_{s \in \mathscr{S}} (\hat{Z}(t) \cdot Rs) \frac{d}{dt} \hat{T}_s(t) \\ &= \sum_{s \in \mathscr{S}} (\hat{Z}(t) \cdot Rs^*) \frac{d}{dt} \hat{T}_s(t) \\ &= (\hat{Z}(t) \cdot Rs^*) \sum_{s \in \mathscr{S}} \frac{d}{dt} \hat{T}_s(t) \\ &= \hat{Z}(t) \cdot Rs^*, \end{aligned}$$

where the first equality follows from (12.34), the second equality follows from (12.47), and the last equality follows from (12.35). This establishes (12.45), proving the lemma. $\qquad \square$

Lemma 12.21. *If (12.26) is satisfied, then the fluid model defined by (12.31) through (12.36) and (12.44) is stable.*

Proof. The proof is almost identical to that of Theorem 9.12. Let $(\hat{D}, \hat{T}, \hat{Z})$ be a fluid model solution. Define the quadratic Lyapunov function

$$g(t) := \hat{Z}(t) \cdot \hat{Z}(t).$$

It follows that

$$
\begin{aligned}
\dot{g}(t) &= 2\hat{Z}(t) \cdot \frac{d}{dt}\hat{Z}(t) \\
&= 2\hat{Z}(t) \cdot \left(\lambda - R\frac{d}{dt}\hat{D}(t)\right) \quad \text{by (12.31)} \\
&= 2\hat{Z}(t) \cdot \lambda - 2\hat{Z}(t) \cdot R\frac{d}{dt}\hat{D}(t) \\
&= 2\hat{Z}(t) \cdot \lambda - 2\max_{s \in \langle \mathscr{S} \rangle} \hat{Z}(t) \cdot Rs \quad \text{by (12.44)} \\
&\leq 2\hat{Z}(t) \cdot \lambda - 2\hat{Z}(t) \cdot R\hat{s} \\
&= -2\hat{Z}(t) \cdot (R\hat{s} - \lambda) \\
&\leq -2\delta|\hat{Z}(t)| = -2\delta\sqrt{g(t)},
\end{aligned}
$$

where $\hat{s}$ is some $s \in \langle \mathscr{S} \rangle$ that satisfies (12.26), and

$$\delta = \min_{i \in \mathscr{I}}(R\hat{s} - \lambda)_i > 0.$$

It then follows from Lemma 8.6 that $\hat{Z}(t) = 0$ for $t \geq \sqrt{g(0)}/2\delta$, proving the fluid model's stability. $\qquad\square$

12.6 Proportional Scheduling with Fixed Routes

In the remainder of this chapter, we consider the special case $J = I$. In this case, there is just one activity or service type to process packets of any given class, so packets in each arrival class visit a fixed sequence of other classes before exiting, and hence are transferred over a fixed sequence of links. Adopting the obvious convention, we assume it is activity i that processes packets of class i, so the matrix B defined in Section 12.1 is the $I \times I$ identity matrix. It will be convenient to define P as the $I \times I$ matrix that has $P_{ij} = 1$ if $R_{ji} = -1$ and $P_{ij} = 0$ otherwise.

That is, $P_{ij} = 1$ if class i packets become class j after completing the one service type available to them. Thus P is a sub-stochastic Markov transition matrix, and part (b) of Assumption 12.1 requires that it be transient; that is, $P^n \to 0$ as $n \to \infty$, which means that all packets exit after traversing a finite number of links (see Theorem F.1 and the discussion that follows it in Appendix F). For future reference, we note that our input-output matrix R has the following form in the fixed-routing case:

$$(12.48) \qquad\qquad R = I - P'.$$

Partitioning packet classes into link-specific groups.   Recall that links are indexed by $k \in \mathcal{K} = \{1,\dots,K\}$, and there is given a $K \times I$ matrix A such that $A_{ki} = 1$ if class i packets traverse link k in the one service type available to them, and $A_{ki} = 0$ otherwise. In a fixed-routing model, the matrix A partitions the potentially large collection of packet classes into a relatively small number of groups according to the link on which they next require a transfer: if $A_{ki} = 1$, we say that *class i belongs to link k*, or equivalently, that the *link designation* for class i packets is k, and we shall denote by $\mathcal{I}(k)$ the set of all such packet classes i ($k \in \mathcal{K}$). Using general language, one may say that a packet's link designation specifies the resource required for its next processing step, while its class designation carries additional information about its processing needs beyond the next step.

Subcritical region with fixed routing.   Suppose that λ lies in the subcritical region Λ, which was defined in general via (12.24). Using (12.48), let

$$\alpha := R^{-1}\lambda = (I + P + P^2 + \dots)'\lambda,$$

interpreting this as the I-vector of *total arrival rates* into the various classes, including both external arrivals and internal transfers. We now define the K-vector

$$(12.49) \qquad\qquad \rho := A\alpha.$$

Because service times are all identically equal to 1 in a packet network, one interprets ρ_k as the long-run average number of packets that link k must transmit per timeslot ($k \in \mathcal{K}$); it is precisely analogous to the load factor ρ_k that was defined for server pool k of a queueing network in

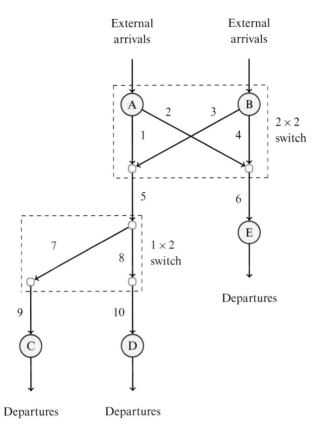

Figure 12.4 A multihop network with two levels of switching.

Section 2.6. Also, it follows immediately from (12.24) that subcriticality (that is, $\lambda \in \Lambda$) implies the following load condition:

$$(12.50) \qquad\qquad \rho < \hat{c} \quad \text{for some } \hat{c} \in \langle \mathcal{C} \rangle.$$

Example with fixed routing. Figure 12.4 pictures a multihop network in which packets arrive at two locations labeled A and B, follow fixed routes, and eventually leave from one of three locations labeled C, D, and E. In this example, we imagine the data flow on all links as unidirectional (this simplifies the discussion of routes), as indicated by the arrowheads in the diagram.

Locations A and B are the input ports of a 2×2 switch, which is represented by a dashed rectangle at the top of the figure. From those two input ports, packets may be transferred to either of the switch's two output ports, and as usual, any such transfer requires one timeslot. The four paths through the 2×2 switch are identified as links 1 through 4 in Figure 12.4, and the constraints on packet transfers over those four links are as stated earlier in (12.11). That is, there are two feasible configurations for the four-link cluster that constitutes the 2×2 switch, each of which makes two transfers available.

The dashed rectangle in the lower-left portion of Figure 12.4 represents a 1×2 switch (that is, a switch with one input port and two output ports), and the two paths through that switch are identified as links 7 and 8. At most, one packet can be transferred per timeslot over either of those two links, and they cannot be active simultaneously. Thus there are two feasible configurations for the two-link cluster that constitutes the 1×2 switch: the system manager can make one transfer available over either link 7 or link 8, but not over both of them simultaneously.

The remaining four links identified in Figure 12.4, numbered 5, 6, 9, and 10, fall into two categories: link 5 connects an output port of one switch to the input port of another, and links 6, 9, and 10 each connect an output port of one switch to a location from which packets exit. They can each transfer one packet per timeslot, and each of them can be scheduled independently from the remainder of the network. Moreover, there is just one feasible configuration for each of those four links, namely, the configuration that makes available the transfer of one packet over the link in question.

It is natural to distinguish $I = 20$ packet classes in this example, as follows. First, packets arriving from external sources fall into six distinct classes, each corresponding to one of two arrival points (A or B) and one of three possible departure points (C, D, or E). Four of those six classes (the ones departing from locations C and D) follow four-stage routes, and hence each of them gives rise to three additional classes corresponding to three stages of partial processing. For example, each new arrival in class AC will transition to a second class after traversing link 1, transition to a third class after traversing link 5, and transition to a fourth class after traversing link 7. There are also two classes of new arrivals (namely, the ones exiting from location E) that follow two-stage routes, and each of them gives rise to one additional class that has completed one of the two required transfers. Thus we arrive at a total of $4 \times 4 + 2 \times 2 = 20$ classes.

An important point for future discussion is that packet classes proliferate rapidly as network size increases. Imagine, for example, that the
system pictured in Figure 12.4 is expanded to have dozens of switches
rather than two, and that each switch is 16×16 rather than 2×2 or
1×2. Such numbers are realistic even for the communication network
within a data center of modest size, and they cause a combinatorial
explosion in the number of classes.

Link clusters. In discussion of the preceding example, the word
"cluster" has been used to mean a collection of links that can be
configured independently of other links. That is, within any given
timeslot, transfers can be scheduled over such a collection of links
without consideration of the transfers scheduled over other links.
Mathematically, this means that the set $\mathscr{C}$ of feasible link configurations
(see Section 12.1) can be written as a Cartesian product

$$(12.51) \qquad \mathscr{C} = \mathscr{C}^1 \times \cdots \times \mathscr{C}^M,$$

with one factor for each cluster. For the example pictured in Figure 12.4,
there are $M = 6$ clusters: one (the 2×2 switch) contains links 1 through
4 and has two feasible configurations; a second one (the 1×2 switch)
contains links 7 and 8 and has two feasible configurations; the other
four each contain a single link and have a single feasible configuration.
Thus there are $2 \times 2 \times 1 \times \cdots \times 1 = 4$ feasible configurations for the full
set of 10 links. They are

$$(1,0,0,1,1,1,1,0,1,1),$$
$$(1,0,0,1,1,1,0,1,1,1),$$
$$(0,1,1,0,1,1,1,0,1,1),$$
$$(0,1,1,0,1,1,0,1,1,1).$$

As we shall see shortly, the product structure allows a decomposition of
certain schedule optimization problems, which may drastically reduce
computational complexity.

Randomized configurations   Recall that $\langle \mathscr{C} \rangle$ denotes the convex hull of
$\mathscr{C}$. Thus each $\hat{c} = (\hat{c}_k) \in \langle \mathscr{C} \rangle$ can be represented in the form

$$(12.52) \qquad \hat{c}_k = \sum_{c \in \mathscr{C}} c_k \pi_c \quad \text{for } k = 1, \cdots, K,$$

where $\pi = (\pi_c, c \in \mathscr{C})$ is a probability distribution. Of course, for any distribution π on $\mathscr{C}$, one can construct a mapping $\sigma : (0,1) \to \mathscr{C}$ such that, for any uniformly distributed random variable U on $(0,1)$,

(12.53) $\mathbb{P}\{\sigma(U) = c\} = \pi_c$ for each $c \in \mathscr{C}$.

Thus, given a $\hat{c} \in \langle\mathscr{C}\rangle$ with representation (12.52), one has

(12.54) $\hat{c} = \mathbb{E}[\sigma(U)]$.

Hereafter we shall refer to $\sigma(U)$ as a *randomized configuration*, and to $\hat{c}$ as its *associated mean vector*.

Walton's random proportional scheduler.   Let $z = (z_1, \ldots, z_I)$ be the vector of class-level packet counts at the beginning of a timeslot. Then $y := Az$ is the corresponding K-vector of *link-level packet counts*. In the remainder of this chapter, we focus on a particular control policy, called the *random proportional scheduler* (RPS), that chooses the schedule for each timeslot based solely on the vector y. To be specific, implementation of the RPS involves four steps in each timeslot, as follows. (a) Choose a point

(12.55) $\hat{c} = \psi(y) \in \langle\mathscr{C}\rangle$,

where $\psi : \mathbb{Z}_+^K \to \langle\mathscr{C}\rangle$ is a function to be specified shortly. (b) Let π be a probability distribution on $\mathscr{C}$ such that (12.52) holds, and let $\sigma : (0,1) \to \mathscr{C}$ be chosen to satisfy (12.53). The link configuration for the timeslot is then taken to be

$$c = \sigma(U),$$

where U is uniformly distributed on $(0,1)$. Thus implementation of the RPS involves randomized choice of a link configuration in each timeslot, but the policy does not specify the complete randomization, only its associated mean vector. (c) The number of packets transferred over link k during the timeslot is

(12.56) $x_k = c_k \wedge y_k$ for $k = 1, \cdots, K$.

That is, the number of packets actually transferred over a given link equals either the link capacity determined by the choice of a configuration, or else the number of packets available for transfer, whichever is smaller. (d) The x_k packets transferred over link k during the times-lot are randomly selected (without replacement) from the y_k packets awaiting transfer over that link ($k = 1, \ldots, K$).

In the obvious way, the uniform random variables used to generate configurations via step (b) in successive periods must be independent of one another and of the network's external arrivals. With this stipulation, even without knowing the function ψ in (12.55), one readily sees that the resulting control policy is Markovian and admissible (see Section 12.1 for the meaning of those terms). Thus $Z = \{Z(\tau) : \tau \in \mathbb{Z}_+\}$ is a DTMC.

The RPS optimization problem.   The terms "proportional scheduler" and "proportional scheduling" originated in the work of Walton (2015), and they refer to use of the following specific function in step (a) in the preceding subsection:

$$(12.57) \qquad \psi(y) = \operatorname{argmax}\left\{\sum_{k=1}^{K} y_k \log(\hat{c}_k) : \hat{c} \in \langle \mathscr{C} \rangle \right\}.$$

For purposes of this book, we have added the modifier "random" to emphasize the use of random selection in step (d), which is a particularly simple method of "disaggregating" the link-level transfer totals determined in step (c). See Section 12.9 for a brief discussion of one alternative to random selection, namely, first-in-first-out processing.

Lemma 10.1 shows that the maximization problem (12.57) is meaningful for real vectors $y \in \mathbb{R}_+^K$, not just integer-valued y. Also, the maximum is achieved (that is, a maximizer exists) for any such y, because the domain $\langle \mathscr{C} \rangle$ is compact and the function to be maximized is continuous. This specific optimization problem is associated with the notion of "proportionally fair" capacity allocations, which was introduced in Section 10.2.

When $y_k > 0$ for *all* $k \in \mathscr{K}$, the objective function in (12.57) is strictly concave in $\hat{c}$, and as a result, $\psi(y)$ is uniquely defined; see part (a) of Lemma 10.1. On the other hand, if $y \in \mathbb{R}_+^K$ and $y_k = 0$ for some $k \in \mathscr{K}$, then $\psi_k(y)$ is *not* uniquely determined by (12.57). Nevertheless, a value for $\psi(y)$ can be selected from among the alternative optima for each $y \in \mathbb{R}_+^K$ so that the function $\psi : \mathbb{R}_+^K \to \langle \mathscr{C} \rangle$ is well defined; see the discussion preceding the statement of Lemma 10.1.

Decomposing the RPS optimization problem. The dimension of the RPS optimization problem (12.57) is nominally K, the number of links in the network under study, which may be prohibitively large. But as noted earlier, in virtually all systems of practical interest, the K links can be partitioned into M noninteracting clusters that are each relatively small, giving us the Cartesian product structure (12.51). In that case, it can be verified that

$$(12.58) \qquad \langle \mathscr{C} \rangle = \langle \mathscr{C}^1 \rangle \times \cdots \times \langle \mathscr{C}^M \rangle.$$

Let us denote by $\mathscr{K}(m)$ the set of links that constitute cluster m, and denote cluster-specific components of a vector $y \in \mathbb{R}_+^K$ in the following, somewhat abusive fashion:

$$y^m = (y_k, k \in \mathscr{K}(m)).$$

Now for $m = 1, \ldots, M$ let

$$(12.59) \qquad \psi^m(y^m) := \mathrm{argmax}\left\{ \sum_{k \in \mathscr{K}(m)} y_k \log(\hat{c}_k) : \hat{c} \in \langle \mathscr{C}^m \rangle \right\}.$$

Remark 12.22. The meaning given to the symbols m and M in the preceding subsection, and again in the paragraphs that follow, is different from the meaning given to them in previous chapters.

Given the product structure (12.58) of the convex set $\langle \mathscr{C} \rangle$, and given the separable character of the objective in (10.4), one sees that the solution $\psi(y)$ for the RPS optimization problem can be gotten by concatenating the solutions $\psi^m(y^m)$ for the cluster-specific optimization problems in (12.59). Again abusing notation slightly, we write this mathematically as

$$(12.60) \qquad \psi(y) = (\psi^1(y^1), \cdots, \psi^M(y^M)).$$

This decomposition is significant for two reasons, both of which have been alluded to earlier. First, the number of decision variables in the mth cluster-specific optimization problem (12.59) is just the number of links in cluster m. When modeling a network of input-queued switches, for example, clusters essentially correspond to individual switches (that is, the links constituting a cluster are the paths through a particular switch), so the RPS optimization problem for a network of switches reduces to numerous smaller problems, each of a size required to schedule a single switch in isolation.

Second, the data y^m that serve as input to the mth cluster-specific optimization problem are simply the numbers of packets currently awaiting transfer over links in cluster m, without any need for further information about the subsequent routing of those packets, nor additional information about packet counts elsewhere in the network.

Contrast with the back-pressure algorithm. It is natural to ask whether, given a Cartesian product structure (12.51), there is a similar decomposition of the back-pressure (BP) optimization problem (12.43). That is, can one decompose the BP optimization problem into separate, cluster-specific problems? The answer is in general negative, but only because the BP algorithm applies to a larger class of problems than does the RPS algorithm.

To be specific, back-pressure control applies to problems with discretionary routing, where there is a many-to-one relationship between processing activities and packet classes. In such a setting, there may exist, for example, a packet class that can be next transmitted over either a link in cluster A or a link in cluster B, and the presence of such discretion prohibits independent scheduling of the two clusters.

If, however, the BP algorithm is applied to a packet network with fixed routing (see the first paragraph of this section for the meaning of that phrase), then a cluster-by-cluster decomposition analogous to (12.60) occurs. As noted earlier, the data involved in BP scheduling of any given cluster involve not only packet counts for classes awaiting transfer over links in that cluster, but also packet counts for "downstream" classes.

12.7 Fluid Limits and Fluid Model under Random Proportional Scheduling

To develop the RPS fluid model, we extend in the following ways the notation used in Section 12.6. First, for $\tau = 1, 2, \ldots$, let $Z(\tau - 1)$ be the I-vector of class-level packet counts at the beginning of timeslot τ, and define $Y(\tau - 1) := AZ(\tau - 1)$. In step (a) of the RPS scheduling algorithm (see Section 12.6), given that $Y(\tau - 1) = y$, we solve the concave optimization problem (12.57) to obtain $\hat{c} = \psi(y)$.

Let $\{U_1, U_2, \ldots\}$ be a sequence of i.i.d. random variables that are uniformly distributed on $(0, 1)$ and are independent of the external arrival process E. For each fixed $y \in \mathbb{Z}_+^K$, let $\sigma(y, \cdot)$ be a map $(0, 1) \to \mathscr{C}$

such that $\mathbb{E}[\sigma(y, U_1)] = \psi(y)$. In timeslot τ, executing step (b) of the RPS algorithm, we choose link configuration

$$(12.61) \qquad c(\tau) = \sigma(Y(\tau - 1), U_\tau).$$

As in Section 12.6, we denote by $\mathscr{I}(k)$ the set of packet classes that belong to link k ($k \in \mathscr{K}$). We denote by $s(\tau)$ the schedule chosen by the RPS algorithm in period τ. (Recall that this is an I-vector of actual packet transfers by class.) Executing steps (c) and (d) of the algorithm, given $c(\tau)$ and $Z(\tau - 1)$, the schedule $s(\tau)$ is chosen as follows: for each link $k \in \mathscr{K}$, we choose $c_k(\tau) \wedge Y_k(\tau - 1)$ packets randomly without replacement from buffers $i \in \mathscr{I}(k)$, denoting by $s_i(\tau)$ the number of packets thereby selected from buffer $i \in \mathscr{I}(k)$. It is convenient to define the conditional expectation $\hat{s}_i(\tau) = \mathbb{E}[s_i(\tau)|Z(\tau - 1)]$ and the constant

$$(12.62) \qquad c_{\max} = \max_{c \in \mathscr{C}} \max_{k \in \mathscr{K}} c_k.$$

One can easily verify that

$$(12.63) \quad \hat{s}_i(\tau) 1_{\{Y_k(\tau-1) > c_{\max}\}}$$
$$= 1_{\{Y_k(\tau-1) > c_{\max}\}} \frac{Z_i(\tau - 1)}{Y_k(\tau - 1)} \psi_k(Y(\tau - 1)) \quad \text{for all } i \in \mathscr{I}(k).$$

We turn now to the task of defining fluid limits under RPS control. In Section 12.3, fluid limit paths were defined on the set Ω_1 of sample paths that satisfy the SLLN for the arrival process, where Ω_1 is defined via (12.29). When the packet network is operated under the random proportional scheduler, additional randomness is introduced due to (i) the random selection of a configuration in (12.61), and (ii) the random selection of packets for transfer in step (d) of the algorithm. Thus, under the random proportional scheduler, fluid limit paths are defined on a set of sample paths $\Omega_2 \cap \Omega_1$, where Ω_1 is as before and Ω_2 satisfies $\mathbb{P}\{\Omega_2\} = 1$. As we shall see shortly (see Theorem 12.27), stability of the fluid limit still implies positive recurrence of the DTMC Z.

The set Ω_2 is defined in Lemma 12.23. In preparation for the lemma's statement, we need to establish the following modified version of the mathematical setup specified in Section 6.3. (The modifications reflect a change from the continuous-time setting of previous chapters to the current slotted-time setting.) First, with attention fixed on a particular admissible Markovian control policy, namely, the random proportional scheduler, we construct on a single probability space $(\Omega, \mathscr{F}, \mathbb{P})$ a complete array of network processes for each initial state z of the DTMC Z.

Second, to indicate the dependence of a constructed variable or process on the initial state, we attach a superscript z to the notation developed in Section 12.1. In particular, for each initial state $z \in \mathbb{Z}_+^I$ and each integer $\tau \geq 1$, let us define

$$\xi_i^z(\tau) := \sum_{m=1}^{\tau} \left(s_i^z(m) - \hat{s}_i^z(m) \right).$$

As usual, we write $\xi_i^z(\tau, \omega)$ to explicitly indicate the dependence of this random variable on $\omega \in \Omega$.

Lemma 12.23. *Let $\{z^\ell : \ell \geq 1\} \subset \mathbb{Z}_+^I$ be a sequence of initial states satisfying $|z^\ell| \geq \ell^2$ for each $\ell \geq 1$, and define*

$$\Omega_2 := \left\{ \omega : \lim_{\ell \to \infty} \frac{\xi^{z^\ell}(|z^\ell|, \omega)}{|z^\ell|} = 0 \right\}.$$

Then

(12.64) $$\mathbb{P}(\Omega_2) = 1.$$

The proof of Lemma 12.23, like that of the following theorem, is delayed until later in this section.

Theorem 12.24. *Consider a packet network operating under the random proportional scheduler. Let $\{z^\ell : \ell \geq 1\} \subset \mathbb{Z}_+^I$ be a sequence of initial states satisfying $|z^\ell| \geq \ell^2$ for each $\ell \geq 1$, and let Ω_2 be as in Lemma 12.23. For each $\omega \in \Omega_1 \cap \Omega_2$, all fluid limits $(\hat{D}, \hat{T}, \hat{Z})$ in (12.30) satisfy the following: for each packet class $i \in \mathscr{I}$,*

(12.65) $$\hat{Z}_i(t) > 0 \text{ implies that } \frac{d}{dt}\hat{D}_i(t) = \frac{\hat{Z}_i(t)}{\hat{Y}_k(t)} \psi_k(\hat{Y}(t)),$$

where $k \in \mathscr{K}$ is the link to which class i belongs, $\hat{Y}(t) = A\hat{Z}(t)$, and for each $y \in \mathbb{R}_+^K$, $\psi(y)$ is a solution to the concave optimization problem (12.57).

Definition 12.25. The *RPS fluid model* consists of (12.31) through (12.36) and (12.65).

Proposition 12.26. *The RPS fluid model is a special case of the PF fluid model formulated in Section 10.4, specifically with the following correspondences: the I packet classes of the RPS fluid model play the role of job classes in the PF fluid model; we identify one demand group for each link, so $L = K$; the partition $\{\mathscr{I}(k), k \in \mathscr{K}\}$ is defined as in Section 12.6;*

and it is $\langle \mathcal{C} \rangle$ *that plays the role of* $\tilde{\mathcal{A}}$ *in the concave optimization problem (10.23).*

Proof. This is a straightforward verification. □

Unlike in Theorem 12.10, where the set Ω_1 does not depend on initial states, the set Ω_2 in Theorem 12.24 does depend on a sequence of initial states. Not surprisingly, though, a version of Theorem 12.10 still holds in the current setting, as follows.

Theorem 12.27. *If the RPS fluid model is stable, then the DTMC Z is positive recurrent.*

Proof. The proof is analogous to that of Theorem 6.2 in the continuous-time setting. Suppose h is a positive integer such that $\hat{Z}(h) = 0$ for each RPS fluid model solution $(\hat{D}, \hat{T}, \hat{Z})$. By Lemma 3.7, it suffices to prove that

$$(12.66) \qquad \lim_{|z| \to \infty} \frac{1}{|z|} \mathbb{E}|Z^z(|z|h)| = 0.$$

Suppose that (12.66) does not hold. Then there exists a sequence $\{z^\ell\} \subset \mathbb{Z}_+^I$ with $|z^\ell| \geq \ell^2$ such that

$$(12.67) \qquad \lim_{\ell \to \infty} \frac{1}{|z^\ell|} \mathbb{E}|Z^{z^\ell}(|z^\ell|h)| > 0.$$

Given this sequence $\{z^\ell\}$, define Ω_2 as in Lemma 12.23. It follows that, for each $\omega \in \Omega_1 \cap \Omega_2$, each fluid limit path in (12.30) is a fluid model solution. Because the fluid model is stable by hypothesis, one has

$$\lim_{\ell \to \infty} \frac{1}{|z^\ell|} |Z^{z^\ell}(|z^\ell|h, \omega)| = 0 \quad \text{for each } \omega \in \Omega_1 \cap \Omega_2.$$

Because $\mathbb{P}(\Omega_1 \cap \Omega_2) = 1$, by invoking a slotted time version of Lemma 6.10, one can argue that

$$\lim_{\ell \to \infty} \frac{1}{|z^\ell|} \mathbb{E}|Z^{z^\ell}(|z^\ell|h)| = 0,$$

contradicting (12.67). □

In the rest of this section, we prove Theorem 12.24 and Lemma 12.23, in that order. In preparation, let us define

$$(12.68) \qquad s_{\max} := \max_{i \in \mathscr{I}} \max_{s \in \mathscr{S}} s_i < \infty.$$

Proof of Theorem 12.24. Assume that $\{z^\ell : \ell \geq 1\} \subset \mathbb{Z}_+^I$ is a sequence of initial states satisfying $|z^\ell| \geq \ell^2$. Fix an $\omega \in \Omega_1 \cap \Omega_2$. Let $(\hat{D}(\cdot), \hat{T}(\cdot), \hat{Z}(\cdot))$ be a fluid limit path in (12.30). Fix a buffer $i \in \mathscr{I}$ and a $t > 0$, and let $k \in \mathscr{K}$ be the link to which class i belongs. Assume that $\epsilon := \hat{Z}_i(t) > 0$. By the continuity of $\hat{Z}_i(\cdot)$, there exists a $\delta_0 \in (0, t)$ such that

$$(12.69) \qquad \min_{u \in [t-\delta_0, t+\delta_0]} \hat{Y}_k(u) \geq \min_{u \in [t-\delta_0, t+\delta_0]} \hat{Z}_i(u) \geq \epsilon/2.$$

We would like to prove that for any $\delta \in (0, \delta_0)$,

$$(12.70) \qquad \hat{D}_i(t+\delta) - \hat{D}_i(t-\delta) = \int_{t-\delta}^{t+\delta} \frac{\hat{Z}_i(u)}{\hat{Y}_k(u)} \psi_k(\hat{Y}(u)) du,$$

from which (12.65) will follow.

Fix a $\delta \in (0, \delta_0)$. To prove (12.70), it follows from (12.5) that for each $\ell \geq 1$,

$$(12.71)$$

$$\hat{D}_i^{z^\ell}(t+\delta) - \hat{D}_i^{z^\ell}(t-\delta)$$

$$= \frac{1}{|z^\ell|} \sum_{\tau \in \left(|z^\ell|(t-\delta), |z^\ell|(t+\delta)\right]} \hat{s}_i(\tau) + \frac{1}{|z^\ell|} \sum_{\tau \in \left(|z^\ell|(t-\delta), |z^\ell|(t+\delta)\right]} \left(s_i(\tau) - \hat{s}_i(\tau)\right)$$

$$= \frac{1}{|z^\ell|} \sum_{\tau \in \left(|z^\ell|(t-\delta), |z^\ell|(t+\delta)\right]} \hat{s}_i(\tau) + \hat{\xi}_i^{z^\ell}(t+\delta) - \hat{\xi}_i^{z^\ell}(t-\delta)$$

$$= \int_{t-\delta}^{t+\delta} \hat{s}_i\left(\lfloor |z^\ell|u \rfloor\right) du + \eta^\ell + \hat{\xi}_i^{z^\ell}(t+\delta) - \hat{\xi}_i^{z^\ell}(t-\delta),$$

where

$$\hat{\xi}_i^{z^\ell}(t) = \frac{1}{|z^\ell|}\xi^{z^\ell}\big(\lfloor |z^\ell| t \rfloor\big), \quad \text{and}$$

$$\eta^\ell = \frac{1}{|z^\ell|}\Big(\lfloor (t+\delta)|z^\ell| \rfloor + 1 - (t+\delta)|z^\ell| \Big)\hat{s}_i\big(\lfloor (t+\delta)|z^\ell| \rfloor\big)$$
$$- \frac{1}{|z^\ell|}\Big(\lfloor (t-\delta)|z^\ell| \rfloor + 1 - (t-\delta)|z^\ell| \Big)\hat{s}_i\big(\lfloor (t-\delta)|z^\ell| \rfloor\big).$$

Because $|\eta^\ell| \leq s_{\max}/|z^\ell|$, where $s_{\max}$ is defined via (12.68), one has

$$(12.72) \qquad\qquad\qquad \lim_{\ell \to \infty} \eta^\ell = 0.$$

Also, because $\omega \in \Omega_2$, we have that

$$(12.73) \qquad\qquad \lim_{\ell \to \infty} \hat{\xi}^{z^\ell}(t+\delta) - \hat{\xi}^{z^\ell}(t-\delta) = 0.$$

Because (12.30) holds, it follows that

$$(12.74) \qquad \big(\hat{D}^{z^\ell}(\cdot,\omega), \hat{T}^{z^\ell}(\cdot,\omega), \hat{Z}^{z^\ell}(\cdot,\omega)\big) \to \big(\hat{D}(\cdot), \hat{T}(\cdot), \hat{Z}(\cdot)\big) \text{ u.o.c.}$$

as $\ell \to \infty$, which implies that

$$(12.75) \qquad \lim_{\ell \to \infty} \Big(\hat{D}_i^{z^\ell}(t+\delta) - \hat{D}_i^{z^\ell}(t-\delta)\Big) = \hat{D}_i(t+\delta) - \hat{D}_i(t-\delta).$$

Recall that we are striving to prove (12.70). Toward that end, it follows from (12.71) through (12.73) and (12.75) that it will suffice to prove the following:

$$(12.76) \qquad \lim_{\ell \to \infty} \int_{t-\delta}^{t+\delta} \hat{s}_i\big(\lfloor |z^\ell| u \rfloor\big)du = \int_{t-\delta}^{t+\delta} \frac{\hat{Z}_i(u)}{\hat{Y}_k(u)}\psi\big(\hat{Y}(u)\big)du.$$

The remainder of this proof is devoted to establishing (12.76). It follows from (12.74) and (12.69) that there exists an integer $\ell_0 > 0$ such that, for $\ell \geq \ell_0$,

$$\min_{u \in [t-\delta, t+\delta]} \hat{Z}_i^{z^\ell}(u) \geq \epsilon/4,$$

and

$$|z^\ell|\epsilon/4 > \max_{c \in \mathscr{C}} c_k.$$

Thus, for any $\ell \geq \ell_0$ and for any timeslot $\tau \in \left(|z^\ell|(t-\delta), |z^\ell|(t+\delta)\right]$, one has

$$c_\ell(\tau) \leq \max_{c \in \mathscr{C}} c_\ell \leq |z^\ell|\epsilon/4 \leq Z_i(\tau) \leq Y_\ell(\tau),$$

where $c_\ell(\tau)$ is defined via (12.61). It follows from (12.63) that for all τ in the indicated interval,

$$\hat{s}_i(\tau) = \frac{Z_i(\tau-1)}{Y_\ell(\tau-1)} \psi_\ell(Y(\tau-1)).$$

Now the homogeneity property (10.8) of $\psi_\ell(\cdot)$ implies that

$$\int_{t-\delta}^{t+\delta} \hat{s}_i\left(\lfloor |z^\ell|u \rfloor\right) du = \int_{t-\delta}^{t+\delta} \frac{\hat{Z}_i^{z^\ell}(u)}{\hat{Y}_k^{z^\ell}(u)} \psi_k\left(\hat{Y}^{z^\ell}(u)\right) du.$$

Also, it follows from (12.30), part (c) of Lemma 10.1, and (12.69) that, for each $u \in (t-\delta, t+\delta)$,

$$\lim_{\ell \to \infty} \frac{\hat{Z}_i^{z^\ell}(u)}{\hat{Y}_\ell^{z^\ell}(u)} \psi_\ell\left(\hat{Y}^{z^\ell}(u)\right) = \frac{\hat{Z}_i(u)}{\hat{Y}_\ell(u)} \psi_\ell\left(\hat{Y}(u)\right).$$

Thus, by the dominated convergence theorem (Corollary B.3),

$$(12.77) \qquad \lim_{\ell \to \infty} \int_{t-\delta}^{t+\delta} \frac{\hat{Z}_i^{z^\ell}(u)}{\hat{Y}_\ell^{\ell}(u)} \psi_\ell\left(\hat{Y}^{z^\ell}(u)\right) du = \int_{t-\delta}^{t+\delta} \frac{\hat{Z}_i(u)}{\hat{Y}_\ell(u)} \psi_\ell(\hat{Y}(u)) du,$$

proving (12.76). □

Proof of Lemma 12.23. It suffices to prove that, for each buffer $i \in \mathscr{I}$,

$$\mathbb{P}\left\{\omega: \lim_{\ell \to \infty} \frac{\xi_i^{z^\ell}(|z^\ell|, \omega)}{|z^\ell|} = 0\right\} = 1.$$

Fix an $i \in \mathscr{I}$. By the definition of $\hat{s}_i(\tau)$ for $\tau \geq 1$, $\{s_i(\tau) - \hat{s}_i(\tau), \tau \geq 1\}$ is a martingale-difference sequence, meaning that

$$\mathbb{E}[s_i(\tau) - \hat{s}_i(\tau)|Z(\tau-1)] = 0 \quad \text{for } \tau \geq 1.$$

Because $|s_i(\tau) - \hat{s}_i(\tau)| \leq s_{max}$, we conclude that $\{s_i(\tau) - \hat{s}_i(\tau), \tau \geq 1\}$ is a bounded martingale-difference sequence, for which the standard SLLN holds; see, for example, theorem 3 in section VII.8 of Feller (1971). Because the martingale-difference sequence $\{s_i(\tau) - \hat{s}_i(\tau), \tau \geq 1\}$

depends on an initial state z^ℓ, we have a family of martingale-difference sequences indexed by $\{z^\ell, \ell \geq 1\}$. To define fluid limit paths, we need a version of the SLLN for this family of sequences.

For arbitrary $\epsilon > 0$, one has

$$(12.78) \quad \mathbb{P}\left\{ \omega : \frac{|\xi_i^{z^\ell}(|z^\ell|, \omega)|}{|z^\ell|} > \epsilon \right\} \leq \frac{\mathbb{E}\left[\xi_i^{z^\ell}(|z^\ell|)\right]^2}{\epsilon^2 |z^\ell|^2} \leq \frac{s_{\max}^2}{\epsilon^2 |z^\ell|} \leq \frac{s_{\max}^2}{\epsilon^2 \ell^2}.$$

The first inequality in (12.78) is an application of Chebyshev's inequality, the last one follows from the condition $|z^\ell| \geq \ell^2$, and the following estimates justify the second one:

$$\mathbb{E}\left[\xi_i^{z^\ell}(|z^\ell|)\right]^2 = \sum_{\tau=1}^{|z^\ell|} \mathbb{E}[s_i(\tau) - \hat{s}_i(\tau)]^2$$

$$+ 2\sum_{\tau < \tau'}^{|z^\ell|} \mathbb{E}\left[\left(s_i(\tau) - \hat{s}_i(\tau)\right)\left(s_i(\tau') - \hat{s}_i(\tau')\right)\right]$$

$$= \sum_{\tau=1}^{|z^\ell|} \mathbb{E}[s_i(\tau) - \hat{s}_i(\tau)]^2 \leq s_{\max}^2 |z^\ell|,$$

where the second equality follows from the fact that

$$\mathbb{E}\left[\left(s_i(\tau) - \hat{s}_i(\tau)\right)\left(s_i(\tau') - \hat{s}_i(\tau')\right)\right]$$

$$= \mathbb{E}\left[\left(s_i(\tau) - \hat{s}_i(\tau)\right)\mathbb{E}\left[\left(s_i(\tau') - \hat{s}_i(\tau')\right)|Z(\tau' - 1)\right]\right] = 0$$

for $\tau < \tau'$, and the final inequality follows from $|s_i(\tau) - \hat{s}_i(\tau)| \leq s_{\max}$ for each $\tau \geq 1$. Because

$$\sum_{\ell=1}^{\infty} \frac{s_{\max}^2}{\epsilon^2 \ell^2} < \infty,$$

the lemma follows from (12.78) and the Borel–Cantelli lemma (see Section B.2). □

12.8 Maximal Stability of Random Proportional Scheduling

It follows from Theorem 12.8 and Theorem 12.28 that the RPS control policy is maximally stable when packet routes are fixed; that is, a packet network with fixed routing is stable under the RPS control policy if there exists *any* Markovian policy under which it is stable.

Theorem 12.28. *Assume that the load condition (12.50) is satisfied. Then (a) the RPS fluid model is stable, and hence (b) the DTMC Z under the RPS control policy is positive recurrent.*

Proof. It has already been established (Proposition 12.26) that the RPS fluid model is a special case of the PF fluid model formulated in Section 10.4, specifically with $\langle \mathcal{C} \rangle$ playing the role of $\mathcal{A}$. Also, our earlier definition (8.30) of the load vector ρ for the PF fluid model, when specialized to the fixed routing model now under discussion, agrees with (12.49). Thus the general stability condition (10.37), when specialized to the RPS fluid model, is equivalent to (12.50), and part (a) of the current theorem is then immediate from Theorem 10.5. Part (b) is immediate from part (a) and Theorem 12.27. □

12.9 Sources and Literature

The slotted-time model formulated in Sections 12.1 and 12.2 is a generalized version of the multihop network model propounded by Tassiulas and Ephremides (1992), incorporating both dynamic routing and order-of-service decisions. As noted earlier in Chapter 9, it was Tassiulas and Ephremides (1992) who introduced and analyzed the max-weight and back-pressure control policies, although the term "back-pressure" only appeared later in work by Tassiulas (1995). Fluid limits in slotted time were first developed in the work of Dai and Prabhakar (2000) on data switch models. Those authors obtained fluid-scaled limits while keeping the initial state *fixed*, a restrictive setup that allowed them to prove results on "rate stability," but not to prove the stronger property of positive recurrence that we obtain here in Theorem 12.27.

The fixed-routing model analyzed in Section 12.6 is essentially the same model considered by Walton (2015), although our notation and terminology differ substantially from his, and also from the notation and terminology of Tassiulas and Ephremides (1992). For example, what we call a "configuration" corresponds roughly to what Tassiulas and Ephremides (1992) call an "activation set," and in our terminology a "schedule" specifies numbers of transfers to be executed in a timeslot *per packet class*, but in Walton (2015) it specifies numbers of *potential* transfers made available *per link*. Also, Walton (2015) uses the term "queue" where we use "link." Some of Walton's terminology is adopted from earlier work on slotted-time "switched network" models by Shah and Wischik (2012).

In making decisions about model formulation and exposition, authors must balance conflicting desires for generality, simplicity, and mathematical precision. The choices we have made in this chapter, particulary those deviating from usage by previous authors, often reflect yet another concern, namely, consistency with the notation, terminology, and modeling style used in earlier parts of this book.

As noted earlier, it was Walton (2015) who introduced what we have called the random proportional scheduling (RPS) policy. He articulated the informational advantages of that policy over the back-pressure algorithm, and proved its maximal stability using an entropy Lyapunov function. The RPS policy uses random selection to "disaggregate" the link-level transmission totals that it determines via proportional fairness logic. An alternative disaggregation approach is to transmit packets queued for any given link on a FIFO basis, which is at least roughly the mechanism used in the management of Internet flows. Analysis of that "FIFO proportional scheduler" is substantially more complex than the RPS analysis, but it has been undertaken recently by Bramson et al. (2017), who prove the maximal stability of the policy in a fixed-routing packet network.

Another example of a "low-information" control policy for packet networks, apart from the two versions of Walton's proportional scheduler mentioned previously, is the queue-proportional rate allocation (QPRA) scheme propounded in recent work by Li and Srikant (2016), details of which will not be presented here.

Appendix A

Selected Topics in Real Analysis

The primary focus of this appendix is real analysis. We review a few key definitions and collect results that are important for this book. Our standard reference is Royden (1988) and its updated edition Royden and Fitzpatrick (2010).

A.1 Absolutely Continuous Functions

Absolutely continuous functions are prominent in the proof of Lemma 8.5, which plays a key role in the the analysis of fluid models. In the following definition, a and b are numbers in $\mathbb{R}$ satisfying $a < b$.

Definition A.1. A function $f : [a,b] \to \mathbb{R}$ is said to be *absolutely continuous* if for each $\epsilon > 0$ there exists a $\delta > 0$ such that

$$\sum_k |f(y_k) - f(x_k)| < \epsilon$$

for any finite sequence of pairwise disjoint intervals $(x_k, y_k) \subset [a,b]$ satisfying

$$\sum_k (y_k - x_k) < \delta.$$

The following is proposition 6.7 of Royden and Fitzpatrick (2010).

Lemma A.2. *If $f : [a,b] \to \mathbb{R}$ is Lipschitz continuous (see Definition 8.1), then f is absolutely continuous on $[a,b]$.*

The following is a generalization of the fundamental theorem of calculus. It is a deep result in real analysis, originally due to H. Lebesgue in 1902. See Fitzpatrick and Hunt (2015) for a modern proof and an account of its history. This result, in combination with Lemma A.2, is used repeatedly in Chapter 8.

Lemma A.3. *Assume that $f : [a,b] \to \mathbb{R}$ is absolutely continuous. Then (i) the derivative of f, denoted $\dot{f}(t)$, exists for almost every t in (a,b); (ii) $\dot{f}$ is Lebesgue integrable on (a,b); and (iii)*

$$\text{(A.1)} \qquad f(t) = f(a) + \int_a^t \dot{f}(u)du \quad for \ t \in [a,b].$$

Proof. Part (i) follows from corollary 5.12 in Royden (1988). Parts (ii) and (iii) follow from the proof of theorem 5.14 in Royden (1988). $\square$

A.2 Sequential Compactness

In this section, we use $\mathbb{N}$ to denote the set $\{1, 2, \ldots\}$ of natural numbers.

Definition A.4. Let (X, ρ) be a metric space with distance function ρ. A set $A \subset X$ is said to be *sequentially compact* if for every infinite sequence $\{x_n : n \in \mathbb{N}\} \subset A$ there is a subsequence $\{n_k : k \in \mathbb{N}\}$ such that

$$\lim_{k \to \infty} \rho(x_{n_k}, x_*) = 0 \quad \text{for some } x_* \in A.$$

In a metric space, a set is sequentially compact if and only if it is compact (see, for example, theorem 9.16 of Royden and Fitzpatrick, 2010), but for purposes of this book, it is convenient to focus on the former concept rather than the latter. The following is a well-known result for the special case of a Euclidean space; see, for example, theorem 9.20 of Royden and Fitzpatrick (2010).

Theorem A.5 (Bolzano–Weierstrass). *Fix an integer $d > 0$. Any bounded and closed set $A \subset \mathbb{R}^d$ is sequentially compact.*

A.3 Path Spaces, u.o.c. Convergence, and Equicontinuity

Let $d > 0$ be a fixed integer. For an interval $[a,b] \subset \mathbb{R}_+$, let $\mathbb{D}([a,b], \mathbb{R}^d)$ be the set of functions $x : [a,b] \to \mathbb{R}^d$ that are right continuous in $[a,b)$ and have left limits in $(a,b]$. Also, let $\mathbb{C}([a,b], \mathbb{R}^d)$ be the subset of continuous functions on $[a,b]$. For a function $x \in \mathbb{D}([a,b], \mathbb{R}^d)$, let

$$\|x\|_{[a,b]} := \sup_{a \le s \le b} |x(s)|.$$

Given a sequence of functions $\{x_n\} \subset \mathbb{C}([a,b], \mathbb{R}^d)$, we say that x_n converges to $x_* \in \mathbb{C}([a,b], \mathbb{R}^d)$ if $\|x_n - x_*\|_{[a,b]} \to 0$ as $n \to \infty$. In this case, x_* is said to be the limit (or less commonly, *limit point*) of the sequence. A set $A \subset \mathbb{C}([a,b], \mathbb{R}^d)$ is said to be closed if it contains all of its limit points.

We denote by $\mathbb{D}^d$ the set of functions $x : \mathbb{R}_+ \to \mathbb{R}^d$ that are right continuous on $\mathbb{R}_+$ and have left limits in $(0, \infty)$, and denote by $\mathbb{C}^d$ the subset of functions in $\mathbb{D}^d$ that are continuous. If $x \in \mathbb{D}^d$ and $0 \leq a < b$, then the restriction of x to $[a, b]$ belongs to $\mathbb{D}([a, b], \mathbb{R}^d)$. For a function $x \in \mathbb{D}^d$ and $t > 0$, we write $\|x\|_t$ as shorthand for $\|x\|_{[0,t]}$.

Definition A.6. A sequence $\{x_n\} \subset \mathbb{D}^d$ is said to converge *uniformly on compact sets*, written $x_n \to x_*$ u.o.c., if there is an $x_* \in \mathbb{D}^d$ such that, for every $t > 0$,

$$\lim_{n \to \infty} \|x_n - x_*\|_t = 0.$$

One can metrize u.o.c. convergence by defining

$$\rho(x, y) = \int_0^\infty e^{-t} \min(1, \|x - y\|_t) \, dt \quad \text{for any } x, y \in \mathbb{D}^d;$$

that is, $x_n \to x_*$ u.o.c. if and only if $\rho(x_n, x_*) \to 0$ as $n \to \infty$. While u.o.c. is a common convergence mode for the space $\mathbb{C}^d$, it is not a common one for $\mathbb{D}^d$. The popular topology on $\mathbb{D}^d$ is the Skorohod J_1-topology; see, for example, Ethier and Kurtz (1986). When x_* is in $\mathbb{C}^d$, it is well known that $x_n \to x_*$ u.o.c. if and only if $x_n \to x_*$ in the Skorohod J_1-topology. In our book, the limit x_* is always in $\mathbb{C}^d$ for some integer $d > 0$, and therefore u.o.c. convergence of functions in $\mathbb{D}^d$ is always equivalent to convergence in the Skorohod J_1-topology.

Definition A.7. Fix an integer $d > 0$ and an interval $[a, b] \subset \mathbb{R}_+$. A set of functions $A \subset \mathbb{C}([a, b], \mathbb{R}^d)$ is said to be *equicontinuous* if the following holds: for every $\epsilon > 0$ there exists a $\delta > 0$ such that

$$\sup_{x \in A} |x(t) - x(s)| < \epsilon$$

for every pair $s, t \in [a, b]$ with $|t - s| < \delta$.

When the set A contains a single function, equicontinuity is simply uniform continuity of that single function. The key in Definition A.7 is that one needs to choose a $\delta > 0$ that works for *all* functions in A. Clearly, if a set of functions A is equicontinuous, then its closure is also equicontinuous.

Theorem A.8 (Arzela–Ascoli). *Fix an integer $d > 0$ and an interval $[a,b] \subset \mathbb{R}_+$. Let $A \subset \mathbb{C}([a,b],\mathbb{R}^d)$ be a closed set of continuous functions. Then A is sequentially compact if and only if (i) the set $\{x(a), x \in A\} \subset \mathbb{R}$ is bounded, and (ii) A is equicontinuous.*

For a proof, see theorem 10.3 of Royden and Fitzpatrick (2010). The following corollary will be used in this book.

Corollary A.9. *Fix an integer $d > 0$. Let $A \subset \mathbb{C}^d$ be a set of continuous functions satisfying (a) $x(0) = 0$ for each $x \in A$, and (b) A is equi-Lipschitz in the following sense: there exists a constant $L > 0$ such that*

$$(A.2) \qquad \sup_{x \in A} |x(t) - x(s)| \le L|t - s|$$

for any $s, t \in \mathbb{R}_+$. Then for any sequence $\{x_n\} \subset A$, there exists a subsequence $\{x_{n_k}\}$ such that $x_{n_k} \to y$ u.o.c. for some $y \in \mathbb{C}^d$.

Proof. We use a standard "diagonal argument." The sequence $\{x_n\}$, restricted to $[0,1]$, is a subset of $\mathbb{C}([0,1],\mathbb{R}^d)$ that satisfies both conditions (i) and (ii) of Theorem A.8. Thus there is a subsequence $\{x_{n_{1,k}}, k \ge 1\}$ such that

$$\lim_{k \to \infty} \|x_{n_{1,k}} - y_1\|_1 = 0$$

for some $y_1 \in \mathbb{C}([0,1],\mathbb{R}^d)$. Assume there is a subsequence $\{x_{n_{m,k}}, k \ge 1\}$ such that

$$(A.3) \qquad \lim_{k \to \infty} \|x_{n_{m,k}} - y_m\|_m = 0$$

for some $y_m \in \mathbb{C}([0,m],\mathbb{R}^d)$. The sequence $\{x_{n_{m,k}}, k \ge 1\}$, restricted to $[0, m+1]$, is a subset of $\mathbb{C}([0,m+1],\mathbb{R}^d)$ that satisfies both conditions (i) and (ii) of Theorem A.8. Thus there is a subsequence $\{x_{n_{m+1,k}}, k \ge 1\} \subset \{x_{n_{m,k}}, k \ge 1\}$ such that

$$\lim_{k \to \infty} \|x_{n_{m+1,k}} - y_{m+1}\|_{m+1} = 0.$$

Because $\{x_{n_{m+1,k}}, k \ge 1\}$ is a subsequence of $\{x_{n_{m,k}}, k \ge 1\}$, we have

$$(A.4) \qquad y_{m+1}(t) = y_m(t) \quad \text{for } t \in [0,m].$$

Thus, by induction, for each integer $m \ge 1$ we have constructed a sequence $\{x_{n_{m,k}}, k \ge 1\}$ and $y_m \in \mathbb{C}([0,m],\mathbb{R}^d)$ such that (A.3) holds. Define

$$y(t) := y_m(t) \quad \text{for } t \in [0, m].$$

It follows that $y(t)$ is well defined for all $t \in \mathbb{R}_+$ and $y \in \mathbb{C}^d$. We now show that

(A.5) $\qquad\qquad x_{n_{k,k}} \to y \quad \text{u.o.c. as } k \to \infty.$

For that argument, fix an $\epsilon > 0$ and a $t > 0$. Let $m > t$ be an integer. There exists an integer $N_m \geq m$ such that

$$\|x_{n_{m,k}} - y\|_m < \epsilon$$

for $k > N_m$. Because $\{n_{k,k}, k \geq m\}$ is a subsequence of $\{n_{m,k}, k \geq m\}$, we have

$$\|x_{n_{k,k}} - y\|_t \leq \|x_{n_{k,k}} - y\|_m < \epsilon$$

for $k > N_m$. $\qquad\qquad\qquad\qquad\qquad\qquad\qquad\qquad\qquad$ $\square$

The following lemma can be found in Dai (1995a).

Lemma A.10. *Let $\{f_n\} \subset \mathbb{D}$ be a sequence of nondecreasing functions on $\mathbb{R}_+$ and $f \in \mathbb{C}$ be a continuous function on $\mathbb{R}_+$. Assume that $f_n(t) \to f(t)$ for all rational $t \geq 0$. Then $f_n \to f$ u.o.c.*

Proof. First, because f_n is nondecreasing and f is continuous, one can check that $f_n(t) \to f(t)$ for every $t \in \mathbb{R}_+$. To see this, suppose that $t \in \mathbb{R}_+$ and t is not rational. Then there exists a rational sequence $t_k > t$ with $t_k \to t$ as $k \to \infty$. Because

$$f_n(t) \leq f_n(t_k)$$

for each $n \geq 1$,

$$\limsup_{n \to \infty} f_n(t) \leq \lim_{n \to \infty} f_n(t_k) = f(t_k).$$

Taking $k \to \infty$ and using the fact that f is continuous, one has

$$\limsup_{n \to \infty} f_n(t) \leq f(t).$$

Similarly, by taking $t_k < t$ with $t_k \to t$, one can show that

$$\liminf_{n \to \infty} f_n(t) \geq f(t).$$

The preceding two displays imply that $\lim_{n \to \infty} f_n(t) = f(t)$.

Next, suppose that f_n does not converge to f uniformly on compact sets. Then there exist $\epsilon > 0$, $t > 0$ and $\{t_{n_l}\}$ such that $t_{n_l} \leq t$ and

$$(A.6) \qquad |f_{n_l}(t_{n_l}) - f(t_{n_l})| \geq \epsilon \qquad \text{for all } l.$$

Because $\{t_{n_l}\}$ is bounded, we may assume that $t_{n_l} \to t_0 \leq t$. Thus for any $\delta > 0$, t_{n_l} eventually is less than $t_0 + \delta$. Hence for l large enough,

$$\begin{aligned}
f_{n_l}(t_{n_l}) - f(t_{n_l}) &\leq f_{n_l}(t_0 + \delta) - f(t_{n_l}) \\
&= f_{n_l}(t_0 + \delta) - f(t_0 + \delta) + f(t_0 + \delta) - f(t_0) + f(t_0) - f(t_{n_l}).
\end{aligned}$$

Therefore,

$$\limsup_{l \to \infty} \left(f_{n_l}(t_{n_l}) - f(t_{n_l}) \right) \leq f(t_0 + \delta) - f(t_0).$$

Because f is continuous and δ is arbitrary, we have

$$\limsup_{l \to \infty} \left(f_{n_l}(t_{n_l}) - f(t_{n_l}) \right) \leq 0.$$

When $t_0 > 0$, one can similarly prove that

$$\liminf_{l \to \infty} \left(f_{n_l}(t_{n_l}) - f(t_{n_l}) \right) \geq 0.$$

When $t_0 = 0$,

$$\liminf_{l \to \infty} \left(f_{n_l}(t_{n_l}) - f(t_{n_l}) \right) \geq \lim_{l \to \infty} \left(f_{n_l}(0) - f(t_{n_l}) \right) = 0.$$

Thus we have

$$\lim_{l \to \infty} \left(f_{n_l}(t_{n_l}) - f(t_{n_l}) \right) = 0,$$

which contradicts (A.6). Hence the lemma is proved. $\qquad\square$

The following lemma is used in the proof of Theorem 11.4.

Lemma A.11. *Suppose that for each $n \in \mathbb{N}$ the functions $f_n, g_n \in \mathbb{D}(\mathbb{R}_+, \mathbb{R})$ are nondecreasing with $f_n(0) = g_n(0) = 0$, and that each pair (f_n, g_n) satisfies*

$$(A.7) \qquad f_n(t) - f_n(s) \leq g_n(t) - g_n(s), \quad 0 \leq s \leq t.$$

Assume that $g_n \to g$ u.o.c. and g is continuous. Then there exists a subsequence $\{n_k\}$ such that $f_{n_k} \to f$ u.o.c., where f is a continuous function.

Proof. By the standard "diagonal argument," there exists a subsequence $\{n_k\}$ such that

$$f(t) := \lim_{k \to \infty} f_{n_k}(t),$$

exists for each rational $t \geq 0$. For each irrational $t > 0$, we define

$$f(t) := \inf_{s > t, s \text{ rational}} f(s).$$

It will be shown that

(A.8) $\qquad 0 \leq f(t) - f(s) \leq g(t) - g(s) \qquad$ for any $0 \leq s < t$.

Equation (A.8) and the continuity of g imply that f is continuous, and it then follows from Lemma A.10 that $f_{n_k} \to f$ u.o.c. as $k \to \infty$.

To prove (A.8), fix $0 \leq s < t$. Pick two sequences of rationals $\{s_\ell\}$ and $\{t_\ell\}$ such that

$$t_\ell > s_\ell, \quad t_\ell \geq t, \quad s_\ell \geq s \quad \text{for each } \ell$$

and

$$\lim_{\ell \to \infty} s_\ell = s, \quad \lim_{\ell \to \infty} f(s_\ell) = f(s), \quad \lim_{\ell \to \infty} t_\ell = t, \quad \lim_{\ell \to \infty} f(t_\ell) = f(t).$$

By (A.7), for each k and each ℓ,

$$f_{n_k}(t_\ell) - f_{n_k}(s_\ell) \leq g_{n_k}(t_\ell) - g_{n_k}(s_\ell).$$

Taking the limit as $k \to \infty$,

$$f(t_\ell) - f(s_\ell) \leq g(t_\ell) - g(s_\ell).$$

Finally, taking the limit as $\ell \to \infty$ gives (A.8). $\qquad \square$

A.4 Dini Derivatives

For a function $f : \mathbb{R}_+ \to \mathbb{R}$, the upper Dini derivative and lower Dini derivative of f at t, denoted $D^+ f(t)$ and $D^- f(t)$, respectively, are defined as follows:

(A.9) $\qquad D^+ f(t) := \limsup_{h \downarrow 0} \frac{f(t+h) - f(t)}{h},$

(A.10) $\qquad D^- f(t) := \liminf_{h \downarrow 0} \frac{f(t+h) - f(t)}{h}.$

The following is used repeatedly in the proofs of Lemmas 10.8 and 10.13.

Lemma A.12. *For any functions $f, f_1, f_2 : \mathbb{R}_+ \to \mathbb{R}$,*

$$(\text{A.11}) \qquad D^-(-f(t)) = -D^+ f(t),$$

$$(\text{A.12}) \qquad D^+(f_1(t) + f_2(t)) \leq D^+ f_1(t) + D^+ f_2(t),$$

$$(\text{A.13}) \qquad D^-(f_1(t) + f_2(t)) \geq D^- f_1(t) + D^- f_2(t).$$

Furthermore, assuming $f : \mathbb{R}_+ \to \mathbb{R}$ is continuous and $g : \mathbb{R} \to \mathbb{R}$ is continuously differentiable and nondecreasing, we have the following:

$$(\text{A.14}) \qquad D^+ g(f(t)) = g'(f(t)) D^+ f(t),$$

$$(\text{A.15}) \qquad D^- g(f(t)) = g'(f(t)) D^- f(t).$$

Proof. Equation (A.11) follows from the definitions (A.9) and (A.10) and the fact that

$$\limsup_{n \to \infty} (-a_n) = -\liminf_{n \to \infty} a_n$$

for any sequence $\{a_n\} \subset \mathbb{R}$. Equation (A.12) follows from the fact that

$$\limsup_n (a_n + b_n) \leq \limsup_n a_n + \limsup_n b_n$$

for any two sequences $\{a_n\}, \{b_n\} \subset \mathbb{R}$. Equation (A.13) follows from (A.11) and (A.12). The proofs of (A.14) and (A.15) are routine and are omitted. $\qquad \square$

Appendix B

Selected Topics in Probability

B.1 Uniform Integrability, Dominated Convergence, and Vitali Convergence

In their most general form, the results presented in this section involve integration over general measure spaces; see, for example, chapter 18 of Royden and Fitzpatrick (2010). The general measure space framework encompasses both the integral versions and the summation versions of the results in question. For example, the proof of Theorem 12.24 utilizes the integral version of the dominated convergence theorem presented in Corollary B.3, whereas the proof of Theorem D.24 utilizes the summation version of the same theorem. For notational simplicity, we choose to state these results in terms of random variables on a probability space. Generalizing results from probability measures to finite measures should be straightforward.

As usual, given a random variable X on a probability space $(\Omega, \mathscr{F}, \mathbb{P})$, we define the expectation of X as the integral

$$\mathbb{E}(X) := \int_{\Omega} X(\omega) \, d\mathbb{P}.$$

Definition B.1. A sequence of random variables $\{X_n : n \in \mathbb{N}\}$ is said to be *uniformly integrable* if

(B.1)
$$\lim_{a \to \infty} \sup_{n \in \mathbb{N}} \mathbb{E}\left(|X_n| 1_{\{|X_n| > a\}}\right) = 0.$$

Because

$$\mathbb{E}\left(|X_n|^{1+\epsilon}\right) \geq a^{\epsilon} \mathbb{E}\left(|X_n| 1_{\{|X_n| > a\}}\right)$$

319

for any $a > 0$ and $\epsilon > 0$, a sufficient condition for (B.1) to hold is

(B.2) $$\sup_{n \in \mathbb{N}} \mathbb{E}\left(|X_n|^{1+\epsilon}\right) < \infty$$

for some $\epsilon > 0$. The following is theorem 4.5.4 of Chung (2001).

Theorem B.2 (Vitali convergence theorem). *Let $\{X_n : n \in \mathbb{N}\}$ be a sequence of random variables on a probability space $(\Omega, \mathcal{F}, \mathbb{P})$. Let X be a random variable on the same probability space. Assume that*

(B.3) $$\lim_{n \to \infty} \mathbb{P}\{|X_n - X| > \epsilon\} = 0$$

for each $\epsilon > 0$. Assume further that the sequence $\{X_n : n \in \mathbb{N}\}$ is uniformly integrable. Then,

(B.4) $$\lim_{n \to \infty} \mathbb{E}|X_n - X| = 0.$$

As a consequence,

(B.5) $$\lim_{n \to \infty} \mathbb{E}(X_n) = \mathbb{E}(X).$$

Corollary B.3 (Dominated convergence theorem). *Consider the same setting as in Theorem B.2, but with the uniform integrability condition replaced by the following:*

(B.6) $$|X_n(\omega)| \le Y(\omega)$$

for all $n \in \mathbb{N}$ and almost all $\omega \in \Omega$, where Y is some nonnegative random variable on the same probability space satisfying

(B.7) $$\mathbb{E}(Y) < \infty.$$

Then (B.4) and (B.5) hold.

Proof. One can verify that conditions (B.6) and (B.7) imply uniformly integrability of the sequence $\{X_n\}$. Thus, (B.4) and (B.5) hold by Theorem B.2. $\qquad\square$

When (B.3) holds, $\{X_n\}$ is said to converge to X in probability. It is known that almost sure convergence implies convergence in probability.

Lemma B.4 (Fatou's lemma). *Let $\{X_n : n \in \mathbb{N}\}$ be a sequence of negative random variables on a probability space $(\Omega, \mathscr{F}, \mathbb{P})$. The following inequality holds*

$$(B.8) \qquad \mathbb{E}\left(\liminf_{n \to \infty} X_n\right) \le \liminf_{n \to \infty} \mathbb{E}(X_n).$$

We conclude this section with the following simple proposition.

Proposition B.5. *Let $E = \{E(t), t \ge 0\}$ be a Poisson process with rate $\lambda > 0$. Fix an $h > 0$. The family of random variables*

$$\left\{\frac{1}{t} E(th), t \ge 1\right\} \qquad \text{is uniformly integrable.}$$

Proof. For any $t \ge 1$,

$$\mathbb{E}\left(\frac{1}{t} E(th)\right)^2 = \frac{1}{t^2}(\lambda th + (\lambda th)^2) = \frac{\lambda h}{t} + (\lambda h)^2 \le \lambda h + (\lambda h)^2.$$

Thus, condition (B.2) is satisfied for the family with $\epsilon = 1$. Therefore, the family of random variables is uniformly integrable. $\qquad \square$

B.2 Borel–Cantelli Lemma

Let $\{A_n : n \ge 1\}$ be a sequence of events in a probability space $(\Omega, \mathscr{F}, \mathbb{P})$. When we speak of the event $\{A_n$ occurs for infinitely many $n\}$, this should be interpreted as shorthand for

$$\cap_{k=1}^{\infty} \cup_{n=k}^{\infty} A_n.$$

The proof of the following lemma can be found on page 77 of Chung (2001).

Lemma B.6 (Borel–Cantelli). *Let $\{A_n : n \ge 1\}$ be a sequence of events in a given probability space. If*

$$(B.9) \qquad \sum_{n=1}^{\infty} \mathbb{P}(A_n) < \infty,$$

then

$$\mathbb{P}\{A_n \text{ occurs for infinitely many } n\} = 0.$$

The following is a corollary of the Borel–Cantelli lemma.

Corollary B.7. *Let $\{X_n : n \geq 1\}$ be a sequence of random variables on a probability space $(\Omega, \mathscr{F}, \mathbb{P})$. Suppose that for each $\epsilon > 0$,*

(B.10)
$$\sum_{n=1}^{\infty} \mathbb{P}\{|X_n| > \epsilon\} < \infty.$$

Then, with probability one, $X_n \to 0$ as $n \to \infty$. That is, $\mathbb{P}\{\omega \in \Omega : \lim X_n(\omega) = 0\} = 1$.

Proof. Let $A = \{\omega \in \Omega : \lim_{n\to\infty} X_n(\omega) = 0\}$. It suffices to prove $\mathbb{P}(A^c) = 0$, which follows from

$$A^c = \cup_{\ell=1}^{\infty} \cap_{k=1}^{\infty} \cup_{n=k}^{\infty} \left\{|X_n| \geq \frac{1}{\ell}\right\}$$

and

$$\mathbb{P}\left(\cap_{k=1}^{\infty} \cup_{n=k}^{\infty} \left\{|X_n| \geq \frac{1}{\ell}\right\}\right) = 0 \quad \text{for each } \ell \geq 1.$$

The latter equality is proved for each $\ell \geq 1$ by applying Lemma B.6 with

$$A_n = \left\{\omega \in \Omega : |X_n(\omega)| \geq \frac{1}{\ell}\right\},$$

observing that (B.9) holds because of (B.10) with $\epsilon = 1/\ell$. $\qquad\square$

B.3 The Maximum of Many i.i.d. Random Variables

Proposition B.8. *Let $X_1, X_2, \ldots$ be i.i.d. random variables such that $\mathbb{P}\{X_1 \geq 0\} = 1$ and $\mathbb{E}(X_1^{1+\epsilon}) < \infty$ for some $\epsilon > 0$. Then*

$$\mathbb{P}\left\{\lim_{n\to\infty} \frac{1}{n} \max_{1\leq k\leq n} X_k = 0\right\} = 1.$$

Proof. By the SLLN for i.i.d. random variables, with probability one,

$$\lim_{n\to\infty} \frac{1}{n} \sum_{k=1}^{n} X_k^{1+\epsilon} = \mathbb{E}\left(X_k^{1+\epsilon}\right),$$

which implies that, with probability one,

(B.11)
$$\lim_{n\to\infty} \frac{1}{n^{1+\epsilon}} \sum_{k=1}^{n} X_k^{1+\epsilon} = 0.$$

Because

$$\max_{1\le k\le n} X_k^{1+\epsilon} \le \sum_{k=1}^{n} X_k^{1+\epsilon},$$

we have from (B.11) that, with probability one,

(B.12)
$$\lim_{n\to\infty} \frac{1}{n^{1+\epsilon}} \max_{1\le k\le n} X_k^{1+\epsilon} = 0.$$

Because the function $y^{1/(1+\epsilon)}$ is continuous in $y \ge 0$, it follows that (B.12) is equivalent to

$$\lim_{n\to\infty} \frac{1}{n} \max_{1\le k\le n} X_k = 0,$$

which proves the lemma. $\qquad\square$

B.4 Propositions Related to Entropy like Functions

In this section, we state and prove some propositions that play a central role in Section 10.5, where an entropy Lyapunov function is used to prove stability of a certain fluid model. Unlike other topics treated in the appendices of this book, these are *not* results that appear in standard reference works. Rather, they are specialized, technical propositions that are essential for our purposes but represent a digression from the main flow of ideas in the body of the book.

We take as given a positive integer I, a vector $\lambda \in \mathbb{R}_+^I$, and an $I \times I$ non negative matrix P. We assume that P has spectral radius less than 1. Thus, by Theorem F.1 in Appendix F, $(I - P')^{-1}$ exists and has the following Neumann expansion:

(B.13)
$$(I - P')^{-1} = \left(\sum_{k=0}^{\infty} P^k\right)'.$$

(Here prime denotes transpose as usual, and the letter I is reused to denote the identity matrix.) Let $\mathscr{I} = \{1,\dots,I\}$, calling $\mathscr{I}$ the set of *classes*. We interpret λ_i as an arrival rate from the external world into class i, and P_{ij} as a transition rate from class i to class j. For each $i \in \mathscr{I}$, let

(B.14)
$$P_{i0} := 1 - \sum_{j \in \mathscr{I}} P_{ij}.$$

Then

(B.15)
$$\sum_{j \in \mathscr{I}_+} P_{ij} = 1 \text{ for each } i \in \mathscr{I},$$

where $\mathscr{I}_+ = \mathscr{I} \cup \{0\}$.

Traffic equations. Let $\alpha := (I - P')^{-1}\lambda$, interpreting α_i as a total arrival rate into class i, including both external arrivals and internal transitions. Then α uniquely solves the system of *traffic equations* (2.36), which we restate here for convenience:

(B.16)
$$\alpha = \lambda + P'\alpha.$$

Throughout this section, we define $\alpha_0 := 1$ and assume λ and P to be such that $\alpha > 0$. That is, by assumption and convention,

(B.17)
$$\alpha_i > 0 \text{ for } i \in \mathscr{I} \quad \text{and} \quad \alpha_0 = 1.$$

Let

(B.18)
$$\mathscr{I}_0 := \{i \in \mathscr{I} : \lambda_i > 0\},$$

calling the elements of this set *input classes*. In words, the following lemma says that $\mathscr{I}_0$ is not empty and each class $i \in \mathscr{I}$ is reachable from an input class $i_0 \in \mathscr{I}_0$.

Lemma B.9. *For each class $i \in \mathscr{I}$, there exist an integer $p \geq 0$ and classes $i_0, i_1, \ldots, i_p$ such that*

(B.19) $\quad i_0 \in \mathscr{I}_0, \quad P_{i_k i_{k+1}} > 0 \quad$ *for* $k = 0, 1, \ldots, p-1, \quad$ *and* $\quad i_p = i$.

Proof. From (B.13) and the definition $\alpha := (I - P')^{-1}\lambda$, we have

$$\alpha_i = \sum_{k=0}^{\infty} \sum_{j \in \mathscr{I}} \lambda_j P_{ji}^k = \sum_{k=0}^{\infty} \sum_{j \in \mathscr{I}_0} \lambda_j P_{ji}^k.$$

Because $\alpha_i > 0$ by assumption, there must then exist an $i_0 \in \mathscr{I}_0$ and $p \geq 0$ such that

$$P_{i_0 i}^p > 0,$$

which implies (B.19). $\qquad\qquad\square$

Let e be the column vector of ones. Then,

$$(\text{B.20}) \qquad \sum_{i \in \mathscr{I}} \lambda_i = e'\lambda = e'(I - P')\alpha = ((I - P)e)'\alpha = \sum_{i \in \mathscr{I}} P_{i0}\alpha_i.$$

The left-hand side of (B.20) is the total input rate to all classes from the external world, and the right-hand side is the total outflow rate from all classes to the external world. The following lemma uses (B.20).

Lemma B.10. *Set $d_0 = 1$ by convention and let $d_1, \ldots, d_I$ be arbitrary. Then*

$$(\text{B.21}) \qquad \sum_{i \in \mathscr{I}} d_i + \sum_{i \in \mathscr{I}} \lambda_i = \sum_{i \in \mathscr{I}} \alpha_i \sum_{j \in \mathscr{I}_+} P_{ij} d_j / \alpha_j + \sum_{i \in \mathscr{I}} d_i \frac{\lambda_i}{\alpha_i}.$$

Proof. Recall that $\alpha_0 = 1$. The first term on the right side of (B.21) is equal to

$$(\text{B.22}) \quad \sum_{i \in \mathscr{I}} \sum_{j \in \mathscr{I}} P_{ij} d_j \frac{\alpha_i}{\alpha_j} + \sum_{i \in \mathscr{I}} P_{i0}\alpha_i$$

$$= \sum_{j \in \mathscr{I}} \frac{d_j}{\alpha_j} \sum_{i \in \mathscr{I}} P_{ij}\alpha_i + \sum_{i \in \mathscr{I}} P_{i0}\alpha_i = \sum_{j \in \mathscr{I}} \frac{d_j}{\alpha_j}(\alpha_j - \lambda_j) + \sum_{i \in \mathscr{I}} P_{i0}\alpha_i$$

$$= \sum_{j \in \mathscr{I}} d_j - \sum_{j \in \mathscr{I}} d_j \frac{\lambda_j}{\alpha_j} + \sum_{i \in \mathscr{I}} P_{i0}\alpha_i,$$

where the second equality follows from the traffic equation (B.16). Then the lemma follows immediately from (B.20). $\qquad\square$

Relative entropy. It is convenient to introduce the notion of relative entropy, which is also called *Kullback–Leibler divergence* and *information divergence*. Let $\mathscr{X}$ be a finite set, and let (p_i) and (q_i) be two probability distributions on $\mathscr{X}$ with $q_i > 0$ for each $i \in \mathscr{X}$. We define the *relative entropy*

$$(B.23) \qquad D(p\|q) := \sum_{i \in \mathscr{X}} p_i \log \left(\frac{p_i}{q_i} \right),$$

with the convention that $p_i \log (p_i/q_i) = 0$ when $p_i = 0$. The following lemma is standard; see, for example, lemma 11.6.1 of Cover and Thomas (2006).

Lemma B.11 (Pinsker's inequality). *Let (p_i) and (q_i) be as before. Then*

$$(B.24) \qquad D(p\|q) \geq \frac{1}{2} \left(\sum_{i \in \mathscr{X}} |p_i - q_i| \right)^2.$$

The following easy generalization of Pinsker's inequality removes the requirement that (p_i) and (q_i) be probability distributions.

Lemma B.12. *Let $(p_i, i \in \mathscr{X})$ and $(q_i, i \in \mathscr{X})$ be two nonnegative functions on a finite set $\mathscr{X}$. Assume that $q_i > 0$ for all $i \in \mathscr{X}$ and*

$$(B.25) \qquad \sum_{i \in \mathscr{X}} p_i = \sum_{i \in \mathscr{X}} q_i.$$

Then the relative entropy function defined by (B.23) satisfies

$$(B.26) \qquad D(p\|q) \geq \frac{1}{2 \sum_{i \in \mathscr{X}} p_i} \left(\sum_{i \in \mathscr{X}} |p_i - q_i| \right)^2.$$

Proof. For each $j \in \mathscr{X}$, define

$$\hat{p}_j := \frac{p_j}{\sum_{i \in \mathscr{X}} p_i} \quad \text{and} \quad \hat{q}_j := \frac{q_j}{\sum_{i \in \mathscr{X}} q_i}.$$

Then $(\hat{p}_j)$ and $(\hat{q}_j)$ satisfy the hypotheses of Lemma B.11, which gives us the following:

$$D(p\|q) = \left(\sum_{i \in \mathscr{X}} p_i \right) D(\hat{p}\|\hat{q})$$

$$\geq \frac{1}{2} \left(\sum_{i \in \mathscr{X}} p_i \right) \left(\sum_{j \in \mathscr{X}} |\hat{p}_j - \hat{q}_j| \right)^2$$

$$= \frac{1}{2 \sum_{i \in \mathscr{X}} p_i} \left(\sum_{j \in \mathscr{X}} |p_j - q_j| \right)^2. \qquad \square$$

Lemma B.13. *Let λ and P be as specified at the beginning of this section. Given $d = (d_j) \in \mathbb{R}^I$, define*

(B.27) $$a := \lambda + P'd \quad \text{and} \quad d_0 = 1.$$

Assume that $d_i > 0$ for $i \in \mathscr{I}$. Then

(B.28) $$\sum_{i \in \mathscr{I}} (d_i - a_i) \log \left(d_i / \alpha_i \right) \geq \frac{1}{2} \frac{1}{\sum_{i \in \mathscr{I}} d_i + \sum_{i \in \mathscr{I}} \lambda_i}$$

$$\left(\sum_{i \in \mathscr{I}} \lambda_i \left| 1 - d_i / \alpha_i \right| + \sum_{i \in \mathscr{I}} \alpha_i \left| d_i / \alpha_i - \sum_{j \in \mathscr{I}_+} P_{ij} d_j / \alpha_j \right| \right)^2.$$

Proof. First,

(B.29) $$\sum_{i \in \mathscr{I}} a_i \log \left(d_i / \alpha_i \right) = \sum_{i \in \mathscr{I}} \left(\lambda_i + \sum_{j \in \mathscr{I}} P_{ji} d_j \right) \log \left(d_i / \alpha_i \right)$$

$$= \sum_{i \in \mathscr{I}} \lambda_i \log \left(d_i / \alpha_i \right) + \sum_{j \in \mathscr{I}} d_j \sum_{i \in \mathscr{I}} P_{ji} \log \left(d_i / \alpha_i \right)$$

$$= \sum_{i \in \mathscr{I}} \lambda_i \log \left(d_i / \alpha_i \right) + \sum_{j \in \mathscr{I}} d_j \sum_{i \in \mathscr{I}_+} P_{ji} \log \left(d_i / \alpha_i \right)$$

$$= \sum_{i \in \mathscr{I}} \lambda_i \log \left(d_i / \alpha_i \right) + \sum_{i \in \mathscr{I}} d_i \sum_{j \in \mathscr{I}_+} P_{ij} \log \left(d_j / \alpha_j \right),$$

where, in the third equality, we have used the convention that $d_0 = 1$ and $\alpha_0 = 1$, and in the last equality we have swapped the dummy variables i and j. Because the function $\log(x)$ is concave for $x \in (0, \infty)$ and (B.15) holds, we have the following for each $i \in \mathscr{I}$:

(B.30) $$\sum_{j \in \mathscr{I}_+} P_{ij} \log \left(d_j / \alpha_j \right) \leq \log \left(\sum_{j \in \mathscr{I}_+} P_{ij} d_j / \alpha_j \right).$$

Therefore, it follows from (B.29) and (B.30) that

(B.31)

$$\sum_{i \in \mathscr{I}} (d_i - a_i) \log \left(d_i / \alpha_i \right) = \sum_{i \in \mathscr{I}} d_i \log \left(d_i / \alpha_i \right) - \sum_{i \in \mathscr{I}} a_i \log \left(d_i / \alpha_i \right)$$

$$\geq \sum_{i \in \mathscr{I}} d_i \left(\log \left(d_i / \alpha_i \right) - \log \left(\sum_{j \in \mathscr{I}_+} P_{ij} d_j / \alpha_j \right) \right) - \sum_{i \in \mathscr{I}} \lambda_i \log \left(d_i / \alpha_i \right)$$

$$= \sum_{i \in \mathscr{I}} d_i \log \left(\frac{d_i}{\alpha_i \sum_{j \in \mathscr{I}_+} P_{ij} d_j / \alpha_j} \right) + \sum_{i \in \mathscr{I}} \lambda_i \log \left(\alpha_i / d_i \right)$$

$$= \sum_{i \in \mathscr{I}} d_i \log \left(\frac{d_i}{\alpha_i \sum_{j \in \mathscr{I}_+} P_{ij} d_j / \alpha_j} \right) + \sum_{i \in \mathscr{I}_0} \lambda_i \log \left(\frac{\lambda_i}{d_i \lambda_i / \alpha_i} \right),$$

where $\mathscr{I}_0$ is defined via (B.18). The right side of (B.31) can be expressed as a relative entropy $D(p \| q)$, as defined in (B.23), where $p = (p_i, i \in \mathscr{I} \cup \mathscr{I}_0)$ and $q = (q_i, i \in \mathscr{I} \cup \mathscr{I}_0)$ are defined as follows:

$$p_i := \begin{cases} d_i & \text{for } i \in \mathscr{I}, \\ \lambda_i & \text{for } i \in \mathscr{I}_0, \end{cases} \quad \text{and} \quad q_i := \begin{cases} \alpha_i \sum_{j \in \mathscr{I}_+} P_{ij} d_j / \alpha_j & \text{for } i \in \mathscr{I}, \\ d_i \lambda_i / \alpha_i & \text{for } i \in \mathscr{I}_0. \end{cases}$$

Lemma B.10 implies that condition (B.25) on p and q is satisfied. The latter condition guarantees that inequality (B.26) holds. It follows from (B.31) that

$$\sum_{i \in \mathscr{I}} (d_i - a_i) \log \left(d_i / \alpha_i \right) = D(p \| q)$$

$$\geq \frac{1}{2} \frac{1}{\sum_{i \in \mathscr{I}} d_i + \sum_{i \in \mathscr{I}_0} \lambda_i} \left(\sum_{i \in \mathscr{I}_0} |\lambda_i - d_i \lambda_i / \alpha_i| \right.$$

$$\left. + \sum_{i \in \mathscr{I}} \left| d_i - \alpha_i \sum_{j \in \mathscr{I}_+} P_{ij} d_j / \alpha_j \right| \right)^2$$

$$= \frac{1}{2} \frac{1}{\sum_{i \in \mathscr{I}} d_i + \sum_{i \in \mathscr{I}} \lambda_i} \left(\sum_{i \in \mathscr{I}} \lambda_i |1 - d_i / \alpha_i| \right.$$

$$\left. + \sum_{i \in \mathscr{I}} \alpha_i \left| d_i / \alpha_i - \sum_{j \in \mathscr{I}_+} P_{ij} d_j / \alpha_j \right| \right)^2,$$

proving the lemma. □

Lemma B.14. *Given a constant $\delta > 0$ and a vector $\alpha \in \mathbb{R}^I$ with all components strictly positive, consider the following optimization problem:*

$$(B.32) \qquad \kappa(\alpha, \delta) \equiv \min_{d \in \mathbb{R}_+^I : |d - e| \geq \delta} \sum_{i \in \mathscr{I}} \alpha_i \left| d_i - \sum_{j \in \mathscr{I}_+} P_{ij} d_j \right|,$$

where $d_0 = 1$ by convention and e is the vector of ones. The optimal objective value $\kappa(\alpha, \delta)$ is strictly positive.

Proof. Let us denote by $f(d)$ the objective function in (B.32), that is,

$$(B.33) \qquad f(d) = \sum_{i \in \mathscr{I}} \alpha_i |d_i - \sum_{j \in \mathscr{I}_+} P_{ij} d_j| \geq 0 \quad \text{for } d \in \mathbb{R}_+^I.$$

The following will be proved in the next few paragraphs: the only $d \in \mathbb{R}_+^I$ giving $f(d) = 0$ is $d = e$. Assuming that claim is valid, we now complete the proof of the lemma. First, the objective function $f(\cdot)$ in (B.32) is continuous, and $f(d) \to \infty$ as $|d| \to \infty$, from which it follows that the minimum in (B.32) is achieved at some point $d^* \in \mathbb{R}_+^I$ with $|d^* - e| \geq \delta$. Because $d^* \neq e$, the claim establishes that $\kappa(\alpha, \delta) = f(d^*) > 0$, proving the lemma.

To prove the claim, observe that $f(d) = 0$ if and only if

$$d_i = \sum_{j \in \mathscr{I}} P_{ij} d_j + P_{i0} \quad \text{for each } i \in \mathscr{I}.$$

In vector form, that is expressed as

$$(I - P)d = P_{\cdot 0},$$

where $P_{\cdot 0}$ is the column vector whose ith component is P_{i0}. Obviously, $d = e$ satisfies this equation, and it follows from (B.13) that $I - P$ is invertible, so the solution is unique. This proves the claim. $\qquad \square$

The following lemma can be proved using elementary calculus. Here we recall the convention (10.46) that $0\log(0) = 0$.

Lemma B.15. *The function*

$$(B.34) \qquad h : x \in \mathbb{R}_+ \to h(x) = x\log(x) \in \mathbb{R}$$

is right continuous at $x = 0$, is strictly negative in $(0, 1)$, and satisfies the following:

$$\inf_{x \geq 0} x\log(x) = -\frac{1}{e}.$$

Lemma B.16. *For any $a, d \in \mathbb{R}$ with $0 \le a \le \hat{a}$ and $0 < d \le \hat{d}$, the following holds:*

$$(B.35) \qquad\qquad (a - d)\log(d) \le \hat{a}\log(1 \vee \hat{d}) + \frac{1}{e}.$$

Proof.   The left side of (B.35) is equal to

$$
\begin{aligned}
(a - d)\log(d) &= a\log(d) - d\log(d) \\
&\le a\log(1 \vee \hat{d}) - d\log(d) \\
&\le a\log(1 \vee \hat{d}) + \frac{1}{e} \\
&\le \hat{a}\log(1 \vee \hat{d}) + \frac{1}{e},
\end{aligned}
$$

where the first inequality follows from the monotonicity of the function $\log(x)$ and the nonnegativity of a, the second inequality follows from Lemma B.15, and the last inequality follows from the nonnegativity of $\log(1 \vee d)$. □

Appendix C

Discrete-Time Markov Chains

In this appendix, we review discrete-time Markov chains (DTMCs). They play a critical role in describing continuous-time Markov chains (CTMCs), which will be reviewed in Appendix D. In addition, DTMCs will be used in Chapter 12 to directly model slotted-time communication systems. Our standard reference is chapter 1 of Norris (1998), which contains proofs for most of the results cited in this chapter. The only exceptions are Sections C.7 and C.8, in which sufficient conditions are established for positive recurrence of a DTMC. Norris (1998) does not cover this topic, but it is central to our book. For that reason, proofs are provided in those two sections.

C.1 Definition of a DTMC

Let $\mathscr{S}$ be a discrete space, either finite or countable. A generic element of $\mathscr{S}$ is denoted by i or j. A sequence of random variables $\{X_k, k \in \mathbb{Z}_+\}$, defined on some probability space $(\Omega, \mathbb{P})$, is called an $\mathscr{S}$-valued *stochastic process* with index set $\mathbb{Z}_+$ if $X_k(\omega) \in \mathscr{S}$ for each $k \in \mathbb{Z}_+$ and each sample path $\omega \in \Omega$. When the context is clear, we denote by X the complete stochastic process $\{X_k, k \in \mathbb{Z}_+\}$. An $\mathscr{S} \times \mathscr{S}$ matrix $P = (P_{ij})$ is said to be a *stochastic matrix* if $P_{ij} \geq 0$ and $\sum_{\ell \in \mathscr{S}} P_{i\ell} = 1$ for each $i, j \in \mathscr{S}$.

Definition C.1. An $\mathscr{S}$-valued stochastic process $X = \{X_k : k \in \mathbb{Z}_+\}$ is said to be a *DTMC* with *state space* $\mathscr{S}$, *transition matrix* $P = (P_{ij})$, and *initial distribution* $\mu = (\mu_i)$ if

(C.1) μ is a distribution on $\mathscr{S}$,

(C.2) $P = (P_{ij})$ is an $|\mathscr{S}| \times |\mathscr{S}|$ stochastic matrix, and

(C.3) $\mathbb{P}\{X_0 = i_0, X_1 = i_1, \dots, X_{k-1} = i_{k-1}, X_k = i_k\} = \mu_{i_0} P_{i_0 i_1} \cdots P_{i_{k-1} i_k}$

for each $k \in \mathbb{Z}_+$ and each $i_\ell \in \mathscr{S}$, $\ell = 0, \dots, k$.

Because $k \in \mathbb{Z}_+$ and $i_\ell \in \mathscr{S}$ are arbitrary, (C.3) fully specifies the distribution of the stochastic process X. Setting $k = 0$ in (C.3), one has $\mathbb{P}\{X_0 = i\} = \mu_i$ for $i \in \mathscr{S}$, which justifies calling μ the initial distribution of X. We denote by $\mathbb{P}_\mu$ the distribution of X corresponding to initial distribution μ. For a state $i \in \mathscr{S}$, we use $\mathbb{P}_i$ as shorthand for $\mathbb{P}_\mu$ when μ concentrates all its mass on state i. Thus one has the identity $\mathbb{P}_i\{X_0 = i\} = 1$. From (C.3), using the definition of conditional probability, one can verify the following.

Proposition C.2. *A DTMC X with transition matrix (P_{ij}) satisfies*

(C.4)
$$\begin{aligned} \mathbb{P}\{X_{k+1} = j \mid X_0 = i_0, \ldots, X_{k-1} = i_{k-1}, X_k = i\} \\ = \mathbb{P}\{X_{k+1} = j \mid X_k = i\} \end{aligned}$$

and

(C.5)
$$\mathbb{P}\{X_{k+1} = j \mid X_k = i\} = P_{ij}$$

for each $k \in \mathbb{Z}_+$ and each $i_0, \ldots, i_{k-1}, i, j \in \mathscr{S}$.

Equation (C.4) expresses the *Markov property* of X: knowing the "current" state X_k, the past history $X_0, \cdots, X_{k-1}$ is irrelevant for predicting the future evolution of X. It is understood that (C.4) holds for each $k \in \mathbb{Z}_+$ and each $i_0, \ldots, i_{k-1}, i, j \in \mathscr{S}$ such that

$$\mathbb{P}\{X_0 = i_0, \ldots, X_{k-1} = i_{k-1}, X_k = i\} > 0.$$

Because $i_0, \ldots, i_{k-1} \in \mathscr{S}$ in (C.4) are arbitrary, sometimes (C.4) is written in the following alternative form:

(C.6) $\quad \mathbb{P}\{X_{k+1} = j \mid X_0, \ldots, X_{k-1}, X_k = i\} = \mathbb{P}\{X_{k+1} = j \mid X_k = i\}.$

Because of (C.5), (P_{ij}) is called the *one-step* transition probability matrix.

C.2 Strong Markov Property

Let X be a DTMC with a countable state space $\mathscr{S}$. We assume that X lives on some probability space $(\Omega, \mathbb{P})$. A function $\tau : \Omega \to \mathbb{Z}_+ \cup \{\infty\}$ is said to be a stopping time with respect to X if, for each integer $k \geq 0$, the set $\{\omega \in \Omega : \tau(\omega) \leq k\}$ can be written as the union of sets having the form

$$\{\omega \in \Omega : X_0 = i_0, X_1 = i_1, \ldots, X_k = i_k\},$$

where $i_0, i_1, \ldots, i_k \in \mathscr{S}$. That is, the occurrence or nonoccurrence of the event $\{\tau \leq k\}$ is determined by the random variables $X_0, X_1, \ldots, X_k$.

Therefore, to check whether a stopping time τ has occurred by time k, one does not need any "future" information about $X_{k+1}, X_{k+2}, \ldots$.

Theorem C.3. *A DTMC X with state space $\mathscr{S}$ satisfies the following strong Markov property: for each stopping time τ, each $k \geq 0$, and each $i, j \in \mathscr{S}$,*

$$\mathbb{P}\{X_{\tau+k} = j | X_0, X_1, \ldots, X_{\tau-1}, X_\tau = i, \tau < \infty\}$$
$$= \mathbb{P}\{X_{\tau+k} = j | X_\tau = i, \tau < \infty\}.$$

C.3 Communicating Classes and Irreducibility

Let X be a DTMC with state space $\mathscr{S}$ and transition matrix P.

Definition C.4. (a) A state $i \in \mathscr{S}$ is said to *reach* state $j \neq i$ if there exists an integer $k > 0$ such that $P_{ij}^k > 0$. (b) States $i \neq j$ are said to *communicate* if state i reaches state j and state j reaches state i. By definition, each state i communicates with itself.

It is easy to check that if state i reaches state j and state j reaches state ℓ, then state i also reaches state ℓ.

Proposition C.5. *The state space $\mathscr{S}$ of a DTMC can be uniquely represented in the form*

$$\mathscr{S} = C_1 \cup \cdots \cup C_K,$$

where $C_1, \ldots, C_K$ are disjoint communicating classes (that is, any two states that belong to the same class C_k communicate, and any two states that belong to two different classes do not communicate).

Definition C.6. A DTMC is said to be *irreducible* if there is exactly one communicating class.

Definition C.7. A communicating class $C \subset \mathscr{S}$ of a DTMC with transition matrix P is said to be *closed* if $P_{ij} = 0$ for each $i \in C$ and each $j \notin C$.

C.4 Recurrence and Invariant Measure

Let X be a DTMC on state space $\mathscr{S}$ with transition matrix P. For a state $i \in \mathscr{S}$, we define

$$\tau_i = \inf\{k \geq 1 : X_k = i\}.$$

In words, τ_i is the first time at which X visits state i. By convention, $\tau_i = \infty$ if X never visits state i. Define $f_i = \mathbb{P}_i\{\tau_i < \infty\}$, the probability that X eventually returns to state i when starting there, and $N_i = \sum_{k=1}^{\infty} 1_{\{X_k=i\}}$, the *total* number of visits to state i (a random variable).

Definition C.8 (Recurrence). State i is said to be *recurrent* if $f_i = 1$. Otherwise, state i is said to be *transient*.

Proposition C.9. *(a) If state i is recurrent, then $\mathbb{P}_i\{N_i = \infty\} = 1$. As a consequence, $\mathbb{E}_i(N_i) = \infty$. (b) If state i is transient, then*

$$(C.7) \qquad\qquad \mathbb{E}_i(N_i) = \frac{1}{1-f_i} < \infty,$$

and as a consequence, $\mathbb{P}_i\{N_i < \infty\} = 1$.

Note that

$$\mathbb{E}_i(N_i) = \mathbb{E}_i\left(\sum_{k=1}^{\infty} 1_{\{X_k=i\}}\right) = \sum_{k=1}^{\infty} \mathbb{E}_i\left(1_{\{X_k=i\}}\right) = \sum_{k=1}^{\infty} \mathbb{P}_i\{X_k = i\} = \sum_{k=1}^{\infty} P_{ii}^k.$$

Thus we have the following corollary.

Corollary C.10. *(a) State i is recurrent if and only if*

$$(C.8) \qquad\qquad \sum_{k=1}^{\infty} P_{ii}^k = \infty.$$

(b) When the state space $\mathscr{S}$ is finite, there is at least one recurrent state.

Proposition C.11. *If state i reaches state j and state i is recurrent, then $\mathbb{P}_j\{\tau_i < \infty\} = 1$. That is, starting in state j, the probability is 1 that X will eventually visit state i. As a consequence, states i and j communicate.*

Corollary C.12 (Solidarity). *Assume two states $i, j \in S$ communicate. Then they are either both recurrent or both transient.*

Definition C.13. A nonzero function $\eta : \mathscr{S} \to \mathbb{R}_+$ is said to be an *invariant measure* for a stochastic matrix (P_{ij}) if, for each state i,

$$(C.9) \qquad\qquad \eta(i) = \sum_{j \in \mathscr{S}} \eta(j) P_{ji}.$$

If the state space $\mathscr{S}$ is countably infinite, an invariant measure η may have infinite total mass, meaning that $\sum_{j \in \mathscr{S}} \eta(j) = \infty$. In that case, η

cannot be normalized to a probability distribution on $\mathscr{S}$. Assuming that state $\ell \in \mathscr{S}$ is recurrent, let us now define

(C.10)
$$\eta^{(\ell)}(i) = \mathbb{E}_\ell \left(\sum_{k=0}^{\tau_\ell - 1} 1_{\{X_k = i\}} \right).$$

That is, $\eta^{(\ell)}(i)$ is the expected number of visits to state i during a "cycle" between two consecutive visits to the recurrent state ℓ.

Lemma C.14. *If state $\ell \in \mathscr{S}$ is recurrent, then $\eta^{(\ell)}(i) < \infty$ for each $i \in \mathscr{S}$, and $\eta^{(\ell)}$ is an invariant measure.*

Theorem C.15. *Let P be the transition matrix of an irreducible, recurrent DTMC. There exists an invariant measure η for P such that $\eta(i) > 0$ for each $i \in \mathscr{S}$, and that invariant measure is unique up to a constant.*

C.5 Positive Recurrence and Stationary Distribution

Definition C.16 (Positive recurrence). For a recurrent state i of a DTMC X, let $m_i = \mathbb{E}_i(\tau_i)$. State i is said to be *positive recurrent* if $m_i < \infty$. Otherwise, state i is said to be *null recurrent*.

Definition C.17 (Stationary distribution). Let X be a DTMC with state space $\mathscr{S}$ and transition matrix P. A probability distribution (or probability mass function) $\pi : \mathscr{S} \to [0, 1]$ is said to be a *stationary distribution* of X if π is an invariant measure for P.

Assume state $\ell \in \mathscr{S}$ is positive recurrent. By Theorem C.14, the function $\eta^{(\ell)}$ defined via (C.10) is an invariant measure for P. Define

$$\pi^{(\ell)}(i) = \frac{1}{m_\ell} \eta^{(\ell)}(i), \quad i \in \mathscr{S}.$$

One can verify that $\sum_{i \in \mathscr{S}} \pi^{(\ell)}(i) = \mathbb{E}_\ell(\tau_\ell)/m_\ell = 1$. Therefore, we have the following lemma.

Lemma C.18. *Assume state $\ell \in \mathscr{S}$ is positive recurrent for a DTMC X. Then $\pi^{(\ell)}$ is a stationary distribution of X.*

Theorem C.19. *Assume the DTMC X is irreducible. The following statements are equivalent. (a) There is a state $\ell \in \mathscr{S}$ that is positive recurrent. (b) Every state is positive recurrent. (c) The DTMC has*

a stationary distribution. When any one of the preceding conditions is
satisfied, the stationary distribution π is unique and is given by

(C.11) $$\pi(i) = \frac{1}{m_i} \quad \text{for } i \in \mathcal{S}.$$

C.6 Convergence to Equilibrium and Strong Law of Large Numbers

Let X be a DTMC with countable state space. The *period* of state i,
denoted $d(i)$, is the largest integer d having the following property:
$\mathbb{P}_i\{X_k = i\} = 0$ for all $k \geq 1$ that are *not* divisible by d. In words, this
means that a return to state i can only occur after a number of periods
that is an integer multiple of $d(i)$. State i is said to be *aperiodic* if
$d(i) = 1$. If one state in a communicating class is aperiodic, then all
states in that class are aperiodic. If the states of an irreducible DTMC
X are aperiodic, then X itself is said to be aperiodic.

Theorem C.20.   *Let $X = \{X_n : n \in \mathbb{Z}_+\}$ be a DTMC that is irreducible,
positive recurrent, and aperiodic, with any initial distribution. Then*

(C.12) $$\lim_{n \to \infty} \mathbb{P}\{X_n = j\} = \pi_j \quad \text{for each state } j,$$

where $\pi = (\pi_j)$ is the unique stationary distribution of the DTMC.

   In the rest of this section, we assume that the DTMC X is irreducible
and positive recurrent, but aperiodicity is not assumed. By Theorem
C.19, there is a unique stationary distribution π. For a nonnegative
function $f \geq 0 : \mathcal{S} \to \mathbb{R}_+$, define

(C.13) $$\pi(f) = \sum_{i \in \mathcal{S}} f(i)\pi(i),$$

which can be infinite. For a general function $f : \mathcal{S} \to \mathbb{R}$, let $f^+(i) = \max(f(i), 0)$ and $f^-(i) = \max(-f(i), 0)$ for each $i \in \mathcal{S}$. If at least one of
$\pi(f^+)$ and $\pi(f^-)$ is finite, then we define

$$\pi(f) = \pi(f^+) - \pi(f^-).$$

Theorem C.21 (SLLN for a DTMC).   *Let $f : \mathcal{S} \to \mathbb{R}$ be a function.
Assume that at least one of $\pi(f^+)$ and $\pi(f^-)$ is finite. Then for each
initial distribution μ one has the following:*

(C.14) $$\mathbb{P}_\mu \left\{ \lim_{N \to \infty} \frac{1}{N} \sum_{k=0}^{N-1} f(X_k) = \pi(f) \right\} = 1.$$

Corollary C.22.   *For each $i, j \in \mathcal{S}$ and each initial distribution μ,*

(C.15)
$$\mathbb{P}_\mu \left\{ \lim_{N \to \infty} \frac{1}{N} \sum_{k=0}^{N-1} 1_{\{X_k=i,\, X_{k+1}=j\}} = \pi(i) P_{ij} \right\} = 1.$$

Proof.   Let $Y_n = (X_n, X_{n+1})$. Then, $Y = \{Y_n : n = 0, 1, \dots\}$ is a DTMC on the countable state space $\mathcal{S} \times \mathcal{S}$. One can verify that Y is irreducible and positive recurrent with stationary distribution given by $\pi(i) P_{ij}$ for each state $(i, j) \in \mathcal{S} \times \mathcal{S}$. Then (C.15) follows from (C.14) for the DTMC Y with

$$f(s, \tilde{s}) = 1_{\{s=i,\, \tilde{s}=j\}} \quad \text{for } s, \tilde{s} \in \mathcal{S}. \qquad \square$$

C.7 A Preliminary Criterion for Positive Recurrence

When the state space of a DTMC is infinite, there is no guarantee that a stationary distribution exists, but the following lemma is a starting point for deriving sufficient conditions; see Section C.8.

Lemma C.23.   *Let X be an irreducible DTMC on a countable state space $\mathcal{S}$. Let $C \subset \mathcal{S}$ be a finite subset and define T_C as the first passage time of X into C, that is,*

(C.16)
$$T_C := \inf\{k \geq 1 : X_k \in C\}.$$

Suppose that

(C.17)
$$\mathbb{E}_x\left(T_C\right) < \infty \quad \text{for each } x \in C.$$

Then X is positive recurrent. As a consequence, it has a unique stationary distribution.

Proof.   Define $T_C^0 = 0$, and for each $k \geq 1$ define

$$T_C^k = \inf\left\{ n > T_C^{k-1} : X_n \in C \right\} = T_C^{k-1} + \inf\left\{ n \geq 1 : X_{T_C^{k-1}+n} \in C \right\}.$$

Then $0 = T_C^0 < T_C^1 < \dots$ is a sequence of stopping times for X. Define $\bar{X}_k = X_{T_C^k}$ for $k \in \mathbb{Z}_+$. By the strong Markov property of X, we have the following: $\bar{X} = \{\bar{X}_k : k \in \mathbb{Z}_+\}$ is a DTMC on the *finite* state space C. In particular,

$$\mathbb{E}\left(X_{T_C^{k+1}} = j | X_{T_C^1} = i_1, X_{T_C^{k-1}} = i_{k-1}, X_{T_C^k} = i\right) = \mathbb{E}\left(X_{T_C^{k+1}} = j | X_{T_C^k} = i\right)$$

for each $k \geq 1$ and each $i_1, \ldots, i_{k-1}, i, j \in C$.

Because X is irreducible, one can check that $\bar{X}$ is also irreducible. Therefore, $\bar{X}$ is positive recurrent and has a unique stationary distribution. In particular,

$$\mathbb{E}_x(\sigma_x) < \infty \quad \text{for each } x \in C,$$

where

$$\sigma_x = \inf\{k \geq 1 : \bar{X}_k \in x\}.$$

Fix an $x \in C$. It follows that $T_x = T_C^{\sigma_x}$. Therefore,

$$\mathbb{E}_x(T_x) = \mathbb{E}_x(T_C^{\sigma_x}) = \mathbb{E}_x\left(\sum_{k=1}^{\sigma_x} Y_k\right)$$

$$= \sum_{k=1}^{\infty} \mathbb{E}_x\left(Y_k 1_{\{k \leq \sigma_x\}}\right)$$

$$\leq m \sum_{k=1}^{\infty} \mathbb{E}_x\left(1_{\{k \leq \sigma_x\}}\right) = m \mathbb{E}_x(\sigma_x) < \infty,$$

where

$$Y_k = T_C^k - T_C^{k-1} = \inf\left\{n \geq 1 : X_{T_C^{k-1}+n} \in C\right\}$$

and we have used the facts that

$$\{k \leq \sigma_x\} = \left\{X_{T_C^1} \neq x, \ldots, X_{T_C^{k-1}} \neq x\right\}$$

and

$$\mathbb{E}\left(Y_k 1_{\{k \leq \sigma_x\}} | X_{T_C^1}, \ldots X_{T_C^{k-1}}\right) = 1_{\{k \leq \sigma_x\}} \mathbb{E}\left(Y_k | X_{T_C^{k-1}}\right)$$

$$\leq m \equiv \max_{x \in C} \mathbb{E}_x(T_C) < \infty. \qquad \square$$

C.8 Foster-Lyapunov Criterion

Let $X = \{X_k, k \in \mathbb{Z}_+\}$ be a DTMC on a countable state space $\mathscr{S}$ with transition matrix P. Suppose that X is irreducible. When $\mathscr{S}$ is finite, Lemma C.14 and Theorem C.15 imply that X is positive recurrent. When $\mathscr{S}$ is infinite, positive recurrence can be proved either by checking condition (C.17) or by finding a stationary distribution π satisfying the balance equations $\pi P = \pi$. Both these approaches can be difficult if not impossible.

The Foster–Lyapunov criterion is the most widely used method to prove positive recurrence of a DTMC. It provides a sufficient condition, involving a so-called *Lyapunov function* V, for (C.17) to hold. This section develops two versions: a standard Foster–Lyapunov criterion (Theorem C.26) and a state-dependent Foster–Lyapunov criterion (Theorem C.27).

Lemma C.24 (Comparison theorem). *Suppose that $V, f, g : \mathscr{S} \to \mathbb{R}_+$ are three non-negative functions satisfying*

$$(C.18) \qquad PV(x) \le V(x) - f(x) + g(x) \quad x \in \mathscr{S}.$$

Then for any stopping time τ and any state $x \in \mathscr{S}$,

$$(C.19) \qquad \mathbb{E}_x\left(\sum_{k=0}^{\tau-1} f(X_k)\right) \le V(x) + \mathbb{E}_x\left(\sum_{k=0}^{\tau-1} g(X_k)\right).$$

Proof. For each integer $k \ge 0$, define $Z_k = V(X_k)$, and for each integer $n \ge 1$, define

$$\tau^n = \min\left\{n, \tau, \inf\{k \ge 0 : Z_k \ge n\}\right\}.$$

Then τ^n is a stopping time with respect to X, and for $k < \tau^n$ one has

$$Z_0 \le n, \quad Z_1 \le n, \dots, Z_k \le n.$$

We first prove that

$$(C.20)$$

$$\mathbb{E}_x(Z_{\tau^n}) = \mathbb{E}_x(Z_0) + \mathbb{E}_x\left(\sum_{k=1}^{\tau_n} \left(\mathbb{E}[Z_k | X_0, X_1, \dots, X_{k-1}] - Z_{k-1}\right)\right),$$

which is known as the *Dynkin formula*. Note that $\mathbb{E}_x(Z_{\tau^n})$ can be infinite, in which case the right side of (C.20) is also infinite. To prove (C.20), we note that

$$Z_{\tau^n} = Z_0 + \sum_{k=1}^{\tau^n}(Z_k - Z_{k-1})$$

$$= Z_0 + \sum_{k=1}^{n} 1_{\{\tau^n \geq k\}}(Z_k - Z_{k-1}).$$

Because τ^n is a stopping time, the occurrence or nonoccurrence of the event $\{\tau^n \geq k\}$ is determined by the random variables $X_0, \ldots, X_{k-1}$, or equivalently, the complementary event $\{\tau^n \leq k-1\}$ is determined by those same random variables. Therefore,

$$\mathbb{E}_x(Z_{\tau^n}) = \mathbb{E}_x(Z_0) + \mathbb{E}_x\left(\sum_{k=1}^{n} \mathbb{E}_x(Z_k - Z_{k-1}|X_0,\ldots,X_{k-1})1_{\{\tau^n \geq k\}}\right)$$

$$= \mathbb{E}_x(Z_0) + \mathbb{E}_x\left(\sum_{k=1}^{\tau^n} \mathbb{E}_x(Z_k|X_0,\ldots,X_{k-1}) - Z_{k-1}\right),$$

proving (C.20).

We now use (C.20) to prove (C.19). By (C.18),

$$\mathbb{E}_x(Z_k|X_0,\ldots,X_{k-1}) = \mathbb{E}_x(V(X_k)|X_0,\ldots,X_{k-1}) = PV(X_{k-1})$$

$$\leq V(X_{k-1}) - f(X_{k-1}) + g(X_{k-1})$$

$$= Z_{k-1} - f(X_{k-1}) + g(X_{k-1})$$

$$\leq Z_{k-1} - f(X_{k-1}) \wedge N + g(X_{k-1}),$$

where $N \geq 1$ is an integer. Noting that $\mathbb{E}_x(Z_0) = V(x)$ by (C.20), one has

$$\mathbb{E}_x(Z_{\tau^n}) \leq V(x) + \mathbb{E}_x\left(\sum_{k=1}^{\tau^n} g(X_{k-1}) - f(X_{k-1}) \wedge N\right),$$

from which

$$\mathbb{E}_x\left(\sum_{k=1}^{\tau^n} f(X_{k-1}) \wedge N\right) \leq V(x) + \mathbb{E}_x\left(\sum_{k=1}^{\tau^n} g(X_{k-1})\right).$$

Taking $N \to \infty$, by the monotone convergence theorem, one has

$$\mathbb{E}_x \left(\sum_{k=1}^{\tau^n} f(X_{k-1}) \right) \leq V(x) + \mathbb{E}_x \left(\sum_{k=1}^{\tau^n} g(X_{k-1}) \right).$$

Letting $n \to \infty$ on the both sides, again by the monotone convergence theorem, one has

$$\mathbb{E}_x \left(\sum_{k=1}^{\tau} f(X_{k-1}) \right) \leq V(x) + \mathbb{E}_x \left(\sum_{k=1}^{\tau} g(X_{k-1}) \right),$$

proving (C.19). □

Throughout the remainder of this section, the definition (C.16) remains in force. That is, we continue to define T_C as the first passage time of X into C.

Lemma C.25. *Suppose there exists a constant $b > 0$, a set $C \subset \mathscr{S}$, and two nonnegative functions $V, f : \mathscr{S} \to \mathbb{R}_+$ such that*

(C.21) $\qquad PV(x) \leq V(x) - f(x) + b1_C(x) \quad x \in \mathscr{S}.$

Then

(C.22) $\qquad \mathbb{E}_x \left(\sum_{k=0}^{T_C-1} f(X_k) \right) \leq V(x) + b \quad x \in \mathscr{S}.$

Proof. Applying Theorem C.24 with $g(x) = 1_C(x)$ and $\tau = T_C$, one has

$$\mathbb{E}_x \left(\sum_{k=0}^{T_c-1} f(X_k) \right) \leq V(x) + b\mathbb{E}_x \left(\sum_{k=0}^{T_C-1} 1_C(X_k) \right).$$

Then (C.22) follows from the fact that $1_C(X_k) = 0$ for $1 \leq k < T_C$. □

The following theorem is known as the Foster–Lyapunov criterion for positive recurrence of a DTMC. Condition (C.23) is known as the *drift condition*.

Theorem C.26 (Foster–Lyapunov criterion). *Assume that X is irreducible. Suppose there exist constants $b > 0$ and $c > 0$, a finite set $C \subset \mathscr{S}$, and a nonnegative function $V : \mathscr{S} \to \mathbb{R}_+$ such that*

(C.23) $PV(x) \le V(x) - c + b 1_C(x) \quad x \in \mathscr{S}.$

Then X is positive recurrent.

Proof.   We prove the theorem by using Lemma C.25. The drift condition (C.23) implies that (C.21) is satisfied with $f(x) = c$. Thus (C.22) holds. With our choice of $f(x)$, (C.22) becomes

$$c \mathbb{E}_x(T_C) \le V(x) + b \quad \text{for each } x \in \mathscr{S},$$

which implies that (C.17) holds. Because the set C is assumed to be finite and X is assumed to be irreducible, Lemma C.23 implies that X is positive recurrent. $\square$

   A key result in this book's development of fluid-based stability theory is Lemma 3.7, and for its proof the standard Foster–Lyapunov criterion is not sufficient. Rather, we need the following *state-dependent* Foster–Lyapunov criterion. This version is a special case of theorem 1 of Foss and Konstantopoulos (2004), whose proof was attributed to Tweedie (1976). This version also appeared in Meyn and Tweedie (1994) for general state space.

Theorem C.27 (State-dependent Foster–Lyapunov criterion). *Assume that the DTMC X is irreducible. Suppose there exists a constant $b > 0$, a finite set $C \subset \mathscr{S}$, a function*

$$n : \mathscr{S} \to \{1, 2, \ldots\},$$

and a nonnegative function $V : \mathscr{S} \to \mathbb{R}_+$ such that

(C.24) $\mathbb{E}_x \Big(V(X_{n(x)}) \Big) \le V(x) - n(x) + b 1_C(x), \quad x \in \mathscr{S}.$

Then X is positive recurrent.

Proof.   Define $\bar{X}_0 = x$, $N(k) = N(k-1) + n(\bar{X}_{k-1})$ and $\bar{X}_k = X_{N(k)}$ for $k \ge 1$. Let $\bar{P}(x, x') = \mathbb{P}\{X_{n(x)} = x' | X_0 = x\}$. Then $\bar{P} = (\bar{P}(x, x'))$ is a stochastic matrix. It follows that $\bar{X} = \{\bar{X}_k : k \in \mathbb{Z}_+\}$ is a DTMC satisfying (C.21) with $f(x) = n(x)$ and with $\bar{P}$ replacing P. Therefore,

(C.25)
$$\mathbb{E}_x \left(\sum_{k=0}^{\overline{T}_C - 1} n(\bar{X}_k) \right) \leq V(x) + b, \quad x \in \mathcal{S},$$

where

$$\overline{T}_C = \inf\{k \geq 1 : \bar{X}_k \in C\}.$$

Because

$$T_C \leq \sum_{k=0}^{\overline{T}_C - 1} n(\bar{X}_k),$$

Equation (C.25) implies that (C.17) holds. Because the set C is assumed to be finite and X is assumed to be irreducible, by Lemma C.23, (C.17) implies that X is positive recurrent. □

Appendix D

Continuous-Time Markov Chains and Phase-Type Distributions

In this appendix, we review continuous-time Markov chains (CTMCs). Like DTMCs, which are reviewed in Appendix C, a CTMC lives in a countable state space $\mathscr{S}$, and it jumps from one state to another. For a DTMC, the time intervals between jumps are all of unit length, but for a CTMC they are of random length, with the mean time to jump depending on the state from which the jump occurs. As a result, it is possible for a CTMC to have infinitely many jumps within a finite time interval. Sufficient conditions will be specified in this appendix to rule out this phenomenon.

The essential data needed to describe a CTMC are contained in its *generator matrix*, from which the chain's time-dependent transition probabilities can be obtained by solving the Kolmogorov backward or forward equations (see Section D.2). Other concepts relating to a CTMC, including its stationary distribution and long-run behavior, are analogous to their DTMC counterparts. Our standard reference is chapters 2 and 3 of Norris (1998).

D.1 Generator Matrices

Let $\mathscr{S}$ be a countable space, and let $\Lambda = (\Lambda_{ij})$ be an $\mathscr{S} \times \mathscr{S}$ matrix.

Definition D.1. Λ is said to be a *generator matrix*, or just *generator* for brevity, if (i) $\Lambda_{ij} \geq 0$ for $i, j \in \mathscr{S}$ with $i \neq j$, and (ii) for each $i \in \mathscr{S}$,

(D.1) $$\Lambda_{ii} = -\sum_{j \neq i} \Lambda_{ij}.$$

Condition (ii) says that each row of Λ is summable, and the row sum is equal to 0. Thus a generator matrix is completely determined by its off-diagonal elements, which will be called *transition rates* (of the corresponding CTMC) in this book; a frequently encountered alternative term is *transition intensities*. It will be convenient to define $\lambda(i) = -\Lambda_{ii} \geq 0$ for $i \in \mathscr{S}$. For reasons explained in the next section, the constant $\lambda(i)$ is called the *exit rate* from state i. Every generator matrix encountered in this book has the property that

(D.2) $$\lambda(i) > 0 \quad \text{for all } i \in \mathscr{S},$$

which is equivalent to saying that the corresponding CTMC has no absorbing states (see Section D.2). Property (D.2) is assumed throughout the remainder of this appendix, and throughout Appendix E and the body of the book as well, whenever reference is made to a "generator matrix."

Given a generator matrix Λ, we define the corresponding *jump matrix* $G = (G_{ij})$ as follows:

(D.3) $$G_{ij} = \begin{cases} \Lambda_{ij}/\lambda(i) & \text{for } j \neq i, \\ 0 & \text{for } j = i. \end{cases}$$

This is a stochastic matrix, so it can be interpreted as the (one-step) transition matrix of a DTMC. For example, when

(D.4) $$\Lambda = \begin{pmatrix} -3 & 2 & 1 \\ 3 & -7 & 4 \\ 2 & 0 & -2 \end{pmatrix},$$

the corresponding jump matrix is

$$G = \begin{pmatrix} 0 & 2/3 & 1/3 \\ 3/7 & 0 & 4/7 \\ 1 & 0 & 0 \end{pmatrix}.$$

Clearly, one can recover a generator matrix from the corresponding jump matrix and the rates $\{\lambda(i), i \in \mathscr{S}\}$.

D.2 Sample Path Construction of a CTMC

Given a countable state space $\mathscr{S}$, an initial distribution ρ on $\mathscr{S}$, and a generator matrix Λ, we now present a sample path construction of a corresponding CTMC $X = \{X(t), t \geq 0\}$. In the CTMC literature, this X is often called the *minimal process* for the triple $(\mathscr{S}, \rho, \Lambda)$.

Construction of a CTMC.   Let G be the jump matrix corresponding to Λ. Let $Y = \{Y_n : n \in \mathbb{Z}_+\}$ be a DTMC on some probability space $(\Omega, \mathscr{F}, \mathbb{P})$ that has state space $\mathscr{S}$, initial distribution ρ, and transition matrix G, and let $\{\tau_k : k \in \mathbb{Z}_+\}$ be a sequence of i.i.d. exponential random variables with mean 1, defined on the same probability space and independent of Y. Define

(D.5) $$X(t) = Y_n \quad \text{for} \quad \sigma_n \le t < \sigma_{n+1},$$

where

(D.6) $$\sigma_0 = 0, \quad \text{and} \quad \sigma_{n+1} = \sigma_n + \tau_n/\lambda(Y_n) \quad \text{for } n \ge 0.$$

Thus the sample path of X is piecewise constant, the sequence of states visited by X is the same as the sequence visited by the jump chain Y, and following a transition into a state i, the occupancy time of X in that state is exponentially distributed with mean $1/\lambda(i)$, which justifies calling $\lambda(i)$ the "exit rate" from state i.

The preceding construction defines $X(t)$ for all $t \in [0, \sigma_\infty)$, where

$$\sigma_\infty = \lim_{n \to \infty} \sigma_n.$$

If $\sigma_\infty = \infty$, then $X(t)$ is defined for all $t \in [0, \infty)$. Otherwise, define

$$X(t) = \Delta \quad \text{for } t \ge \sigma_\infty,$$

where Δ is an additional state outside of $\mathscr{S}$. In the latter case, X lives in the augmented state space $\overline{\mathscr{S}} \equiv \mathscr{S} \cup \{\Delta\}$, and X has infinitely many jumps in the finite interval $[0, \sigma_\infty]$.

Definition D.2.   The generator matrix Λ is said to be *nonexplosive* if

(D.7) $$\mathbb{P}_i\{\sigma_\infty = \infty\} = 1$$

for each $i \in \mathscr{S}$, where $\mathbb{P}_i(A) = \mathbb{P}(A \mid X(0) = i)$ for $A \in \mathscr{F}$.

One can check that the condition

(D.8) $$\sup_{i \in \mathscr{S}} \lambda(i) < \infty$$

is sufficient for Λ to be nonexplosive. Another sufficient condition is that the DTMC Y be recurrent; see theorem 2.7.1 of Norris (1998). All of the generator matrices encountered in this book satisfy (D.8), but there are some commonly encountered CTMC models, most notably queueing models with customer abandonment, for which (D.8) fails.

Definition D.3. A generator matrix Λ will be called *regular* in this book if it satisfies (D.2) and is nonexplosive.

All generator matrices are assumed to be regular hereafter, so there will be no need to deal with augmented state spaces. The following theorem is standard. For a proof, see, for example, theorem 2.8.4 of Norris (1998).

Theorem D.4. *The stochastic process $X = \{X(t), t \geq 0\}$ constructed in (D.5) satisfies the following. (i) X has right-continuous sample paths. (ii) $\mathbb{P}\{X(0) = i\} = \rho(i)$ for $i \in \mathscr{S}$. (iii) For each integer $k \geq 0$, each $t > 0$ and $h > 0$, each $i_0, \ldots, i_k, i, j \in \mathscr{S}$, and each $0 \leq t_0 < \cdots < t_k < h$,*

$$(D.9) \qquad \mathbb{P}\{X(t+h) = j | X(t_0) = i_0, \ldots, X(t_k) = i_k, X(h) = i\}$$
$$= \mathbb{P}\{X(t+h) = j | X(h) = i\}.$$

(iv) For each $i, j \in \mathscr{S}$ and each $t > 0$,

$$(D.10) \qquad P_{ij}(t) := \mathbb{P}\{X(t+h) = j | X(h) = i\}$$

is independent of $h \geq 0$. Furthermore, the family of matrices $\{P(t) = (P_{ij}(t)), t \geq 0\}$ is the unique solution to the Kolmogorov backward equation

$$(D.11) \qquad P'(t) = \Lambda P(t) \quad \text{for } t \geq 0, \quad P(0) = I,$$

where $P'(t)$ is the matrix of componentwise derivatives of $P(\cdot)$ at t.

Remark D.5. (a) Property (iii) says that X satisfies the Markov property, analogous to (C.3) in the DTMC case. (b) From (D.11), the matrix-valued function $P(t) = (P_{ij}(t))$ defined via (D.10) satisfies

$$(D.12) \qquad P'(0) = \Lambda,$$

where the derivative is interpreted as the right derivative. By the Markov property, one can check that $\{P(t), t \geq 0\}$ satisfies the Chapman–Kolmogorov equation

$$(D.13) \qquad P(t+s) = P(t)P(s) \quad \text{for } t, s \geq 0.$$

Therefore, $\{P(t), t \geq 0\}$ forms a *semigroup*. For a semigroup $\{P(t), t \geq 0\}$, conditions (D.11) and (D.12) are equivalent. Condition (D.12) states that Λ *generates* the semigroup $\{P(t), t \geq 0\}$. (c) Given our nonexplosive assumption, the solution $\{P(t), t \geq 0\}$ to the Kolmogorov backward equation is unique, and $P(t)$ is a stochastic matrix for each $t \geq 0$. When

Λ is explosive, the matrix-valued function $P(t)$ defined by (D.10) is sub-stochastic but not necessarily stochastic. In that case, $\{P(t), t \geq 0\}$ is still *a* solution to the Kolmogorov backward equation (more particularly, it is called the *minimal solution*), but the solution is no longer unique. (d) When $\mathscr{S}$ is finite, (D.8) is satisfied and therefore Λ is nonexplosive. The unique solution $P(t)$ to the Kolmogorov equation is given by

$$(D.14) \qquad P(t) = e^{t\Lambda} \equiv \sum_{k=0}^{\infty} \frac{t^k}{k!} \Lambda^k,$$

where the right side is well defined; see section 2.10 of Norris (1998). (e) One can replace the backward equation (D.11) with the Kolmogorov forward equation

$$P'(t) = P(t)\Lambda \quad \text{for } t \geq 0, \qquad P(0) = I,$$

which is often more difficult to work with than the backward equation.

Definition D.6. Let $\mathscr{S}$ be a countable space, ρ a distribution on $\mathscr{S}$, and Λ a (regular) $\mathscr{S} \times \mathscr{S}$ generator matrix. A stochastic process $X = \{X(t), t \geq 0\}$ taking values in $\mathscr{S}$ is said to be *a CTMC with state space $\mathscr{S}$, initial distribution ρ, and generator Λ* if conditions (i) through (iv) in Theorem D.4 are satisfied.

Two CTMCs with the same state space, initial distribution, and generator are equal in distribution. Whenever a CTMC X is mentioned, without loss of generality we envision X to be constructed as in (D.5). We call $Y = \{Y_n, n \in \mathbb{Z}_+\}$ the corresponding *jump chain*.

D.3 Transience, Recurrence, Invariant Measures, and Irreducibility

Let $X = \{X(t), t \geq 0\}$ be a CTMC with (regular) generator Λ and jump chain $Y = \{Y_n, n \in \mathbb{Z}_+\}$, defined on some probability space $(\Omega, \mathscr{F}, \mathbb{P})$. For a state $i \in \mathscr{S}$ and each $\omega \in \Omega$, let $T_i(\omega)$ be the *first passage time* of X to state i, meaning that

$$(D.15) \qquad T_i(\omega) = \inf\{t \geq \sigma_1 : X(t, \omega) = i\},$$

where σ_1 is the first jump time of X, as defined in (D.6). Also, let

$$(D.16) \qquad N_i(\omega) = \inf\{n \geq 1 : Y_n(\omega) = i\}$$

be the first passage time of the jump chain to i.

Definition D.7. State $i \in \mathscr{S}$ is said to be *recurrent* for X if $\mathbb{P}_i\{T_i < \infty\} = 1$, and to be *transient* for X if it is not recurrent.

Because

$$\{\omega : N_i(\omega) < \infty\} = \{\omega : T_i(\omega) < \infty\},$$

the recurrence or transience of X is the same as that of the jump chain Y. In the following definition, we envision η as a row vector, so the product on the left side of (D.17) is well defined.

Definition D.8. Let $\eta = (\eta(i), i \in \mathscr{S})$ be a measure (that is, a non-negative vector, not necessarily a probability distribution) on $\mathscr{S}$. The measure η is said to be *invariant* for Λ if

$$(D.17) \qquad\qquad \eta \Lambda = 0.$$

The following lemma says that there is a one-to-one correspondence between invariant measures of the CTMC and invariant measures of its jump chain. The proof is a straightforward algebraic verification.

Lemma D.9. *Let Λ be a (regular) generator matrix with jump matrix G, and let η be a measure on $\mathscr{S}$. Define*

$$(D.18) \qquad\qquad \mu(i) = \lambda(i)\eta(i) \quad \text{for } i \in \mathscr{S}.$$

Then η is invariant for Λ if and only if $\mu G = \mu$.

Definition D.10. We say state j is *reachable* from state i by the CTMC X, written $i \to j$, if j is reachable from state i by the DTMC Y.

Given this definition, the notions of *communicating class, closed class*, and *irreducibility* for the CTMC X (or equivalently, for its generator Λ) are inherited in the obvious way from the jump chain Y (or equivalently, from the jump matrix G). Lemma D.9 and Theorem C.15 then give the following; see also theorem 3.5.2 of Norris (1998).

Theorem D.11. *Suppose that Λ is irreducible and recurrent. Then Λ has an invariant measure η that is unique up to a constant.*

D.4 Strong Markov Property

To state the strong Markov property for a continuous-time Markov chain, we first give the definition of a stopping time. Here $X = \{X(t), t \geq 0\}$ is the regular CTMC constructed in (D.5).

Definition D.12.   A continuous random variable $T : \Omega \to \mathbb{R}^+ \cup \{\infty\}$ is said to be a *stopping time* with respect to X if, for each $t \in \mathbb{R}_+$, the occurrence or nonoccurrence of the event $\{T \leq t\}$ is completely determined by the random variables $\{X(s), 0 \leq s \leq t\}$.

The first passage time $T_i(\omega)$ defined in (D.15) is an example of a stopping time. The following is proved as theorem 6.5.4 of Norris (1998).

Theorem D.13 (Strong Markov property).   *Assume the CTMC $X = \{X(t), t \geq 0\}$ is regular with generator matrix G, state space $\mathscr{S}$, and initial distribution $\rho = (\rho(i))$. Let T be a stopping time with respect to X. Then, conditional on $\{T < \infty\}$ and $X(T) = i$, the process $\{X(T+t), t \geq 0\}$ is a CTMC with generator matrix G and initial distribution δ_i, where δ_i is the distribution on the state space $\mathscr{S}$ that concentrates all its mass on state i. Furthermore, conditional on $\{T < \infty\}$ and $X(T) = i$, the processes $\big(X(T+t), t \geq 0\big)$ and $(X(t), t \leq T)$ are independent.*

Assume that i is a recurrent state for the CTMC X, and that $X(0) = i$. Define $T_i^{(0)} = 0$, $\sigma_i^{(0)} = \inf\{t \geq 0 : X(t) \neq i\}$, and for $\ell = 1, 2, \ldots$

$$T_i^{(\ell)} = \inf\left\{t \geq \sigma_i^{(\ell-1)} : X(t) = i\right\},$$

$$\sigma_i^{(\ell)} = \inf\left\{t \geq T_i^{(\ell)} : X(t) \neq i\right\}.$$

In words, $T_i^{(\ell)}$ is the time of the ℓth entry into state i. For each $\ell = 1, 2, \ldots$, define

(D.19) $$C^{(\ell)} = \left\{X(t), T_i^{(\ell-1)} \leq t \leq T^{(\ell)}\right\},$$

which we call the ℓth i-cycle of the CTMC. From Theorem D.13, we have the following corollary.

Corollary D.14.   *Assume that i is a recurrent state of a regular CTMC X. Then*

$$\left(T_i^{(\ell)} - T_i^{(\ell-1)}, \quad C^{(\ell)}\right), \ell = 1, 2, \ldots$$

is an i.i.d. sequence.

D.5 Positive Recurrence and Stationary Distributions

Let $X = \{X(t), t \geq 0\}$ be a CTMC on state space $\mathscr{S}$ with (regular) generator Λ and jump chain $Y = \{Y_n, n \in \mathbb{Z}_+\}$.

Definition D.15. A state i is said to be *positive recurrent* for X if $\mathbb{E}_i(T_i) < \infty$, where T_i is the first passage time of X to state i, defined via (D.15).

Definition D.16. A probability distribution π on $\mathscr{S}$ is said to be a *stationary distribution* for X if it is invariant for Λ.

Theorem D.17. *If Λ is irreducible, then the following are equivalent: (i) every state is positive recurrent; (ii) some state i is positive recurrent; and (iii) there exists a stationary distribution for X. When these equivalent conditions are satisfied, the stationary distribution π is unique and is given by*

$$\text{(D.20)} \qquad \pi(i) = \frac{1}{\lambda(i)\mathbb{E}_i(T_i)} \quad \text{for } i \in \mathscr{S}.$$

In the proof of this theorem, the main idea is to relate the following two occupancy measures on $\mathscr{S}$: first, η is defined for the CTMC X via

$$\text{(D.21)} \qquad \eta(j) = \mathbb{E}_i \int_0^{T_i} 1_{\{X(s)=j\}} ds, \quad j \in \mathscr{S},$$

and then μ is defined for the jump chain Y via

$$\text{(D.22)} \qquad \mu(j) = \mathbb{E}_i \sum_{n=0}^{N_i-1} 1_{\{Y_n=j\}}, \quad j \in \mathscr{S},$$

where N_i is the first passage time for the jump chain as defined in (D.16), and i is assumed to be positive recurrent for X.

To express these definitions in words, let us define an *i-cycle of X* as an interval between two consecutive times at which X enters state i. Then $\eta(j)$ is the expected amount of time that X spends in state j during an *i*-cycle. The number of visits made by X to state j during the cycle is $\sum_{n=0}^{N_i-1} 1_{\{Y_n=j\}}$, and the expected duration of each visit is $1/\lambda(j)$. Thus, one should have

$$\text{(D.23)} \qquad \eta(j) = \mu(j)/\lambda(j), \quad j \in \mathscr{S}.$$

From Lemma C.14, the measure μ defined by (D.22) is an invariant measure for Y. By Lemma D.9 and (D.23), the measure η defined by

(D.21) is an invariant measure for X. For the details of the proof of this theorem, see theorem 3.5.3 of Norris (1998).

Theorem D.18. *Suppose that the CTMC X is irreducible and π is a stationary distribution of X. If X has initial distribution π, then $X(t)$ has distribution π for each $t > 0$. Furthermore, X is a stationary process.*

To prove Theorem D.18, it suffices to show that $\pi P(s) = \pi$ for each $s > 0$. When $\mathcal{S}$ is finite, one can use the Kolmogorov backward equation to write

$$(D.24) \qquad \frac{d}{ds}\pi P(s) = \pi P'(s) = \pi \Lambda P(s) = 0.$$

Thus $\pi P(s)$ is independent of s and is equal to $\pi P(0) = \pi$. When $\mathcal{S}$ is infinite, however, the first equality in (D.24) is not justified, and a completely different proof is needed. Given the assumed irreducibility, one can use the explicit expression in (D.21) to complete the proof, using the strong Markov property of X. See the proof of theorem 3.5.5 of Norris (1998) for details.

Definition D.19. Fix a *step size* $h > 0$. For each $n \in \mathbb{Z}_+$, define

$$(D.25) \qquad\qquad Z_n = X(nh).$$

Then $Z = \{Z_n, n \in \mathbb{Z}_+\}$ is a DTMC, which we call a *skeleton chain* of the CTMC X.

Lemma D.20. *Assume the CTMC X is irreducible. Then each skeleton chain Z is irreducible and aperiodic.*

Proof. We first prove that $i \to j$ implies $P_{ij}(t) > 0$ for *all* $t > 0$. If $\Lambda_{ij} > 0$, then

$$P_{ij}(t) \geq \mathbb{P}_i\{\tau_0/\lambda(Y_0) \leq t, Y_1 = j, \tau_1/\lambda(Y_1) > t\}$$
$$= \left(1 - e^{-\lambda(i)t}\right)\frac{\Lambda_{ij}}{\lambda(i)}e^{-\lambda(j)t} > 0$$

for all $t > 0$. Assume $i \to j$. There exists an integer $k > 0$ and states $i_0 = i, i_1, \ldots, i_{k-1}, i_k = j$ such that $\Lambda_{i_\ell i_{\ell+1}} > 0$ for $\ell = 0, \ldots, k - 1$. Therefore,

$$P_{ij}(t) \geq P_{i_0 i_1}(t/k) \cdots P_{i_{k-1} i_k}(t/k) > 0$$

for all $t > 0$. Suppose that $h > 0$ is the step size used to define the skeleton chain Z. Because $P_{ij}(h) > 0$ for each pair of $i, j \in \mathcal{S}$, the skeleton chain Z is irreducible and aperiodic. $\qquad\square$

Remark D.21.   In the first part of the proof of the lemma, the conclusion is much stronger than $P_{ij}(t) > 0$ for *some* $t > 0$. The analogous conclusion does *not* hold in a discrete-time setting. As a consequence of this lemma, there is no need to introduce the concept of periodicity in the continuous-time setting.

Theorem D.22.   *Assume that the CTMC X is irreducible, and fix a step size $h > 0$. Then the following two statements are equivalent.*

(i) The CTMC X is positive recurrent.
(ii) The skeleton DTMC Z is positive recurrent.

Proof.   *(i) to (ii).* Assume X is positive recurrent. Then X has a unique stationary distribution π. By Theorem D.18, π is also a stationary distribution of the skeleton chain Z. Thus Z is positive recurrent.

 (ii) to (i). Assume Z is positive recurrent. Fix a state $i \in \mathscr{S}$. Let T_i^Z be the first passage time of Z to i. Then $\mathbb{E}_i(T_i^Z) < \infty$. Because $T_i \leq h T_i^Z$, one has $\mathbb{E}_i(T_i) < \infty$, proving that X is positive recurrent. $\square$

It follows from Lemma D.20 and Theorem D.22 that to prove the positive recurrence of a CTMC, it is sufficient to prove the positive recurrence of a skeleton DTMC Z, to which one can apply the Foster–Lyapunov criteria (see Section C.8 of Appendix C).

D.6 Convergence to Equilibrium and Strong Law of Large Numbers

The main results of this section are Theorems D.24 and D.25. The former shows that for an irreducible CTMC, positive recurrence is equivalent to the existence of a limit distribution. The latter states a strong law of large numbers (SLLN) for an irreducible and positive recurrent CTMC. The proof of the latter theorem is similar to the one for a DTMC. The following preliminary result is from lemma 3.6.1 of Norris (1998). It establishes the uniform continuity of $P_{ij}(t)$ as a function t.

Lemma D.23.   *Let $P(t) = (P_{ij}(t))$ be the family of time-dependent transition matrices for a CTMC with generator matrix Λ. Then for all $t, h \geq 0$,*

$$|P_{ij}(t+h) - P_{ij}(t)| \leq 1 - e^{-\lambda(i)h}.$$

Theorem D.24.   *Let X be an irreducible CTMC on state space $\mathscr{S}$ with initial distribution ρ.*

(i) If X is positive recurrent, then

$$\lim_{t \to \infty} \mathbb{P}\{X(t) = j\} = \pi(j) > 0 \quad \text{for each } j \in \mathscr{S},$$

where $\pi = (\pi(i))$ is the unique stationary distribution of X.
(ii) If X is not positive recurrent, then

$$\lim_{t \to \infty} \mathbb{P}\{X(t) = j\} = 0 \quad \text{for each } j \in \mathscr{S}.$$

Proof.   *(i)*. By Theorem D.17, X has a unique stationary distribution $\pi = (\pi(i), i \in \mathscr{S})$ with $\pi(i) > 0$ given by (D.20). Because

$$\mathbb{P}\{X(t) = j\} = \sum_{i \in \mathscr{S}} \rho(i)\mathbb{P}_i\{X(t) = j\},$$

it suffices by the dominated convergence theorem (Theorem B.3) to prove that

(D.26) $$\lim_{t \to \infty} \mathbb{P}_i\{X(t) = j\} = \lim_{t \to \infty} P_{ij}(t) = \pi(j)$$

for each $i \in \mathscr{S}$. We now prove (D.26). Fix an $h > 0$ and let $Z = \{Z_n, n \in \mathbb{Z}_+\}$ be the corresponding skeleton chain. By Lemma D.20, the DTMC Z is irreducible and aperiodic, and by Theorem D.18, π is the stationary distribution of Z. Therefore, Theorem C.20 for a DTMC implies that

(D.27) $$\lim_{n \to \infty} P_{ij}(nh) = \pi(j) \quad \text{for each } j \in \mathscr{S}.$$

We now complete the proof of (D.26) by using the uniform continuity of $P_{ij}(t)$, which was established in Lemma D.23. Fix a state $i \in \mathscr{S}$. Given an $\epsilon > 0$, one can find an $h > 0$ such that

$$1 - e^{-\lambda(i)s} \le \epsilon/2 \quad \text{for } 0 \le s \le h.$$

Next, by (D.27) there exists an integer $N > 0$ such that

$$|P_{ij}(nh) - \pi(j)| \le \epsilon/2 \quad \text{for } n \ge N.$$

For $t \ge Nh$, we have $nh \le t < (n+1)h$ for some $n \ge N$. Thus, by Lemma D.23,

$$|P_{ij}(t) - \pi(j)| \le |P_{ij}(t) - P_{ij}(nh)| + |P_{ij}(nh) - \pi(j)| \le \epsilon,$$

proving (D.26).

(ii). When X is transient, the proof is immediate. When X is null recurrent, the proof follows from step 3 of theorem 1.8.5 in Norris (1998). □

The proof of the following SLLN follows that of Theorem C.21 for DTMCs.

Theorem D.25. *Let X be an irreducible CTMC with (regular) generator matrix Λ and initial distribution ρ. Then*

$$\mathbb{P}\left\{\lim_{t\to\infty}\frac{1}{t}\int_0^t 1_{\{X(s)=i\}}ds = \frac{1}{\mathbb{E}_i(T_i)\lambda(i)} \quad \text{for all } i \in \mathscr{S}\right\} = 1,$$

where T_i is the first passage time of X to state i. Moreover, when X is positive recurrent with stationary distribution $\pi = (\pi(i))$, one has the following for any bounded function $f : \mathscr{S} \to \mathbb{R}$:

$$\mathbb{P}\left\{\lim_{t\to\infty}\frac{1}{t}\int_0^t f(X(s))ds = \bar{f}\right\} = 1,$$

where $\bar{f} = \sum_{j\in\mathscr{S}}\pi(j)f(j)$.

D.7 Uniformization and the Foster–Lyapunov Criterion

Let X be a CTMC with countable state space $\mathscr{S}$ and generator matrix Λ. Assume its transition rates are bounded, that is,

(D.28) $$\sup_{i\in\mathscr{S}}\lambda(i) < \infty.$$

Under assumption (D.28), the CTMC X has the sample path representation

(D.29) $$X(t) = Z_{N(t)} \quad \text{for } t \geq 0,$$

where $Z = \{Z_n, n \in \mathbb{Z}_+\}$ is a DTMC on $\mathscr{S}$, $N = \{N(t), t \geq 0\}$ is a Poisson process, and N and Z are independent. The pair (N, Z) is not unique. Indeed, for any

$$\eta \geq \sup_{i\in\mathscr{S}}\lambda(i),$$

define

(D.30)
$$P_{ij} = \begin{cases} \lambda_{ij}/\eta & \text{for } i \neq j, \\ 1 - \lambda(i)/\eta & \text{for } i = j. \end{cases}$$

Then $P = (P_{ij})$ is a stochastic matrix on $\mathcal{S}$. Let Z be a DTMC on $\mathcal{S}$ with transition matrix P, and let N be an independent Poisson process with rate η. One can show that the continuous-time process $\{Z_{N(t)}, t \geq 0\}$ is a CTMC with generator matrix Λ. Therefore, the representation (D.29) holds.

Recall that the jump chain Y is also a DTMC whose transition matrix is given by the jump matrix (D.3) with zero diagonal entries. Thus, whenever Y jumps, it lands in a new state that is different from the current one. In contrast, the transition matrix (D.30) for the DTMC Z may have positive diagonal elements. Therefore, when Z jumps, it may land back in the current state with positive probability.

In the sample path construction of the CTMC X in (D.29), we have a sequence of intervals over which X remains constant, and the value of X at the end of the nth such interval is set equal to Z_n. The salient feature of the construction is that the expected duration of interval $n+1$, given the observed state Z_n, is independent of Z_n. That is, the expected duration of any given transition interval is a single constant, *uniformly* across all current states. Therefore, the representation (N, Z) in (D.29) is known as a *uniformization representation* of X. From (D.29), we have that, for any $t \geq 0$ and $i, j \in \mathcal{S}$,

(D.31)
$$P_{ij}(t) = e^{-\eta t} \sum_{n=0}^{\infty} \frac{(\eta t)^n}{n!} P_{ij}^n,$$

which provides a stable numerical algorithm to compute transient probabilities of the CTMC. From (D.31), one can verify that any stationary distribution of Z is a stationary distribution of X.

Theorem D.26. *Let X be a CTMC on a countable state space $\mathcal{S}$. Assume that X is irreducible and positive recurrent with unique stationary distribution π. For each $i, j \in \mathcal{S}$ with $i \neq j$, let $J(i, j, t)$ be the number of jumps from state i to state j in $(0, t]$. Then,*

(D.32)
$$\mathbb{P}_\mu \left\{ \lim_{t \to \infty} \frac{1}{t} J(i, j, t) = \pi(i) \Lambda_{ij} \right\} = 1$$

for each initial distribution μ.

Proof.   From (D.29), one has that

$$(D.33) \qquad J(i,j,t) = \sum_{k=0}^{N(t)} 1_{\{Z_k=i,\, Z_{k+1}=j\}}.$$

By Corollary C.22,

$$(D.34) \qquad \mathbb{P}_\mu\left\{ \lim_{n\to\infty} \frac{1}{n} \sum_{k=0}^{n} 1_{\{Z_k=i,\, Z_{k+1}=j\}} = \pi(i) P_{ij} \right\} = 1.$$

Now the SLLN for a Poisson process gives

$$(D.35) \qquad \mathbb{P}\left\{ \lim_{t\to\infty} \frac{1}{t} N(t) = \eta \right\} = 1.$$

The SLLN (D.32) follows from (D.33), (D.34), (D.35), and the fact that $\eta P_{ij} = \Lambda_{ij}$.   □

Theorem D.27 (Foster–Lyapunov criterion for CTMC).   *Let X be an irreducible CTMC that has countable state space $\mathscr{S}$ and satisfies* (D.28). *Assume there exist constants $b > 0$ and $c > 0$, a finite set $C \subset \mathscr{S}$, and a nonnegative function $V : \mathscr{S} \to \mathbb{R}_+$ such that*

$$(D.36) \qquad \Lambda V(i) \le -c + b 1_C(i), \quad i \in \mathscr{S}.$$

Then X is positive recurrent with a unique stationary distribution.

Proof.   Let (N,Z) be a uniformization representation of X, where Z has transition probability matrix P of the form (D.30). Then Z is also irreducible. Spelling out condition (D.36) gives

$$\sum_{j\neq i} \lambda_{ij} V(j) - \lambda(i) V(i) \le -c + b 1_C(i), \quad i \in \mathscr{S},$$

which is equivalent to

$$P V(i) \le V(i) - c/\eta + b/\eta\, 1_C(i), \quad i \in \mathscr{S}.$$

Therefore, given that the CTMC X satisfies (D.36), the DTMC Z satisfies the Foster–Lyapunov drift condition (C.23) with the same finite set C, the same function V, and constants $b/\eta > 0$ and $c/\eta > 0$. By and Theorems C.19 and C.26, Z is positive recurrent and has a unique stationary distribution. Therefore, X has a unique stationary distribution and is positive recurrent, proving the theorem.   □

D.8 Phase-Type Distributions

This section introduces the important family of *phase-type* distributions, which are used in this book to model the service time distributions of a stochastic processing network (SPN), enabling a discrete-state Markov representation of the SPN (see Chapter 3). An important modeling consideration is that *any* distribution on the positive half-line can be approximated to an arbitrary degree of accuracy, in a certain standard sense, by a phase-type distribution; see section III.4 of Asmussen (2003) for a proof and discussion of this *phase-type approximation theorem*. We begin with a few simple examples of phase-type distributions, followed by Definition D.28.

Let v be a positive random variable representing a service time. When v has an exponential distribution with rate parameter $r > 0$, denoted $\exp(r)$, it has density function

$$f(x) = re^{-rx}, \quad x \geq 0,$$

and mean $1/r$. When v has an Erlang distribution with shape parameter 2 and rate parameter r, denoted $\mathrm{Erlang}(2, r)$, it has density function

$$f(x) = r^2 xe^{-rx}, \quad x \geq 0,$$

and mean $2/r$. In this latter case, v can be represented as

$$v = v_1 + v_2,$$

where v_1 and v_2 are independent random variables, each distributed $\exp(r)$.

The following is an alternative construction of an $\mathrm{Erlang}(2, r)$ random variable. Let $Y = \{Y(t), t \geq 0\}$ be a Markov chain with state space $\{1, 2, 3\}$ and jump matrix

$$\begin{pmatrix} 0 & 1 & 0 \\ 0 & 0 & 1 \\ 0 & 0 & 1 \end{pmatrix}.$$

From the structure of this jump matrix, one sees that states 1 and 2 are transient, while state 3 is absorbing. Let the exit rate in each transient state i be r $(i = 1, 2)$, meaning that each excursion in a transient state is distributed $\exp(r)$. Finally, assume that $Y(0) = 1$ with probability 1. Combining all this information, one sees that the random variable

(D.37) $$v = \inf\{t \geq 0 : Y(t) = 3\}$$

is the sum of two i.i.d. random variables, each distributed $\exp(r)$, and hence it has the desired Erlang$(2,r)$ distribution. Because v represents a service time in the applications covered by this book, we call the process Y that generates v a *service regulating Markov chain (SRMC)*.

This representation of the random variable v can be generalized to the following scenario. (a) The state space of the service regulating Markov chain Y is $\mathscr{S} \cup \{\delta\}$, where $\mathscr{S}$ is a finite set of d "service phases" and δ is a "terminal phase." (b) The structure of the SRMC jump matrix is such that phases in $\mathscr{S}$ are transient and δ is absorbing. Hereafter we denote by P the $d \times d$ jump matrix for transitions within $\mathscr{S}$, so P is substochastic and transient. (c) Exit rates for the transient states, denoted $\{\gamma(s), s \in \mathscr{S}\}$, are strictly positive but otherwise arbitrary. (d) The initial state $Y(0)$ is random, with a distribution $\{p(s), s \in \mathscr{S}\}$ concentrated on the transient states. With this more general scenario, let

(D.38) $$v = \inf\{t \geq 0 : Y(t) = \delta\}.$$

Definition D.28. The random variable v defined by (D.38) is said to have a *phase-type distribution* with phase space $\mathscr{S}$ and parameters (p, γ, P).

It is well known that the cumulative distribution function of v is given by

(D.39) $$\mathbb{P}\{v \leq x\} = 1 - e' \exp(-Rx)p \quad \text{for } x \geq 0,$$

where e denotes a vector of ones and

$$R = (I - P')\text{diag}(\gamma),$$

and the matrix exponential is defined as usual via

(D.40) $$\exp(-Rx) = \sum_{n=0}^{\infty} \frac{1}{n!}(-Rx)^n.$$

See, for example, section 2.4 of Latouche and Ramaswami (1999). From (D.39) it follows that

(D.41) $$\mathbb{E}[v] = e' R^{-1} p.$$

Definition D.29. A d-phase *hyperexponential distribution* with parameters (p, γ), denoted $H_d(p, \gamma)$, is a phase-type distribution with phase space $\mathscr{S} = \{1, \ldots, d\}$ and parameters (p, γ, P), where $P(i, \delta) = 1$ for all $i = 1, \ldots, d$.

That is, if v is distributed $H_d(p,\gamma)$, then with probability $p(1)$ it is distributed $\exp(\gamma(1)),\ldots$, and with probability $p(d)$ it is distributed $\exp(\gamma(d))$. For example, if $d = 2, p = (0.2,0.8)$ and $\gamma = (1,0.1)$, then with 20% probability the service time v is exponentially distributed with mean 1, and with 80% probability it is exponentially distributed with mean 10.

D.9 Joint Phase-Type Distributions

The previous section has described a mechanism for generating a non-negative random variable v as the absorption time of a CTMC. That mechanism can be generalized in the following way to generate a jointly distributed pair of random variables (v,φ), where v is nonnegative and φ takes values in a finite space $\Delta = \{\delta^1,\ldots,\delta^r\}$. The application we have in mind is that where v is a service time and φ is the associated output vector (see Section 2.1).

In the generalized setting, the service regulating Markov chain $Y = \{Y(t), t \geq 0\}$ has state space $\mathscr{S} \cup \Delta$, where $\mathscr{S}$ is a finite set of "service phases." Phases in $\mathscr{S}$ are transient and those in Δ are absorbing. Again the data of Y include an initial distribution p over the transient service phases, and a vector γ of exit rates from the transient service phases. Thus each excursion in a transient state s is distributed $\exp(\gamma(s))$. Again we denote by P the jump matrix of Y. Generalizing (D.37), let

$$(D.42) \qquad v = \inf\{t \geq 0 : Y(t) \in \Delta\} \quad \text{and} \quad \varphi = Y(v).$$

That is, we set v equal to the time at which Y reaches an absorbing phase, and select a value for φ according to which absorbing phase is reached. Use of the term "output space" in the following definition reflects our intended application to SPN modeling.

Definition D.30.   The random variables (v,φ) defined by (D.42) are said to have a *joint phase-type distribution* with phase space $\mathscr{S}$, output space Δ, and parameters (p,γ,P).

Given the phase-type approximation theorem mentioned in the first paragraph of the previous section, it is a straightforward exercise to show that any joint distribution on $\mathbb{R}_+ \times \Delta$ can be approximated to any desired degree of accuracy (again in a standard sense) by a joint phase-type distribution.

For all the SPN models considered in this book, service times and their associated output vectors are assumed to have a joint phase-type distribution (see Section 2.2). To achieve a Markov representation of the SPN, one can imagine each service as being generated by its own separate SRMC, with possibly random initial state, random transition times, etc. The evolution of that SRMC begins at the moment when the service is initiated, and ends at the moment of absorption. Its phase at times between initiation and termination can be incorporated in the state description for the system as a whole. Typically, however, we do not imagine such detailed phase information to be observable by the system manager, and hence we consider only control policies based on higher-level system state characteristics, such as buffer contents and service counts.

Definition D.30 identifies the parameters of a joint phase type distribution as (p, γ, P), but for SPN modeling it is often convenient to replace γ and P with the following pair of *transition rate functions F* and *f*: let

$$(D.43) \qquad F(s, \tilde{s}) = \gamma(s) P(s, \tilde{s}) \quad \text{for} \quad s, \tilde{s} \in \mathscr{S},$$

and

$$(D.44) \qquad f(s, \delta) = \gamma(s) P(s, \delta) \quad \text{for} \quad s \in \mathscr{S}, \delta \in \Delta.$$

Thus $F(s, \tilde{s})$ is the rate at which the service regulating Markov chain makes "silent" transitions (so called because they are not observed by the system manager) from s to $\tilde{s}$, and $f(s, \delta)$ is the rate at which it makes transitions from service phase s that terminate the service and trigger output vector δ.

Appendix E

Markovian Arrival Processes

Consider a system with external arrivals of I different classes. The multiple arrival streams can be modeled by a sequence of random pairs $\{(\tau(n), \delta(n)), n = 1, 2, \ldots\}$, where $\tau(1), \tau(2), \ldots$ are *interarrival times* taking values in $(0, \infty)$, and $\delta(1), \delta(2), \ldots$ are *arrival batches* taking values in a finite space $\Delta \subset \mathbb{Z}_+^I$. The corresponding I-dimensional external arrival process $E = \{E(t), t \geq 0\}$ is defined as follows for $t \geq 0$:

$$E(t) = \sum_{n=1}^{\eta(t)} \delta(n) \quad \text{where} \quad \eta(t) = \max\{n \geq 1 : \tau(1) + \ldots + \tau(n) \leq t\}.$$

We interpret $\tau(1) + \cdots + \tau(n)$ as the time of the nth *arrival event*, $\delta_i(n)$ as the number of class i arrivals occurring at that time, $\eta(t)$ as the number of arrival events that occur during the time interval $[0, t]$, and $E_i(t)$ as the number of class i arrivals during $[0, t]$.

Except for the very mild requirement that Δ be a finite set, the arrival process described in the previous paragraph is perfectly general. In the language of stochastic process theory, it is a *marked point process (MPP)* with *mark space* Δ. A familiar special case is that where Δ consists simply of the unit vectors $e^1, \ldots, e^I$, meaning that all arrival batches are singletons, and a still more special case is that where components of E are independent Poisson processes. In general, however, MPP models may incorporate clustered arrivals, correlations between arrivals of different classes, and serially correlated arrivals.

This appendix introduces the notion of a *Markovian arrival process (MArP)*, for which Asmussen (2003), chapter XI, is our standard reference. When a MArP is used to model external arrivals to a processing network, Markov representation of the network is facilitated (see Section 4.1), but such arrival processes are still very general: Asmussen and Koole (1993) have shown that any marked point process can be approximated to an arbitrary degree of accuracy, in a certain standard sense, by a MArP; this *MArP approximation theorem* is a process-level analog of the phase-type approximation theorem cited in Section D.8.

E.1 General Definition and Examples

The central element of a MArP model is a continuous-time Markov chain $Y = \{Y(t), t \geq 0\}$ with state space $\mathscr{S} = \{1, 2, \ldots, d\}$, hereafter called the *arrivals-regulating Markov chain (ARMC)*. To avoid potential confusion with the notation for arrival classes, states of Y will be denoted by s and $\tilde{s}$ in this appendix, rather than i and j. We denote by $p = (p(s) : s \in \mathscr{S})$ the initial distribution of Y, and by $\Lambda = (\Lambda(s, \tilde{s}) : s, \tilde{s} \in \mathscr{S})$ its generator matrix (see Section D.1). To avoid trivial complications, the jump matrix P derived from Λ is assumed to be irreducible. The triple $(\mathscr{S}, p, \Lambda)$ will be referred to later as *ARMC data*.

We take as given a finite batch space Δ as before, the elements of which will be denoted $\delta^1, \ldots, \delta^r$. For each $s \in \mathscr{S}$ and $k = 1, \ldots, r$, there is given an arrival rate $\beta_k(s) \geq 0$, and for each pair of states $s, \tilde{s} \in \mathscr{S}, s \neq \tilde{s}$, there is given a subprobability distribution $q_1(s, \tilde{s}), \ldots, q_r(s, \tilde{s})$, meaning that $q_k(s, \tilde{s}) \geq 0$ for all k and $\sum_{k=1}^r q_k(s, \tilde{s}) \leq 1$. As a matter of convention, let $q_k(s, s) = 0$ for $k = 1, \ldots, r$ and $s \in \mathscr{S}$. The pair (β, q) will be referred to later as *arrivals data*.

With these data in hand, one can describe the MArP verbally as follows. First, whenever Y is in state s, batches of composition $\delta^1, \ldots, \delta^r$ arrive according to independent Poisson processes at rates $\beta_1(s), \ldots, \beta_r(s)$, respectively, independent of all previous history $(s \in \mathscr{S})$. Second, each time Y makes a transition from state s to state $\tilde{s} \neq s$, there is a probability $q_1(s, \tilde{s})$ that the transition triggers an arrival batch of composition $\delta^1, \ldots$, and a probability $q_r(s, \tilde{s})$ that it triggers an arrival batch of composition δ^r, independent of all previous history $(s, \tilde{s} \in \mathscr{S})$. Thus one can identify three distinct types of events in the evolution of a MArP: *spontaneous* batch arrivals of various compositions, which occur without a transition of the underlying ARMC; *triggered* batch arrivals of various compositions, which are accompanied by a simultaneous transition of the underlying ARMC; and *silent transitions* of the underlying ARMC, which are not accompanied by any arrivals.

To write out the MArP in explicit mathematical terms, we take as given a probability space on which are defined the following primitive stochastic elements: the ARMC Y; a family of independent Poisson processes $\xi^1,\ldots,\xi^r$, each with unit arrival rate, also independent of Y; and a family of i.i.d. discrete random variables $\{\zeta(n,s,\tilde{s}),n=1,2,\ldots\}$, indexed by $s,\tilde{s}\in\mathscr{S}$, taking values in the set $\{0,1,\ldots,r\}$, independent of one another and of Y and of the Poisson processes $\xi^1,\ldots,\xi^r$. More specifically, the random variables $\zeta(n,s,\tilde{s})$ are distributed as follows: for $n=1,2,\ldots$ and $s,\tilde{s}\in\mathscr{S}$, we have

$$\mathbb{P}\{\zeta(n,s,\tilde{s})=k\}=q_k(s,\tilde{s})\quad\text{for }k=1,\ldots,r$$

and

$$\mathbb{P}\{\zeta(n,s,\tilde{s})=0\}=1-\sum_{k=1}^{r}q_k(s,\tilde{s}).$$

Our goal now is to express the external arrival process E in terms of these probabilistic primitives, not only in the interest of mathematical precision, but also to facilitate the proof of certain key MArP properties (see Section E.2). Toward that end, let $J(s,\tilde{s},t)$ be the number of jumps that Y makes during the time interval $[0,t]$ from state s to state $\tilde{s}$, with the convention that $J(s,s,t)=0$ for all $s\in\mathscr{S}$. Now for $s,\tilde{s}\in\mathscr{S}$ and $t\geq 0$, let

$$T(s,t)=\int_0^t 1_{\{Y(u)=s\}}\,du$$

and

$$J_k(s,\tilde{s},t)=\sum_{n=1}^{J(s,\tilde{s},t)}1_{\{\zeta(n,s,\tilde{s})=k\}}\quad\text{for }k=1,\ldots,r.$$

In words, $T(s,t)$ is the amount of time that Y spends in state s up to time t, and $J_k(s,\tilde{s},t)$ is the number of batch arrivals with configuration δ^k that are triggered by ARMC transitions from s to $\tilde{s}$ up to time t. We can now represent as follows the *total* number of batch arrivals with configuration δ^k that occur up to time t:

$$\text{(E.1)}\qquad A_k(t)=\sum_{s\in\mathscr{S}}\xi^k(\beta_k(s)T(s,t))+\sum_{s,\tilde{s}\in\mathscr{S}}J_k(s,\tilde{s},t),$$

with the first term on the right representing spontaneous arrivals and the second one representing triggered arrivals. (With our conventions, the second sum on the right side of (E.1) is effectively over $s \neq \tilde{s}$.) Finally, one has the obvious relationship

$$(\text{E.2}) \qquad E(t) = \sum_{k=1}^{r} \delta^k A_k(t), \quad t \geq 0.$$

Definition E.1. The I-dimensional stochastic process E defined by (E.2) is a MArP with batch space Δ, with ARMC data $(\mathscr{S}, p, \Lambda)$, and with arrivals data (β, q).

Example E.2. Consider a renewal arrival process whose interarrival times are distributed Erlang$(2, \gamma)$. This can be represented as a MArP with $I = 1$ and $\Delta = \{1\}$, meaning that all arrivals are singletons and they are of just one class. The ARMC has state space $\mathscr{S} = \{1, 2\}$, initial distribution $p = (1, 0)$, and generator matrix

$$(\text{E.3}) \qquad \Lambda = \begin{pmatrix} -\gamma & \gamma \\ \gamma & -\gamma \end{pmatrix}.$$

That is, the ARMC $Y = \{Y(t), t \geq 0\}$ occupies states $1, 2, 1, 2, \ldots$ deterministically in that order, and the occupancy time of each state is distributed $\exp(\gamma)$. The arrivals data are $\beta(1) = 0, \beta(2) = 0, q(1, 2) = 0$, and $q(2, 1) = 1$. (Because we have $r = 1$ in this example, subscripts are omitted in specifying the arrivals data β and q.) That is, there are no spontaneous arrivals, a transition of the ARMC from state 1 to state 2 never triggers an arrival, and a transition from state 2 to state 1 always triggers an arrival.

Example E.3. Generalizing Example E.2, consider a MArP with $I = 1$ and $\Delta = \{1\}$, allowing the generator matrix of the underlying ARMC to be arbitrary except for the irreducibility constraint previously imposed. Suppose there is one distinguished state $\hat{s} \in \mathscr{S}$ such that $q(s, \hat{s}) = 1$ for all $s \in \mathscr{S}, s \neq \hat{s}$, and $q(s, \tilde{s}) = 0$ for $s, \tilde{s} \in \mathscr{S}$ and $\tilde{s} \neq \hat{s}$; that is, an arrival is triggered each time the ARMC enters the distinguished state, and there are no other triggered arrivals. Also, assume that $\beta(\cdot) \equiv 0$, so there are no spontaneous arrivals, and $p(\hat{s}) = 1$, meaning that the ARMC begins in the distinguished state $\hat{s}$. The corresponding one-dimensional arrival process E is a renewal process whose interarrival times have a phase-type distribution.

Remark E.4. If the description of Example E.3 is modified to allow a general initial distribution p, then the arrival process E is a *delayed* renewal process, meaning that the initial arrival time may have a distribution different from that of the i.i.d. interarrival times that follow.

Example E.5. Again we consider a MArP with $I = 1$ and $\Delta = \{1\}$. The generator matrix of the ARMC is arbitrary except for irreducibility. If $q(\cdot, \cdot) \equiv 0$, then the one-dimensional arrival process E is called a *Markov-modulated Poisson process (MMPP)*. In words, an MMPP is a Poisson process whose arrival rate is determined by the state of an underlying CTMC.

Remark E.6. Many realistic phenomena, such as bursty arrivals or nonstationary Poisson arrivals with a cyclic or periodic arrival rate, can be represented or closely approximated by an MMPP, but in the last section of their paper, Asmussen and Koole (1993) provide an example that proves the following, stated for simplicity in the standard setting with $I = 1$ and $\Delta = \{1\}$: the approximation theorem cited at the beginning of this appendix no longer holds if one replaces the class of MArPs with the smaller class of MMPPs.

E.2 SLLN and Uniform Integrability

The following proposition documents the two facts about MArPs that one needs for purposes of fluid-based analysis. The strong law of large numbers established in part (a) is cited at virtually every point in this book where a MArP is mentioned. The uniform integrability established in part (b) is needed to prove Theorem 11.5, which extends Theorem 6.2 (fluid limit stability implies SPN stability) to a family of network models in which external inputs follow a MArP.

Proposition E.7. *Let $E = \{E(t), t \geq 0\}$ be a MArP and fix a constant $h \geq 0$. (a) The following SLLN holds:*

$$(\text{E.4}) \qquad \mathbb{P}\left\{ \lim_{t \to \infty} \frac{1}{t} E(ht) = \lambda h \right\} = 1,$$

where

$$(\text{E.5}) \qquad \lambda = \sum_{k=1}^{r} \delta^k \theta_k,$$

$$(\text{E.6}) \qquad \theta_k = \sum_{s \in \mathscr{S}} \pi(s)\beta_k(s) + \sum_{s, \tilde{s} \in \mathscr{S}} \pi(s)\Lambda_{s\tilde{s}} q(s, \tilde{s}),$$

and π is the unique stationary distribution of the arrivals-regulating Markov chain Y. (b) The family of random variables

(E.7) $$\left\{\frac{1}{t}E(th), t \geq 1\right\} \quad \text{is uniformly integrable.}$$

Proof. (a). In light of (E.2), it suffices to prove that for each $k = 1,\ldots,r$,

(E.8) $$\mathbb{P}\left\{\lim_{t\to\infty}\frac{1}{t}A_k(ht) = \theta_k h\right\} = 1,$$

where θ_k is defined via (E.6). The SLLN (E.8) follows immediately from (E.9) through (E.12), the justification for which will be explained shortly:

(E.9) $$\mathbb{P}\left\{\lim_{t\to\infty}\frac{1}{t}\xi^k(t) = 1\right\} = 1,$$

(E.10) $$\mathbb{P}\left\{\lim_{t\to\infty}\frac{1}{t}T(s,t) = \pi(s)\right\} = 1 \quad \text{for each } s \in \mathscr{S},$$

(E.11) $$\mathbb{P}\left\{\lim_{N\to\infty}\frac{1}{N}\sum_{n=1}^{N}1_{\{\zeta(n,s,\tilde{s})=k\}} = q_k(s,\tilde{s})\right\} = 1 \quad \text{for each } s,\tilde{s} \in \mathscr{S},$$

(E.12) $$\mathbb{P}\left\{\lim_{t\to\infty}\frac{1}{t}J(s,\tilde{s},t) = \pi(s)\Lambda(s,\tilde{s})\right\} = 1 \quad \text{for each } s,\tilde{s} \in \mathscr{S}.$$

Here (E.9) is the SLLN for a Poisson process, (E.10) follows from Theorem D.25, (E.11) is the SLLN for i.i.d. random variables, and (E.12) follows from Theorem D.26.

 (b) By (E.2), it suffices to prove that for each $k = 1,\ldots,r$,

(E.13) $$\left\{\frac{1}{t}A_k(th), t \geq 1\right\} \quad \text{is uniformly integrable.}$$

Using the representation (D.33) for $J(s,\tilde{s},t)$, one has

(E.14) $$J(s,\tilde{s},t) \leq N(t)$$

for each $t \geq 0$, where N is the Poisson process appearing in the representation (D.29) of the CTMC Y. From (E.1) and (E.14), one has

$$A_k(t) \leq |\mathscr{S}|\xi^k(Bt) + |\mathscr{S}|^2 N(t),$$

where $B = \max_{s\in\mathscr{S}}\beta(s)$. Now (E.13) follows from Proposition B.5. $\quad\square$

Corollary E.8.   *For each $h \geq 0$,*

(E.15)
$$\lim_{t \to \infty} \frac{1}{t} \mathbb{E}\Big[E(ht) \Big] = \lambda h,$$

where λ is given by formula (E.5).

Proof.   This follows from (E.4), (E.7), and theorem 4.5.4 of Chung (2001).   □

Appendix F

Convergent Square Matrices

Let Q be a $d \times d$ real matrix. We denote by $\rho(Q)$ the spectral radius of Q, that is, $\rho(Q) = \max_{1 \le i \le d} |\lambda_i|$, where $\lambda_1, \ldots, \lambda_d$ are the eigenvalues of Q. Also, Q is said to be *convergent* if $Q^k \to 0$ as $k \to \infty$. The following standard result is subsumed in theorems 4 and 5 of Isaacson and Keller (2012), which are stated and proved on pages 14 and 15 of that book. It is true without any further restrictions, but in this book it will be applied only in contexts where Q is nonnegative, such as Section 9.1.

Theorem F.1. *The following three statements are equivalent:*
(a) Q is convergent;
(b) $\rho(Q) < 1$; and
(c) $(I - Q)$ is nonsingular with

$$\text{(F.1)} \qquad (I - Q)^{-1} = \sum_{k=0}^{\infty} Q^k.$$

The sum on the right side of (F.1) is sometimes called a *Neumann series*, or the *Neumann expansion* of $(I - Q)^{-1}$. If Q is substochastic and transient (those terms are defined in the "Guide to Notation and Terminology," page xix), then Q can be interpreted as the transition probability matrix of a transient discrete-time Markov chain X with state space $\mathcal{S} = \{1, \ldots, d\}$. In that case, the Neumann expansion can be interpreted as follows. First, denoting by Q_{ij}^k the (i, j)th element of Q^k, we interpret Q_{ij}^k as the probability that X occupies state j in period k, given that $X = i$ initially. Equivalently stated, Q_{ij}^k is the expected number of visits to state j occurring in period k, and thus the (i, j)th element of the sum on the right side of (F.1) is the expected *total* number of visits to state j, given that $X = i$ initially.

References

Andrews, Matthew, Baruch Awerbuch, Antonio Fernández, Tom Leighton, Zhiyong Liu, and Jon Kleinberg (2001). Universal-stability results and performance bounds for greedy contention-resolution protocols. *Journal of the ACM* **48**, 39–69. MR1867275, Crossref (Cit. on p. 192)

Asmussen, Søren (2003). *Applied Probability and Queues*. Second Edition. Vol. 51. Applications of Mathematics. Stochastic Modelling and Applied Probability. Springer-Verlag, New York, NY. MR1978607, Crossref (Cit. on pp. 106, 130, 358, 362)

Asmussen, Søren and Ger Koole (1993). Marked point processes as limits of Markovian arrival streams. *Journal of Applied Probability* **30**, 365–372. MR1212668, Crossref (Cit. on pp. 362, 366)

Baskett, Forest, K. Mani Chandy, Richard R. Muntz, and Fernando G. Palacios (1975). Open, closed, and mixed networks of queues with different classes of customers. *Journal of the Association for Computing Machinery* **22**, 248–260. MR365749, Crossref (Cit. on pp. 53, 77, 106)

Berman, Abraham and Robert J. Plemmons (1994). *Nonnegative Matrices in the Mathematical Sciences*. Vol. 9. Classics in Applied Mathematics. Revised reprint of the 1979 original. Society for Industrial and Applied Mathematics (SIAM), Philadelphia, PA. MR1298430, Crossref (Cit. on p. 217)

Bertsekas, Dimitris and Robert G. Gallager (1992). *Data Networks*. Prentice Hall, Englewood Cliffs, NJ. Crossref (Cit. on p. 106)

Bertsimas, Dimitris, David Gamarnik, and John N. Tsitsiklis (1996). Stability conditions for multiclass fluid queueing networks. *IEEE Transactions on Automatic Control* **41**, 1618–1631; correction: **42**, 128, 1997. MR1419686, Crossref (Cit. on p. 193)

Bertsimas, Dimitris and John Tsitsiklis (1997). *Introduction to Linear Optimization*. Vol. 6. Athena Scientific Series in Optimization and Neural Computation. Athena Scientific, Nashua, NH. Crossref (Cit. on p. 196)

Billingsley, Patrick (1999). *Convergence of Probability Measures*. Second Edition. Wiley, New York, NY. MR1700749, Crossref (Cit. on p. 147)

Borodin, Allan, Jon Kleinberg, Prabhakar Raghavan, Madhu Sudan, and David P. Williamson (2001). Adversarial queuing theory. *Journal of the ACM* **48**, 13–38. MR1867274, Crossref (Cit. on p. 192)

Borovkov, A. A. (1986). Limit theorems for queueing networks. *Theory of Probability and Its Applications* **31**, 413–427. MR0866868, Crossref (Cit. on p. 78)

Botvich, D. D. and A. A. Zamyatin (1994). Ergodicity of a conservative communications network. *Discrete Mathematics and Applications* **4**, 479–493. MR1310907, Crossref (Cit. on p. 192)

Bramson, Maury (1994). Instability of FIFO queueing networks. *The Annals of Applied Probability* **4**, 414–431. MR1272733, Crossref (Cit. on pp. 28, 159)

Bramson, Maury (1996a). Convergence to equilibria for fluid models of FIFO queueing networks. *Queueing Systems. Theory and Applications* **22**, 5–45. MR1393404, Crossref (Cit. on pp. xvii, 159)

Bramson, Maury (1996b). Convergence to equilibria for fluid models of head-of-the-line proportional processor sharing queueing networks. *Queueing Systems. Theory and Applications* **23**, 1–26. MR1433762, Crossref (Cit. on pp. 106, 250)

Bramson, Maury (1998). Stability of two families of queueing networks and a discussion of fluid limits. *Queueing Systems. Theory and Applications* **28**, 7–31. MR1628469, Crossref (Cit. on pp. 145, 146, 192, 193)

Bramson, Maury (2008). *Stability of Queueing Networks*. Vol. 1950. Lecture Notes in Mathematics. Lectures from the 36th Probability Summer School Held in Saint-Flour, July 2–15, 2006. Springer, Berlin, Germany. MR2445100, Crossref (Cit. on pp. xvii, 77, 145, 146)

Bramson, Maury, Bernardo D'Auria, and Neil Walton (2017). Proportional switching in first-in, first-out networks. *Operations Research* **65**, 496–513. MR3647853, Crossref (Cit. on pp. 192, 310)

Bramson, Maury. and R. J. Williams (2003). Two workload properties for Brownian networks. *Queueing Systems. Theory and Applications* **45**, 191–221. MR2024178, Crossref (Cit. on p. 53)

Chen, Hong (1995). Fluid approximations and stability of multiclass queueing networks: work-conserving disciplines. *The Annals of Applied Probability* **5**, 637–665. MR1359823, Crossref (Cit. on pp. 145, 146, 193)

Chen, Hong and Avi Mandelbaum (1991). Discrete flow networks: bottleneck analysis and fluid approximations. *Mathematics of Operations Research* **16**, 408–446. MR1106809, Crossref (Cit. on p. 146)

Chen, Hong and Hanqin Zhang (1997). Stability of multiclass queueing networks under FIFO service discipline. *Mathematics of Operations Research* **22**, 691–725. MR1467392, Crossref (Cit. on p. 159)

Chen, Hong and Hanqin Zhang (2000). Stability of multiclass queueing networks under priority service disciplines. *Operations Research* **48**, 26–37. MR1753222, Crossref (Cit. on p. 192)

Chung, Kai Lai (2001). *A Course in Probability Theory*. Third Edition. Academic Press, Inc., San Diego, CA. MR1796326, Crossref (Cit. on pp. 320, 321, 368)

Cover, Thomas M. and Joy A. Thomas (2006). *Elements of Information Theory*. Second Edition. Wiley-Interscience [John Wiley & Sons], Hoboken, NJ. MR2239987, Crossref (Cit. on p. 326)

Cruz, Rene L. (1991). A calculus for network delay, Part II: Network analysis. *IEEE Transactions on Information Theory* **37**, 132–141. MR1087891, Crossref (Cit. on p. 192)

Dai, J. G. (1995a). On positive Harris recurrence of multiclass queueing networks: a unified approach via fluid limit models. *The Annals of Applied Probability* **5**, 49–77. MR1325041, Crossref (Cit. on pp. 77, 78, 145, 146, 147, 159, 192, 315)

Dai, J. G. (1995b). Stability of open multiclass queueing networks via fluid models. *Stochastic Networks.* Ed. by F. Kelly and R. J. Williams. Vol. 71. The IMA Volumes in Mathematics and Its Applications. Springer, New York, NY, 71–90. MR1381005, Crossref (Cit. on p. 192)

Dai, J. G. (1996). A fluid limit model criterion for instability of multiclass queueing networks. *The Annals of Applied Probability* **6**, 751–757. MR1410113, Crossref (Cit. on p. 146)

Dai, J. G., J. J. Hasenbein, and J. H. Vande Vate (1999). Stability of a three-station fluid network. *Queueing Systems. Theory and Applications* **33**, 293–325. MR1742573, Crossref (Cit. on p. 193)

Dai, J. G. and Wuqin Lin (2005). Maximum pressure policies in stochastic processing networks. *Operations Research* **53**, 197–218. MR2131925, Crossref (Cit. on pp. 53, 123, 218)

Dai, J. G. and Sean P. Meyn (1995). Stability and convergence of moments for multiclass queueing networks via fluid limit models. *IEEE Transactions on Automatic Control* **40**, 1889–1904. MR1358006, Crossref (Cit. on p. 146)

Dai, J. G. and B. Prabhakar (2000). The throughput of data switches with and without speedup. *IEEE INFOCOM 2000, Nineteenth Annual Joint Conference of the IEEE Computer and Communications Societies*, 556–564. Crossref (Cit. on p. 309)

Dai, J. G. and J. H. Vande Vate (2000). The stability of two-station multitype fluid networks. *Operations Research* **48**, 721–744. MR1792776, Crossref (Cit. on pp. 184, 190, 192, 193)

Dai, J. G. and Y. Wang (1993). Nonexistence of Brownian models for certain multiclass queueing networks. *Queueing Systems. Theory and Applications* **13**, 41–46. MR1218843, Crossref (Cit. on p. 28)

Dai, J. G. and G. Weiss (1996). Stability and instability of fluid models for reentrant lines. *Mathematics of Operations Research* **21**, 115–134. MR1385870, Crossref (Cit. on p. 192)

Dantzig, George B. (1998). *Linear Programming and Extensions.* Corrected Edition. Princeton Landmarks in Mathematics. Princeton University Press, Princeton, NJ. MR1658673, Crossref (Cit. on p. 4)

Davis, M. H. A. (1984). Piecewise-deterministic Markov processes: a general class of nondiffusion stochastic models. *Journal of the Royal Statistical Society. Series B.* **46**, 353–388. MR0790622, Crossref (Cit. on p. 76)

De Veciana, G., Tae-Jin Lee, and T. Konstantopoulos (2001). Stability and performance analysis of networks supporting elastic services. *IEEE/ACM Transactions on Networking* **9**, 2–14. Crossref (Cit. on p. 250)

Dean, Jeffrey and Sanjay Ghemawat (2008). MapReduce: simplified data processing on large clusters. *Communications of the ACM* **51**, 107–113. Crossref (Cit. on p. 267)

Down, D. and S. P. Meyn (1994). Piecewise linear test functions for stability of queueing networks. *Proceedings of 1994 33rd IEEE Conference on Decision and Control.* Vol. 3, 2069–2074. Crossref (Cit. on p. 192)

Dupuis, Paul and Ruth J. Williams (1994). Lyapunov functions for semi-martingale reflecting Brownian motions. *The Annals of Probability* **22**, 680–702. MR1288127, Crossref (Cit. on p. 146)

Ethier, Stewart N. and Thomas G. Kurtz (1986). *Markov Processes: Characterization and Convergence*. Wiley Series in Probability and Mathematical Statistics. John Wiley & Sons, Inc., New York, NY. MR838085, Crossref (Cit. on pp. 147, 313)

Feller, William (1971). *An Introduction to Probability Theory and Its Applications*. Vol. II. Second Edition. John Wiley & Sons, Inc., New York–London–Sydney. MR0270403, Crossref (Cit. on p. 307)

Fitzpatrick, Patrick M. and Brian R. Hunt (2015). Absolute continuity of a function and uniform integrability of its divided differences. *American Mathematical Monthly* **122**, 362–366. MR3343072, Crossref (Cit. on p. 311)

Foss, Serguei G. (1991). Ergodicity of queuing networks. *Siberian Mathematical Journal* **32**, 690–705. Crossref (Cit. on p. 78)

Foss, Serguei G. and Takis Konstantopoulos (2004). An overview of some stochastic stability methods. *Journal of the Operations Research Society of Japan* **47**, 275–303. MR2174067, Crossref (Cit. on p. 342)

Foss, Serguei G. and Artyom Kovalevskii (1999). A stability criterion via fluid limits and its application to a polling system. *Queueing Systems. Theory and Applications* **32**, 131–168. MR1720552, Crossref (Cit. on p. 147)

Gale, David (1960). *The Theory of Linear Economic Models*. McGraw-Hill Book Co., Inc., New York–Toronto–London. MR0115801, Crossref (Cit. on p. 217)

Gamarnik, David (2000). Using fluid models to prove stability of adversarial queueing networks. *IEEE Transactions on Automatic Control* **45**, 741–746. MR1764845, Crossref (Cit. on p. 192)

Gamarnik, David (2003). Stability of adaptive and nonadaptive packet routing policies in adversarial queueing networks. *SIAM Journal on Computing* **32**, 371–385. MR1969395, Crossref (Cit. on p. 192)

Gamarnik, David and John J. Hasenbein (2005). Instability in stochastic and fluid queueing networks. *The Annals of Applied Probability* **15**, 1652–1690. MR2152240, Crossref (Cit. on p. 146)

Goel, Ashish (2001). Stability of networks and protocols in the adversarial queueing model for packet routing. *Networks* **37**, 219–224. MR1837200, Crossref (Cit. on p. 192)

Harrison, J. Michael (1973). Assembly-like queues. *Journal of Applied Probability* **10**, 354–367. MR0356276, Crossref (Cit. on p. 123)

Harrison, J. Michael (1988). Brownian models of queueing networks with heterogeneous customer populations. *Stochastic Differential Systems, Stochastic Control Theory and Applications*. Ed. by W. Fleming and P. L. Lions. Vol. 10. The IMA Volumes in Mathematics and Its Applications. Springer, New York, NY, 147–186. MR0934722, Crossref (Cit. on pp. 27, 53)

Harrison, J. Michael (1998). Heavy traffic analysis of a system with parallel servers: asymptotic optimality of discrete-review policies. *The Annals of Applied Probability* **8**, 822–848. MR1627791, Crossref (Cit. on p. 28)

Harrison, J. Michael (2000). Brownian models of open processing networks: canonical representation of workload. *The Annals of Applied Probability* **10**,

75–103; corrections: **13**, 390–393 (2003), and **16**, 1703–1732 (2006). MR1765204, Crossref (Cit. on pp. 27, 53, 123)

Harrison, J. Michael (2002). Stochastic networks and activity analysis. *Analytic Methods in Applied Probability: in memory of Fridrikh Karpelevich*. Ed. by Yu. M. Suhov. Vol. 207. American Mathematical Society Translations: Series 2. American Mathematical Society, Providence, RI, 53–76. MR1992207, Crossref (Cit. on p. 28)

Harrison, J. Michael (2003). A broader view of Brownian networks. *The Annals of Applied Probability* **13**, 1119–1150. MR1994047, Crossref (Cit. on pp. 27, 53)

Harrison, J. Michael (2013). *Brownian Models of Performance and Control*. Cambridge University Press, Cambridge, UK. MR3157450, Crossref (Cit. on p. 130)

Harrison, J. Michael and Marcel J. López (1999). Heavy traffic resource pooling in parallel-server systems. *Queueing Systems. Theory and Applications* **33**, 339–368. MR1742575, Crossref (Cit. on p. 106)

Harrison, J. Michael and Viên Nguyen (1993). Brownian models of multiclass queueing networks: current status and open problems. *Queueing Systems. Theory and Applications* **13**, 5–40. MR1218842, Crossref (Cit. on p. 159)

Harrison, J. Michael and Lawrence M. Wein (1989). Scheduling networks of queues: heavy traffic analysis of a simple open network. *Queueing Systems. Theory and Applications* **5**, 265–279. MR1030470, Crossref (Cit. on p. 28)

Hasenbein, John J. (1997). Necessary conditions for global stability of multiclass queueing networks. *Operations Research Letters* **21**, 87–94. MR1482493, Crossref (Cit. on p. 193)

Humes Jr., Carlos (1994). A regulator stabilization technique: Kumar–Seidman revisited. *IEEE Transactions on Automatic Control* **39**, 191–196. MR1258702, Crossref (Cit. on p. 193)

Isaacson, Eugene and Herbert B. Keller (2012). *Analysis of Numerical Methods*. Dover Books on Mathematics. Dover Publications, Mineola, NY. Crossref (Cit. on p. 369)

Jackson, J. R. (1957). Networks of waiting lines. *Operations Research* **5**, 518–521. MR0093061, Crossref (Cit. on p. 77)

Kang, W. N., F. P. Kelly, N. H. Lee, and R. J. Williams (2009). State space collapse and diffusion approximation for a network operating under a fair bandwidth sharing policy. *The Annals of Applied Probability* **19**, 1719–1780. MR2569806, Crossref (Cit. on pp. 43, 106, 250)

Kannan, R. and Carole K. Krueger (1996). *Advanced Analysis on the Real Line*. Universitext. Springer-Verlag, New York, NY. MR1390758, Crossref (Cit. on p. 165)

Kelly, Frank P. (1975). Networks of queues with customers of different types. *Journal of Applied Probability* **12**, 542–554. MR0388571, Crossref (Cit. on p. 78)

Kelly, Frank P. (1979). *Reversibility and Stochastic Networks*. Wiley Series in Probability and Mathematical Statistics. John Wiley & Sons, Ltd., Chichester, UK. MR554920, Crossref (Cit. on pp. 50, 52, 53, 78, 106)

Kelly, Frank P. (1997). Charging and rate control for elastic traffic. *European Transactions on Telecommunications* **8**, 33–37. Crossref (Cit. on p. 250)

Kelly, Frank P., A. K. Maulloo, and D. K. H. Tan (1998). Rate control for communication networks: shadow prices, proportional fairness and stability. *Journal of Operations Research Society* **49**, 237–252. Crossref (Cit. on p. 250)

Kleinrock, Leonard (1967). Time-shared systems: a theoretical treatment. *Journal of the Association for Computing Machinery* **14**, 242–261. MR273994, Crossref (Cit. on pp. 46, 87, 95)

Koehler, Gary J., Andrew B. Whinston, and Gordon P. Wright (1975). *Optimization over Leontief Substitution Systems*. North-Holland Publishing Co., Amsterdam, Netherlands–Oxford, UK; American Elsevier Publishing Co., Inc., New York, NY. MR0416562, Crossref (Cit. on p. 217)

Koopmans, T. C., editor (1951). *Activity Analysis of Production and Allocation*. John Wiley & Sons, New York, NY. Crossref (Cit. on p. 3)

Kumar, P. R. (1993). Re-entrant lines. *Queueing Systems. Theory and Applications* **13**, 87–110. MR1218845, Crossref (Cit. on p. 24)

Latouche, G. and V. Ramaswami (1999). *Introduction to Matrix Analytic Methods in Stochastic Modeling*. ASA-SIAM Series on Statistics and Applied Probability. Society for Industrial and Applied Mathematics (SIAM), Philadelphia, PA. Crossref (Cit. on p. 359)

Laws, C. N. and G. M. Louth (1990). Dynamic scheduling of a four-station queueing networks. *Probability in the Engineering and Informational Sciences* **4**, 131–156. Crossref (Cit. on p. 106)

Li, Bin and R. Srikant (2016). Queue-proportional rate allocation with per-link information in multihop wireless networks. *Queueing Systems. Theory and Applications* **83**, 329–359; correction: **84**, 203–210, (2016). MR3528494, Crossref (Cit. on p. 310)

Lu, S. H. and P. R. Kumar (1991). Distributed scheduling based on due dates and buffer priorities. *IEEE Transactions on Automatic Control* **36**, 1406–1416. Crossref (Cit. on p. 28)

Massoulié, Laurent (2007). Structural properties of proportional fairness: stability and insensitivity. *The Annals of Applied Probability* **17**, 809–839. MR2326233, Crossref (Cit. on pp. 106, 165, 192, 250)

Massoulié, Laurent and J. W. Roberts (2000). Bandwidth sharing and admission control for elastic traffic. *Telecommunication Systems* **15**, 185–201. Crossref (Cit. on pp. 106, 229, 250)

Massoulié, Laurent and Kuang Xu (2018). On the capacity of information processing systems. *Operations Research* **66**, 568–586. MR3782815, Crossref (Cit. on p. xvii)

McKeown, N., A. Mekkittikul, V. Anantharam, and J. Walrand (1999). Achieving 100% throughput in an input-queued switch. *IEEE Transactions on Communications* **47**, 1260–1267. Crossref (Cit. on p. 218)

Meyn, Sean P. (1995). Transience of multiclass queueing networks via fluid limit models. *The Annals of Applied Probability* **5**, 946–957. MR1384361, Crossref (Cit. on p. 146)

Meyn, Sean P. and Douglas Down (1994). Stability of generalized Jackson networks. *The Annals of Applied Probability* **4**, 124–148. MR1258176, Crossref (Cit. on pp. 14, 78)

Meyn, Sean P. and R. L. Tweedie (1994). State-dependent criteria for convergence of Markov chains. *The Annals of Applied Probability* **4**, 149–168. MR1258177, Crossref (Cit. on p. 342)

Meyn, Sean and Richard L. Tweedie (2009). *Markov Chains and Stochastic Stability*. Second Edition. Cambridge University Press, Cambridge, UK. MR2509253, Crossref (Cit. on p. 78)

Mo, J. and J. Walrand (2000). Fair end-to-end window-based congestion control. *IEEE-ACM Transactions on Networking* **8**, 556–567. Crossref (Cit. on p. 250)

Neely, M. J., E. Modiano, and C. E. Rohrs (2005). Dynamic power allocation and routing for time-varying wireless networks. *IEEE Journal on Selected Areas in Communications* **23**, 89–103. Crossref (Cit. on p. 217)

Newell, G. F. (1982). *Applications of Queueing Theory*. Second Edition. Monographs on Statistics and Applied Probability. Chapman & Hall, London, UK. MR737381, Crossref (Cit. on p. 146)

Norris, J. R. (1998). *Markov Chains*. Vol. 2. Cambridge Series in Statistical and Probabilistic Mathematics. Reprint of 1997 original. Cambridge University Press, Cambridge, UK. MR1600720, Crossref (Cit. on pp. 331, 344, 346, 347, 348, 349, 350, 352, 353, 355)

Pedarsani, Ramtin., Jean Walrand, and Yuan Zhong (2014). Scheduling tasks with precedence constraints on multiple servers. *52nd Annual Allerton Conference on Communication, Control, and Computing*, 1196–1203. Crossref (Cit. on p. 106)

Pulhaskii, A. A. and A. N. Rybko (2000). Nonergodicity of queueing networks when their fluid models are unstable. *Problems of Information Transmission* **36**, 23–41. MR1746007, Crossref (Cit. on p. 146)

Rockafellar, R. Tyrrell and Roger J.-B. Wets (1998). *Variational Analysis*. Vol. 317. Grundlehren der Mathematischen Wissenschaften [Fundamental Principles of Mathematical Sciences]. Springer-Verlag, Berlin, Germany. MR1491362, Crossref (Cit. on p. 224)

Royden, H. L. (1988). *Real Analysis*. Third Edition. Macmillan Publishing Company, New York, NY. MR1013117 (Cit. on pp. 311, 312)

Royden, H. L. and P. M. Fitzpatrick (2010). *Real Analysis*. Fourth Edition. Pearson, London, UK. Crossref (Cit. on pp. 161, 311, 312, 314, 319)

Rybko, A. N. and A. L. Stolyar (1992). Ergodicity of stochastic processes describing the operation of open queueing networks. *Problems of Information Transmission* **28**, 199–220. MR1189331, Crossref (Cit. on pp. 20, 28, 146)

Serfozo, Richard (1999). *Introduction to Stochastic Networks*. Vol. 44. Applications of Mathematics. Springer-Verlag, New York, NY. MR1704237, Crossref (Cit. on pp. 78, 106)

Serfozo, Richard (2009). *Basics of Applied Stochastic Processes*. Probability and Its Applications. Springer-Verlag, Berlin, Germany. MR2484222, Crossref (Cit. on p. 140)

Shah, Devavrat and Damon Wischik (2012). Switched networks with maximum weight policies: fluid approximation and multiplicative state space collapse. *The Annals of Applied Probability* **22**, 70–127. MR2932543, Crossref (Cit. on p. 310)

Sigman, Karl (1990). The stability of open queueing networks. *Stochastic Processes and Their Applications* **35**, 11–25. MR1062580, Crossref (Cit. on p. 78)

Srikant, R. and Lei Ying (2014). *Communication Networks: An Optimization, Control and Stochastic Networks Perspective.* Cambridge University Press, Cambridge, UK. MR3202391, Crossref (Cit. on pp. 225, 250)

Stolyar, A. L. (1995). On the stability of multiclass queueing networks: a relaxed sufficient condition via limiting fluid processes. *Markov Processes and Related Fields* **1**, 491–512. MR1403094, Crossref (Cit. on pp. 146, 147)

Tassiulas, Leandros (1995). Adaptive back-pressure congestion control based on local information. *IEEE Transactions on Automatic Control* **40**, 236–250. MR1312890, Crossref (Cit. on pp. 217, 218, 309)

Tassiulas, Leandros and Partha P. Bhattacharya (2000). Allocation of interdependent resources for maximal throughput. *Communications in Statistics. Stochastic Models* **16**, 27–48. MR1742516, Crossref (Cit. on p. 218)

Tassiulas, Leandros and Anthony Ephremides (1992). Stability properties of constrained queueing systems and scheduling policies for maximum throughput in multihop radio networks. *IEEE Transactions on Automatic Control* **37**, 1936–1948. MR1200609, Crossref (Cit. on pp. 17, 217, 218, 309)

Tassiulas, Leandros and L. Georgiadis (1994). Any work-conserving policy stabilizes the ring with spatial reuse. *Proceedings of INFOCOM '94 Conference on Computer Communications*, 66–70. Crossref (Cit. on p. 192)

Tweedie, R. L. (1976). Criteria for classifying general Markov chains. *Advances in Applied Probability* **8**, 737–771. MR0451409, Crossref (Cit. on p. 342)

Vlasiou, Maria, Jiheng Zhang, and Bert Zwart (2014). Insensitivity of proportional fairness in critically loaded bandwidth sharing networks. *arXiv e-prints*. arXiv: 1411.4841 [math.PR]. Crossref (Cit. on p. 251)

Walton, Neil (2015). Concave switching in single-hop and multihop networks. *Queueing Systems. Theory and Applications* **81**, 265–299. MR3401872, Crossref (Cit. on pp. xvii, 250, 299, 309, 310)

Wang, Weina, Siva Theja Maguluri, R. Srikant, and Lei Ying (2017). Heavy-traffic delay insensitivity in connection-level models of data transfer with proportionally fair bandwidth sharing. *SIGMETRICS Performance Evaluation Review* **45**, 232–245. Crossref (Cit. on p. 251)

Wang, Weina, K. Zhu, L. Ying, J. Tan, and L. Zhang (2014). Maptask scheduling in MapReduce with data locality: throughput and heavy-traffic optimality. *IEEE/ACM Transactions on Networking* **24**, 190–203. Crossref (Cit. on p. 267)

Wein, Lawrence M. (1991). Brownian networks with discretionary routing. *Operations Research* **39**, 322–340. Crossref (Cit. on p. 28)

Xie, Q. and Y. Lu (2015). Priority algorithm for near-data scheduling: throughput and heavy-traffic optimality. *IEEE INFOCOM 2015, IEEE Conference on Computer Communications*, 963–972. Crossref (Cit. on p. 267)

Xie, Q., A. Yekkehkhany, and Y. Lu (2016). Scheduling with multi-level data locality: throughput and heavy-traffic optimality. *IEEE INFOCOM 2016, The 35th Annual IEEE International Conference on Computer Communications*, 1–9. Crossref (Cit. on p. 267)

Ye, Heng-Qing and David D. Yao (2012). A stochastic network under proportional fair resource control-diffusion limit with multiple bottlenecks. *Operations Research* **60**, 716–738. MR2960540, Crossref (Cit. on p. 250)

Index